The History of

# AMERICAN DEEP SUBMERSIBLE OPERATIONS

1775–1995

*by Will Forman*

BEST PUBLISHING COMPANY

Cover Design: Susan Joiner
Text Layout and Design: Linda Longnaker
Editor: Andy Nordhoff
Production Supervisor: Wes Flynn



ISBN: 0-941332-72-1
Library of Congress catalog card number: 98-89161

Composed, printed, and bound in the United States of America

Best Publishing Company
2355 North Steves Boulevard
P.O. Box 30100
Flagstaff, Arizona 86003-0100 USA

# Preface

The first generation of American deep submersibles was built in the 1960s and 1970s. There was no wet NASA or the presidential mandate with nationally supported funding which NASA enjoyed. The publicity at the time for deep submergence was almost negligible, partly because of the attention and funding given to the space program and the VietNam War.

To date, most of the books about deep submersibles have been compiled by people who were not closely involved in the business. Some library sources present third or fourth-hand compilations of the data. For example, Archimedien screws have been incorrectly shown as the propellers on the 1776 *Turtle* submersible since the late 1800s. In actuallity, the paddle ends of oars were used to form propellers and it is indicated clearly in the available literature of the period.

Many of us who pioneered in the business feel that some of the early events need to be told from first-hand experience. Following the suggestions made by several colleagues, the mistakes and accidents are included and described in detail.

Early in 1964, Dr. Robert Frosch, then Assistant Secretary of the Navy for Research and Development, referred publicly to the *Deep Jeep* submersible as "the first American built deep submersible." Its operating depth of 2,000' was over three times deeper than anything else built in the U.S. at the time. Stories about *Deep Jeep*, the acrylic submersible *Kumukahi,* and the glass submersible *Deep View* are included in the chapters to follow.

In 1967, the Deep Submersible Pilots Association (DSPA) was formed and as of 1995, had about 100 members (1/3 reside in Southern California and approximately 10% overseas). Its monthly newsletter promotes safety through information exchange and many members meet regularly at monthly luncheons held in San Diego. As the Membership Committee Chairman for over twenty years, I have maintained my acquaintance with most of the submersible pilots from the early days of deep submergence and, with the enthusiastic cooperation of many of them, have obtained and reviewed their dive records which has helped immensely in retrieving the information required for this book. For purposes of definition, the Deep Submersible Pilots Constitution states: "A Deep Submersible shall refer to a manned vehicle which is capable of sustained operations to a depth of 200 meters (656.2'). It shall not refer to a combatant military submarine or a craft which is surface-controlled to the exclusion of independent pilot control." Some of the submersibles included herein do not meet the above criteria, but contribute to the general theme enough to merit inclusion. The time frame depicted in this book mainly includes the 60s through the 90s. However, there are earlier entries to establish a historical context and to give credit to those on whose shoulders we stood.

The chapters are listed in approximate order of dates of launch. Some parts of the chapters may read similar to dive logs. On the one hand, I have attempted to ease this burden by including a chart whereby all generic chronologies can be viewed at a glance. Even so, some dive log-like treatment was felt to be essential to present as complete a historical reference as the data could provide. Where these may seem repetitious at first, the reader who studies them and the reader who skims them will both be rewarded. By faithfully representing the operational chronologies, trends emerge in the life of each submersible from initial trials and tentative usage on to a particular vessel's or series of vessels ultimate fulfillment of its potential — however small or great.

I have written this book because no one else has done so. It is difficult at this juncture to predict where and to what extent manned, deep submersible development will follow. It is certain, however, that the 1960s through to 1995 were seminal, vigorous, and significant years. The record needs preservation, not just as a segment of marine history, but as a foundation for future submersible designers, operators, and client/users. No doubt there will be a resurgence of these interests some day and I hope the future participants have every success with their efforts.

In addition to the DSPA sources already mentioned, the chapters that follow are the result of extensive research in San Diego's public library system, numerous articles retrieved from the Naval Institute Proceedings, my personal library and those of generous friends: Dr. Andy Rechnitzer, Doug Privitt, Jay Jefferies, John Pritzlaff, Kevin Hardy, Bob Worthington, Neil Collins, and Bill Walker. Lieutenant Ed Lancaster (USN) was indispensable during the gathering of data for the Navy deep submersible sections. The sources of information and data used are summarized in the bibliography. Of these, Frank Busby's submersible directories and the Half Fathom Philately Society monthly newsletter were most frequently consulted, but only listed as chapter references when used as primary source material. Frequently, both the Half Fathom as well as Busby's other directories, and the reference collections of colleagues were used to substantiate the primary information sources. A few specific references to source material are included in the text but, in general, the bibliography represents such a multitude of sources about the same general information that individual references are not used. The bibliography contains 312 entries including over 60 personal interviews or information obtained through correspondence. In particular, Larry Shumaker, Tap Pryor, and Bill Walker made strong suggestions as to content and clarity. Dan Baits provided the initial editing that made the text easier to read. The final editing by Tap Pryor probably only represented a minute percentage of difference to the total text. However, like the small percentage variation in DNA between a chimpanzee and a human, his efforts made a world of difference to the final result.

# Foreword

By Don Walsh
President
International Maritime Incorporated
Myrtle Point, Oregon

I cannot think of a more qualified person to write this book. Will Forman's professional involvement with the oceans virtually spans the contemporary history of man's exploration and use of the deep ocean.

We first met nearly three decades ago. A former naval aviator, he was an undersea engineer-inventor and man of strong opinions about his design and engineering projects. Will was fortunate to land in a Navy research laboratory that understood, and exploited, that peculiar breed of non-conformist. His first Navy lab employer was just as unorthodox as his career: the Naval Ordnance Test Station at China Lake in the Southern California desert. The Laboratory's mission was aircraft weapons, Will's was to build submersibles. And that's what he did.

Nevertheless, over the rest of his career, both the Navy and Will profited from their mutually creative relationship. Each needed the other.

At the time we met, I was a naval lieutenant and had just left the Navy Electronics Laboratory in San Diego, where I completed three and a half years as the first Officer-in-Charge of the Navy's Bathyscaph *Trieste*. During this time we had piloted *Trieste* to the deepest place in the world ocean. Understandably, the U.S. Navy was getting a lot of attention for its pioneering work in the deep ocean. But those of us in the *Trieste* Program knew how publicity without a real program behind it could be very perishable, a passing thing.

We needed to sell the Navy on its future in the deep oceans. All of us in the *Trieste* Program looked for sponsors and colleagues within the Navy. Will was one of these, and was already working on design and development of one of the first manned deep submersible to be built in the United States. Work began on his two-man *Deep Jeep* out in the desert in 1961, just a year after *Trieste* had made the deep dive. Once built, he became the *Deep Jeep's* pilot and thereby, one of the first deep submersible pilots in the world.

From this time onwards, Will was involved in just about every deep submersible program the Navy had. Eventually, he moved to the Navy Laboratory in San Diego, where the organization's missions were more ocean-oriented. Here he worked on the use of massive acrylic structures which resulted in the world's first all acrylic pressure hull (i.e., one large window). Will also designed, built, and tested *Deep View*, the first manned submersible to use massive glass structures for the pressure hull. He was also *Deep View's* first pilot.

Will also kept busy in the professional world outside the Navy and was one of the founders of the Deep Submersible Pilots Association in the mid-1960s. DSPA was set up to help develop training and safety standards for deep submersible operations. Many of these standards have been adopted throughout the world. Will has been active in DSPA since the beginning, and continues today as an officer.

In the 1960s and 70s the U.S. Navy was doing almost all the real 'frontier work' on deep submersibles in the world. Will and a handful of colleagues, gave many papers and presentations in the U.S., and abroad, to groups eager to know about these technologies and operations. By the late 1970s, more than 200 manned deep submersibles had been built worldwide. Most of the fundamental engineering concepts were based on work done by the U.S. Navy's research laboratories. Will Forman was right in the middle of much of this work.

After his retirement from Civil Service, Will kept right on going.

He designed and built a man-powered submersible for the "Submarine Races" competition where there's a high premium for design and propulsion efficiency. He also completed the design for a radical new 30 passenger tourist submarine which will have an all-transparent acrylic hull. The hull sections were completed, but lack of funding from his sponsor has put this project on the shelf for now. But there are other projects ahead. Will dreams and designs. In his backyard in San Diego, he tests models of his latest projects in his swimming pool.

In sum, he's not only been a witness to history, but also a major participant from the earliest days of deep submergence engineering and operations in the U.S. An engineer, inventor, and submersible pilot; he's also a great story teller. Indeed who else could tell this story better than Will Forman?

I'm glad he wrote this book. This kind of accounting of where we have been is necessary if we are to decide where we are to go. And that future is not an idle or theoretical question. Manned submersibles will continue to be important for exploring the oceans and for doing certain types of work tasks there. Will Forman has helped to shape the future.

Don Walsh

# TABLE OF CONTENTS

# *The Beginnings*

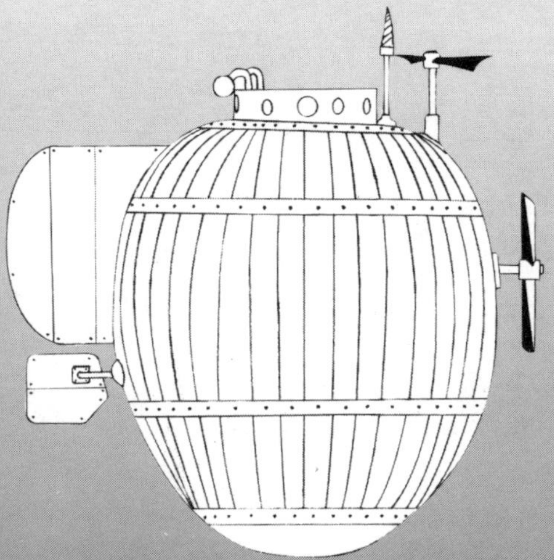

**TURTLE ...**

**NAUTILUS...**

**HUNLEY...**

**ALLIGATOR...**

**INTELLIGENT WHALE...**

**HOLLAND SUBS...**

**LAKE SUBS...**

**BEEBE'S BATHYSPHERE...**

**BARTON'S BENTHOSCOPE...**

***1775–1954***

# The Beginnings

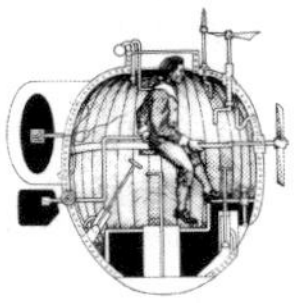

## Turtle (1775)

The first submersible built in America was a tiny, depth limited, homemade vessel constructed by two patriotic brothers in an attempt to repel the Goliath-like British fleet during the American Revolutionary War.

A decade before the Revolutionary War, the average colonist was earning good wages, land was cheap, and personal liberties were free of unnecessary legal and social restrictions. To pay for the French and Indian wars in America and her wars in Europe, England began imposing unfair taxes and forced the garrisoning of her Red Coats in Colonial homes to enforce them. The situation culminated on 19 April 1775 when a Colonial Militia fired on the Red Coats at Lexington and Concord.

That same summer, David Bushnell was a 35 year old senior at Yale University in New Haven, Connecticut. When David's father died and his mother remarried, he sold his inherited share of the property to his brother as a means of entering Yale. In New Haven, his extracurricular activities included the experimenting with underwater explosions for several years and the 2 ounce underwater mine he exploded in 1772 may have been the first of its kind in history. By 1774 he had completed plans for a 150 lb. mine with clockwork timers for delayed firing. It was his intention to blow up and sink the many unwanted

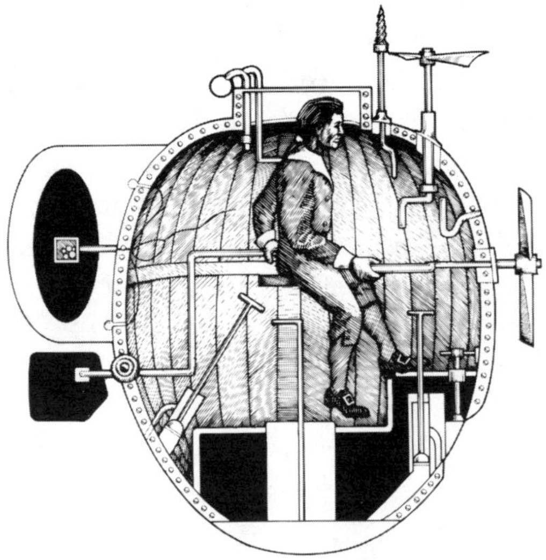

***Name:*** *Turtle* ***Builder:*** *D. Bushnell*
***Length:*** *7'6"* ***Maximum Depth Rating:*** *< 100'*
***Capacity:*** *1 Man* ***Year Launched:*** *1776*
***Description:*** *Cross Section of the Turtle*
Courtesy Naval Undersea Museum Keyport, WA

British ships in the Boston Harbor, but he needed a method of delivery. At spring break, he and his brother Ezra began secretly building a "Sub-Marine" which David designed and called "*Turtle*" because it looked like it was made of two turtle shells clamped together.

Bushnell's *Turtle* was 7'-6" long, 8' high and 4' wide. The construction technique used was similar to that used for making kegs and barrels. Oak planks were laid up vertically to make up the sides of the hull. The vertical hull planks narrowed and were held together at the top and bottom by hoops of iron which were forced over the ends to keep the corked and tarred joints between the planks tight and waterproof. Internally, a strong beam

*Through spies, the British were also told of Bushnell's "machine to blow up ships" – which may have had a good propaganda affect.*

was used to spread and bow the sides apart and keep them from collapsing inward under pressure. At the top of the wooden hull was a brass conning tower and hatch, which looked like "a flat crowned hat with a broad brim." Two small port holes faced forward. In addition, there was one on each side, one aft and one on top. A pair of oar tips 4" wide by 12" long made up the two bladed hand operated propeller with a horizontal axis that was mounted in front of the hull for fore and aft propulsion, depending on the direction of rotation. A smaller, hand operated propeller of similar construction was mounted with a vertical axis located just forward of the conning tower for vertical thrust. A rudder, located behind the pilot outside the hull, was operated with the left hand by a tiller.

A fixed ballast of 900 lbs. of lead was located in the keel. Two hundred lbs. of additional ballast could be lowered 40' or 50' by a hand crank as an anchor or released in an emergency. A foot valve near the floor was used to let water into a ballast tank while a hand pump provided a means of pumping the water overboard. When submerged but near to the surface, two snorkel tubes with check valves provided air. When fully submerged, there was enough air inside the hull for thirty minutes. An 18" long glass tube open at the bottom to the outside water and sealed at the top functioned as a depth gage by measuring the size of the compressed air column in the tube. At the top of the conning tower was an auger for securing a mine to an enemy's wooden hull. It was designed to be screwed into a ship's hull from inside the submersible. The mine attached to the auger would then remain as *Turtle* slipped away from the released auger, with the timer actuated to start a count down.

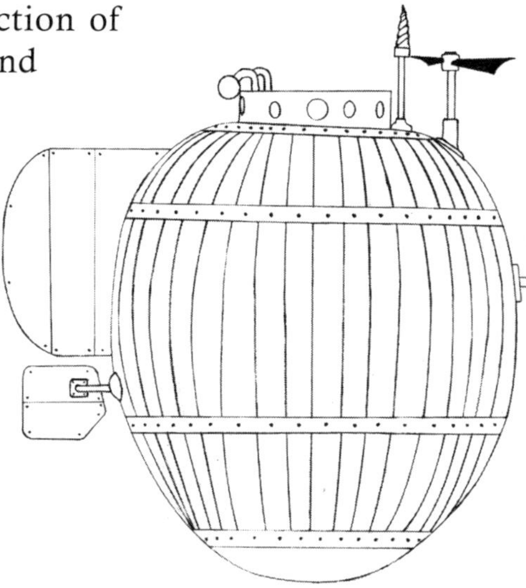

*Turtle Side View*

David Bushnell's 1775 class graduated early due to the war, so he and his brother started testing *Turtle* immediately. David, though sickly, took the first dive at the mouth of the Connecticut River, but after an exhaustive trial decided that his brother Ezra would be a stronger pilot. Ezra practiced using the auger on mock ship hulls till he was proficient. Ben Franklin on his way to visit George Washington at Cambridge stopped by the project and was given a tour of it, which he subsequently reported favorably to Washington. Through spies, the British were also told of Bushnell's "machine to blow up ships" – which may have had a good propaganda affect. Washington, on a trip south to defend Manhattan, stopped by Saybrook, Connecticut where *Turtle* was kept, and was given a tour of the sub on 11 April 1776.

In an effort to defend New York harbor, the *Turtle* was placed on a rented sloop and sailed there. However, Ezra Bushnell came down with a severe fever just as an attack was being readied against the English fleet. He hastily trained Sgt. Ezra Lee, who set off in the sub toward *H.M.S. Eagle*. After some difficulty in navigation, Sgt. Lee got into position, but couldn't set the auger. He discovered too late that he had been trying to screw it into the iron rudder hinge. Daylight caught up and he had to leave. By mistake he headed for the enemy shore. Discovering his error, he released the mine with a 30 minute time delay which scared the enemy away, allowing his friends to pull him to their shore. Upon the exploding of the mine all of the ships in the harbor pulled up anchor and drifted to sea. In time, the British ships returned and *Turtle* was given one more chance to attack, but it too was unsuccessful. A few days later, the sloop with *Turtle* aboard was sunk. The sub was recovered and hidden from the enemy. Short on funds, the Bushnells couldn't repair it and *Turtle* was lost to history.

David Bushnell experimented some more with floating mines and one pair was set alongside *HMS Cerebus*. When spotted, the mines were brought on board and blew up, sinking the vessel with great propaganda impact as well. Another set of Bushnell's mines blew up a barge near Philadelphia on 5 January 1778, terrorizing the blockade located there.

Broke and with no more subs or mines, David Bushnell joined the army as a captain. He was captured and, being unknown, swapped in a prisoner exchange and discharged at war's end. At Jefferson's request, he sent complete details about *Turtle* which Jefferson read to the American Philosophical Society. These were published in 1799.

Shortly after the revolutionary war Bushnell disappeared, but when a Dr. Bush died in Georgia in 1826 and his will was read, it turned out that he was David Bushnell. He had changed his name and profession and lived in Georgia for over 30 years.

Both George Washington and Thomas Jefferson had publicly praised the man and his inventions. Thomas Edison called him "the father of submarine warfare." John Holland called *Turtle* "the most perfect thing of its kind constructed before 1880."

## Nautilus

Robert Fulton left home at age 17 to study art, specializing in miniatures under the patronage of Ben Franklin, who sent him to England at age 21 for tutoring by a famous English artist in 1786. After seven years, he was an accomplished artist whose work was accepted by the Royal Academy. Bored with art, he gave it up and studied canal construction for two years. He devised and patented a new method for replacing water locks, but could not sell these to the English. He went to France for a short visit with a friend, but had an idea to build a submarine and stayed seven years.

Shortly after his arrival in Paris in 1795 and after some months of persistence, he convinced the French Directory to fund the building of his submarine *Nautilus* for "lessening the power of the British fleet," that remained France's periodic enemy.

Fulton described the geometry of the 16'-9" long by 4'-9" wide copper pressure hull as an imperfect ellipsoid with an interior volume of 212 cu. ft. A three or four man crew operated the sub, using a hand crank to turn the shaft of a propeller at the stern. Like *Turtle*, a steel auger was used to attach a mine to an enemy ship before pulling away to allow a timer to set off the charge. Initially, a fan-shaped sail was used for surface transport; later, a 12' pivoting mast with a main sail and jib was employed. A lead weight of 360 lbs. that could be lowered by hand-winch on a 130' line was used as an anchor.

Five years later, *Nautilus* was first tested manned in the Seine river on 13 June 1800. After a number of test dives, several significant innovations were added: a revolving periscope, a surface float with an air tube connected to the sub for extended operations while submerged, a vertically oriented propeller located near the bow for depth control plus several spare propellers with different configurations. A 3' diameter, two bladed propeller was found to be more efficient than either a four bladed one or an Archemedian screw.

On 12 September 1800, Fulton and a crew of three left the port of La Havre near the mouth of the Seine to attempt a daring voyage to cross the Bay of Seine to Cherbourg, a distance of seventy miles in the open sea. For three days, the four men traveled across the stormy waters on the southern boundary of the English Channel beyond sight of land. Several times, the crew remained submerged for six hour periods to avoid the stormy surface seaway, using the tube and float for air. The continuously stormy weather finally forced them ashore near the small town of Gowan where the sub was left unattended all winter.

Due to favorable reports about the Bay of Seine operations by influential friends, Fulton was able to meet Napoleon and was granted 10,000 francs for refurbishment and additional tests. The electrolysis damage between the copper hull and the steel interior components took two months to repair or replace. Fulton added a small window and a 1 cu. ft. copper sphere containing 200 cu. ft. of compressed air which allowed him and three companions to stay submerged, using candles for light, for six hours without the use of the tube and surface float.

The French revolutionary war was still active in early 1801. Trailing a contact mine containing 20 lbs. of gunpowder behind *Nautilus*, Fulton submerged, dragged it into a sloop and blew it 100 ' in the air. The event was reported to Napoleon by the Admiral assigned to witness it, along with a request for another grant to build a new and better sub.. Disgusted with waiting and anxious that *Nautilus* not be copied and the plans offered for sale to others, Fulton disassembled her on 20 April 1801. He remained in France a little over two years more trying to sell his plans. Finally, in 1803, Napoleon sent word that "he wanted never to hear about his submarine again." Apparently an indefatigable salesman, Fulton then proposed his steamboat idea to Napoleon, but this too was ignored.

However, the English, who declared war against France in 1803, knew about *Nautilus* and made a secret offer to Fulton.

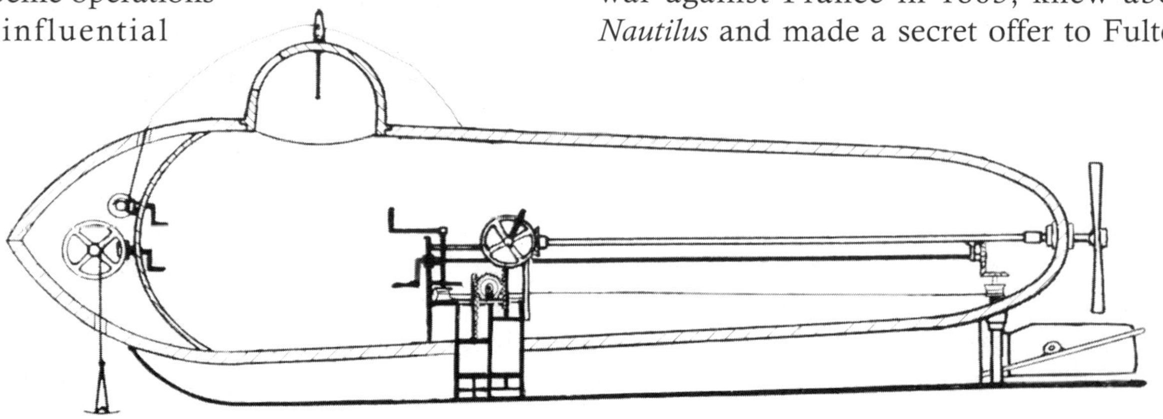

***Name:*** *Nautilus* ***Builder:*** *R. Fulton* ***Length:*** *16' 9"* ***Maximum Depth Rating:*** *< 125'*
***Capacity:*** *3-4 Man* ***Year Launched:*** *1800* ***Description:*** *Cross Section*

He accepted and departed for England. On 20 July 1804, he received a contract to build a new sub for a fee of 200 lbs. sterling a month, plus free materials. He was also offered a percentage bonus for any French tonnage sunk.

At that time, Napoleon had 120,000 men and 3,000 launches ready to invade England whenever the weather cleared. Fulton and a crew took advantage of the delay and floated two mines connected by a line from upwind of the French fleet and blew the French *Dorthea* into smithereens. Fulton was paid. But, for the loss of *Dorthea* and other causes, the invasion fizzled. However, a disagreement over further payments angered him, so he left England for the U.S.A.

After 20 years away from home, the 42 year old artist/inventor/engineer/adventurer/ salesman arrived in New York harbor on 13 December 1806. With a population of 80,000, New York was still the nation's capitol. Still humiliated by the memory of defeat by an ungrateful colony, the British fleet had begun stopping American ships at gunpoint, taking off English sailors who had signed with the Americans. In the process, they were not especially particular and kidnapped many American sailors as well. One of the eastern newspapers announced Fulton's return and, as his exploits with mines and submarines in England and France were well known in America, welcomed him home as a needed addition to the country in its defense of harbors and coastlines by means of submarine systems.

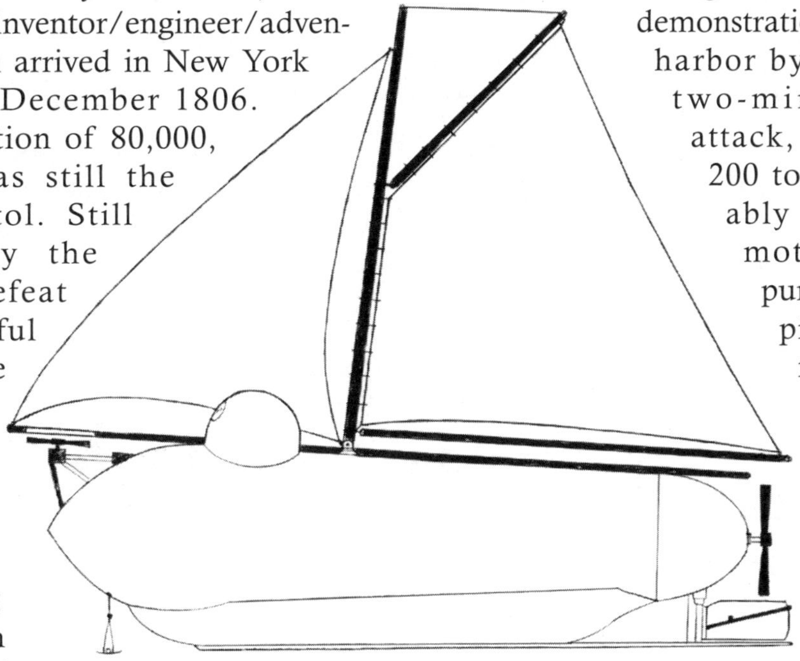

*Nautilus, 1801, Side View with Sail*

Indeed, Fulton had only been back six months when, on 22 June 1807, *H.M.S. Leopard* (from whose small boats he had helped blow up *Dorthea*) intercepted the *U.S.S. Chesapeake* a few miles off Virginia Capes. After a brief exchange of gunfire, *Chesapeake* hove to and four seamen were removed by a British boarding party. The public and press were outraged and demanded revenge, even war, but as yet, the U.S. had nothing afloat to match British seapower. Fulton's response — through friends — to Secretary of Navy Dearborn was to offer to blow up British warships with mines as he had done to the French. President Jefferson asked the Secretary to arrange and report on a demonstration in New York harbor by Fulton of his two-mines-on-a-line attack, this time on a 200 ton brig, presumably selected from mothballs for the purpose. The firing pins failed in the first try. In the second, the mines missed the target and exploded 100 yards off. A third attempt blew up the target, breaking it in two.

Fulton wrote an extended report to Jefferson suggesting improvements to his boat launched system. Jefferson replied on 16 August "Your letter of July 28 came to hand just as I was leaving Washington. It has not sooner been in my power to acknowledge it...I consider your torpedoes as a very valuable means of the defense of harbors [but] not solely to be relied on." Jefferson went on to show much more interest in ship-sinking by submarines and suggested the development of a submarine fleet along with an entire corps of "Submarine Engineers" to operate them.

Clearly not one to sit on his hands — or his laurels — Fulton did not respond to the Presidential request, having in the brief period since the demonstration, turned his attention to steamboat design and, of course, went on to fame and fortune with his historic development and voyage of the steamboat *Claremont*.

Apparently, however, he never lost entirely his love for submersibles, with assets from the *Claremont*, Fulton began constructing an 82' long semi-submersible called the *Mute*, but died in February 1815 while working in poor health. The extent of Fulton's work in the development of submarines did not come to light until 1920 when his papers were purchased from a private collection by his biographer, William Parsons.

## Hunley

During the American Civil War, the Union Navy set a blockade around the Confederate States. Out of the many Confederate efforts to combat the blockade, came one of the first submarine casualties of war. Named after its designer Horace Hunley, it was made from a 4' diameter by 25' long boiler tank. The tank was cut lengthwise into two half round shells 25' long. Twelve inch wide by 25' long plates were added at the cut line of the two halves to form an oval cross section 5' high by 4' wide. A "round end wedge" was added at each end of the tank to enclose the ends and form a pressure hull. Flat plates that fit the oval and reached from the floor almost to the ceiling at each end, served as tank ends for ballast and pitch trim control. The fact that a gap was left at the top of the tank plates contributed to some of the sub's casualties that were to follow. A hand-operated pump for draining and a valve for flooding were provided for each ballast tank. The sub

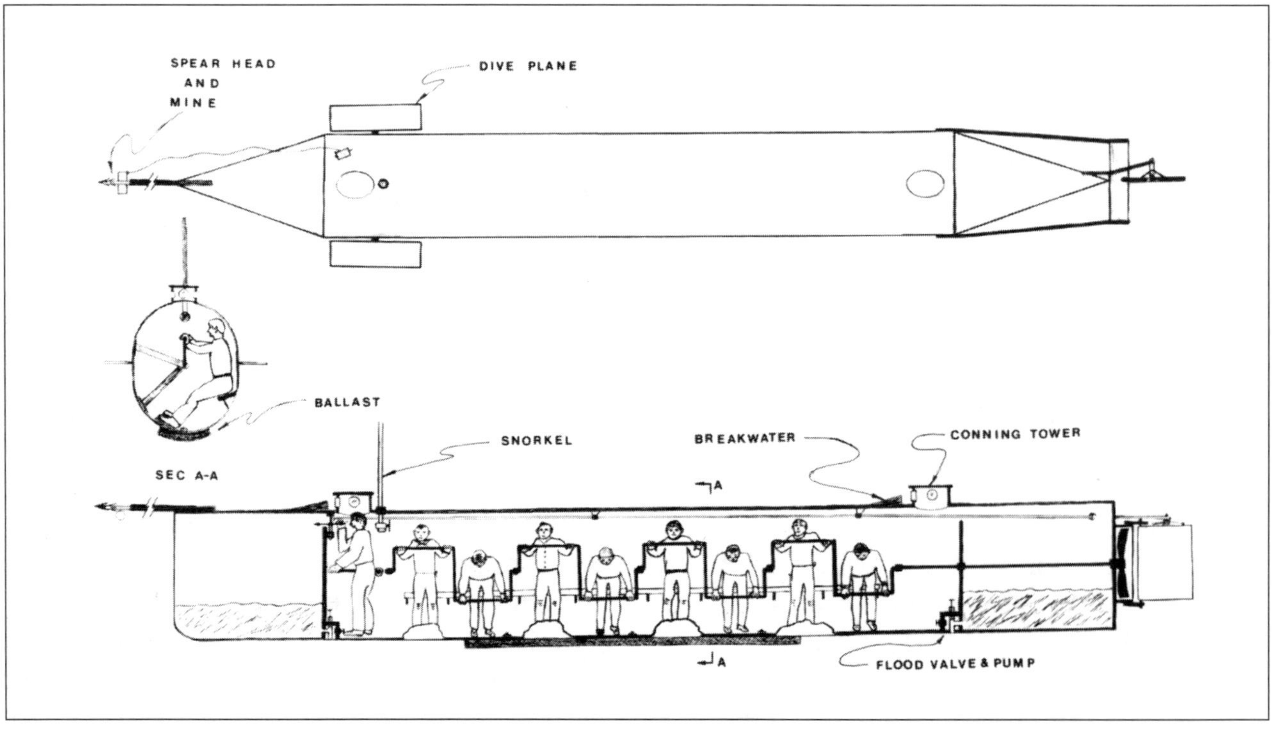


***Name:*** *Hunley* ***Builder:*** *H. Hunley* ***Length:*** *40'* ***Capacity:*** *9 Man*
***Maximun Depth Rating:*** *< 75'* ***Year Launched:*** *1862* ***Description:*** *Top and Side View*

was 40' in length including the propeller and rudder. Small conning towers, with forward and side viewports and hatches were located at each end of the sub. One was forward for the captain/pilot and one was aft for the ballast control officer who also helped with the hand cranking for propulsion. A 4' long snorkel pipe was located just aft of the forward conning tower. Below the snorkel on each side of the hull were the 6' long dive planes operated by the captain/ pilot. Aft of the stern ballast tank was the shrouded 2 1/2' diameter, 3 bladed propeller positioned in front of a 2' long rudder controlled by the pilot from his forward position.

*Hunley, Artist's 1861 Sketch*
Courtesy of the Museum of the Confederacy, Richmond, VA

A crew of eight men – including ballast control officer – propelled the sub at speeds of up to 4 mph by hand crank from a semi-standing position with their backs against the hull. An external cast-iron ballast weight at the bottom of the hull could be unscrewed from inside and dropped in an emergency. The captain/ pilot near the forward conning tower steered by a ship's wheel, adjusted the forward ballast by pump or valve and hand operated the dive planes, guided by the candle-lit compass and glass tube depth gage as well as viewport sightings.

During the last dive in the first test series at Mobile, Alabama all nine of the crew drowned. The sub was raised, put on two flat cars and moved to Charleston, South Carolina. In a second trial there, while surfaced with the hatches still open, a ship's wake flooded the sub and only the Captain, Lt. Payne, got out alive. A third tragedy, the second at Charleston, occurred when the Navy ship that *Hunley* was tied up to took off without notice. With *Hunley* out of trim and the hatches open, she sank. Only Payne and two others got out alive. Payne never dived *Hunley* again and presumably looked for better work in another part of the war.

One of the escapees was the ballast control officer, Lt. Hasker, who was at the aft conning tower when the sub went down. Forty years later, Hasker described to Simon Lake, founder of the Lake Torpedo Boat Company, how he escaped by waiting near the hatch for the sub to flood and the air and water pressure to equalize. He then opened the hatch in forty-two feet of water and made what may have been the first submarine escape by a crewman. Lake was instrumental in building some of the first U.S. Navy submarines in the early 1900s.

The next test dive was on 15 October 1863, and was made near Charleston with both Horace Hunley and the builder Thomas Parks aboard along with a crew of sevenfrom Mobile Alabama. Since *Hunley*'s ballast tanks were open to the crew section, all nine were drowned when the forward ballast tank flood valve broke off. On 17 February 1864, the fifth crew of *Hunley* under Lt. George Dixon attacked and sank the Union's 16-gun steam sloop, *Housatonic.* using a mine on the end of its bow sprit. A year after the Civil War divers located the sub near the wreck of the *Housatonic*, unflooded with a suffocated crew. It was not feasable to recover it so they left it in place. This final event brought the total number of *Hunley* casualties to forty-one. Only five of the *Housatonic* crew were killed. *Hunley's* sinking of the *Housatonic* was the first time an enemy ship was sunk by a submarine in wartime.

In May of 1995, a crew hired by the author Clive Cussler located the hull of *Hunley* several hundred feet from where *Housatonic* sank. Custodial care has been assigned by the U.S. Government to the State of South Carolina which intends to recover it. However, the estimated several million dollar cost has slowed the progress.

## Alligator

In 1859, a Frenchman by the name of DeVilleroi moved from his homeland to Philadelphia and began building a submarine. He remained unnoticed in America until shortly after the Civil War began in April 1861. DeVilleroi had been recognized for his submarine work for the French Navy 30 years previously when he built a 10' long sub that was hand propelled. It had been used for diving and maneuvering demonstrations for political and military leaders in the Seine river, near Paris, in 1835 and later during the Crimean war in 1854, without finding funding either time.

Soon after his arrival in the U.S., he interested a wealthy Philadelphian in funding a submarine for salvaging valuable cargoes from the many wrecks that lay sunken in American coastal waters. The submarine, named *The Propeller*, was built and tested over a two year period near Philadelphia. Two systems on the sub were unique: a hatchway from which extravehicular trips by the crew could be made under water and a system for supplying "manufactured air" which permitted the crew to stay submerged for up to three hours.

On 17 May 1861, just 35 days after the Civil War broke out, two young men were arrested as spies for towing the salvage submarine into the Philadelphia harbor. A newspaper article described the submarine as "a 20' long cylinder with tapered ends having a propeller at the aft end, appearing overall as a large cigar...near the forward end was a hatchway...Two tiers of glass bullseyes existed along each side of the submarine affording light to the inside." The article further described the hand crank propulsion, the rudder and dive planes, pumps for ballasting and a large lead pig for fixed ballast in the keel.

In due time, the innocent young men were released. The newspaper coverage brought DeVilleroi's sub to the attention of the Commanding Officer of the Philadelphia Navy Yard who reported a description to the Secretary of the Navy. After *Propeller* went through some sea trials, a contract was given to DeVilleroi for $14,000 on 1 November 1861 for a larger version to be built in just 40 days. The new submarine was launched on 30 April 1862, propelled by a total of 16 oars all located at the vertical centerline of the hull, eight on each side. "Each (oar) was constructed on the principle of the webbed foot of a duck, the blade being formed in two pieces, opening and closing like a book." The crew consisted of 16 men, plus the commander. The new sub was named *Alligator* (and, later, nicknamed *Lincoln's Alligator*). Its length was 47'-0" with an oval cross section 4'-6"wide by 6'-0" high. Dive planes were located fore and aft, a hatch and observation dome were located topside near the bow and another hatch was located in the bottom portion of the bow for underwater excursions.

*Alligator* was sent up the James River to City Point, near Richmond, with 20 barrels of powder to support McClellan's land forces, but was returned to the Washington Navy Yard due to the shallowness of the river and belated fear that the enemy would capture it. A screw propeller was installed which increased its maximum speed to 4 kts. *Alligator* was next ordered to Port Royal, South Carolina to aid in the taking of Charleston. On 18 March, President Abraham Lincoln observed the submarine during its pre-cruise operations. A month later, on the way to Port Royal under tow in a violent storm, one of two hawsers snapped. The towing ship lost control, making it necessary to set the unmanned sub adrift. It sank soon afterwards, on 2 April 1863.

## Intellegent Whale

Oliver Halstead sold a submarine to the Navy in the 1870s called the *Intelligent Whale*. This sub had a tangled career. Shortly after the loss of the *Alligator*, Scovel Meriam designed the *Intelligent Whale* for Union purposes. Construction of the submersible was initiated by Augustus Price and Cornelius Bushnell, but before completion, it was sold in 1864 to the American Submarine Company, founded by Halstead. After sale and completion of construction and some unfortunate trials, the sub was inactive. Years of litigation to determine ownership followed.

The submarine was 28'-8" in overall length, had a beam diameter of 7' and a

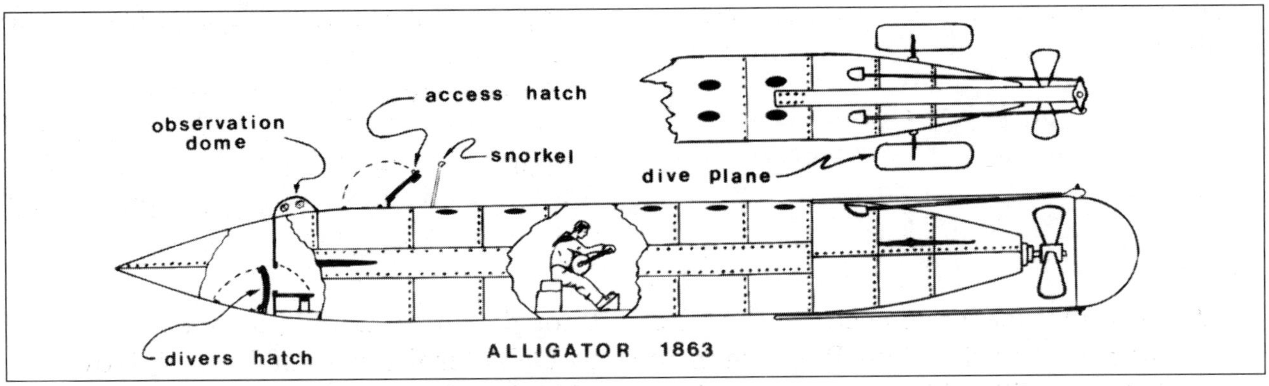


***Name:*** *Alligator* ***Builder:*** *DeVilleroi* ***Length:*** *47'* ***Capacity:*** *17 Man*
***Maximun Depth Rating:*** *< 75'* ***Year Launched:*** *1862* ***Description:*** *Side View with Cutaways*
Courtesy of U.S. Naval Historical Center, Washington, D.C.

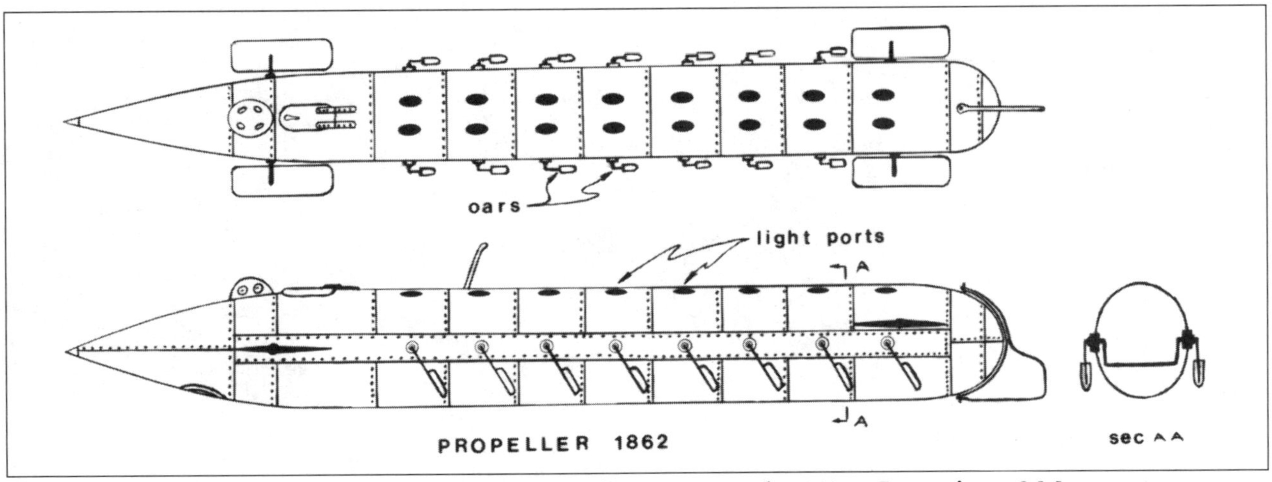


***Name:*** *Propeller* ***Builder:*** *DeVilleroi* ***Length:*** *20'* ***Capacity:*** *9 Man*
***Maximun Depth Rating:*** *< 75'* ***Year Launched:*** *1861* ***Description:*** *Top and Side View with Cutaways*
Courtesy of U.S. Naval Historical Center, Washington, D.C.

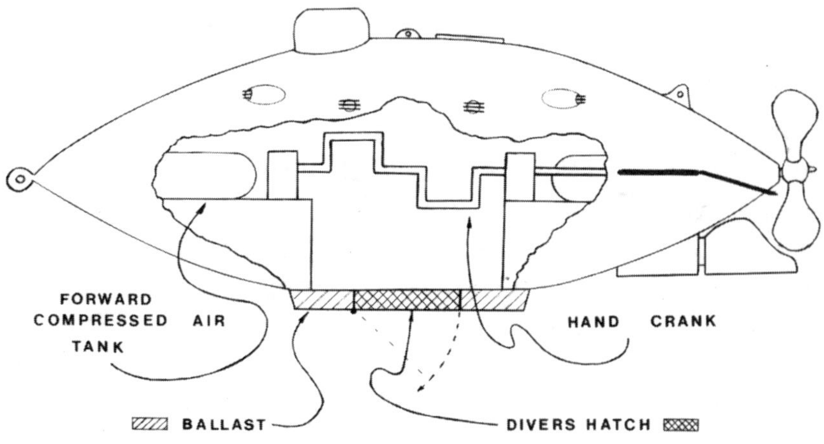


***Name:*** *Intelligent Whale* ***Builder:*** *American Submarine Co.*
***Length:*** *28'8"* ***Capacity:*** *13 Man* ***Maximun Depth Rating:*** *< 150'*
***Year Launched:*** *1868* ***Description:*** *Side View with Cutaways*

height of 9'. Compressed air and hand pumps were used for removing ballast water from on board, and flood valves were used for adding ballast. The internal air volume was considered adequate for ten hours for the normal compliment of six, but could hold up to thirteen crew members for limited periods of time. The pressure hull was fabricated of $1/2$" thick boiler iron. The hand cranked vessel had horizontal and vertical control surfaces at the stern for submerged maneuvering at up to 4 kts.

During the sub's first sea trials in the late 1860s, Army General Thomas W. Sweeney with two others made a pier side test in 16' of water. After equalizing the interior pressure to the outside water pressure, the general in a divers suit exited through a hatch, placed an explosive charge on a nearby barge and returned to the sub. He set off the charge remotely by lanyard and the scow blew up and sunk.

As a result of recommendations based on the test, the Navy offered to buy the sub with partial payments to be made if and when various sea trials were satisfactorily performed. On 27 May 1870, the Navy took possession of the sub with a downpayment of $12,050. It was just as well for Halstead that he was paid in advance. During the acceptance test, the sub flooded, but all of the crew got out. No further payments were made and the sub lay neglected in the Brooklyn Navy Yard for almost 90 years, until exhibited briefly at the Navy Museum in the Washington Navy Yard in 1960. It was in the museum's storage building in 1995.

*Intelligent Whale*
Courtesy of the U.S. Naval Historical Center, Washington, D.C.

# John Holland's Many Submersibles

In 1888, 18 years after turning its back on *Intelligent Whale*, the Navy published an invitation for design bids for a submarine. The two main entries were from a Swede named Nordenfeldt and a forty-seven year old Irish immigrant named John Holland. Nordenfeldt had built a steam-powered submarine in England that would build up a head of steam at the surface and dive briefly until it lost steam pressure and propulsion power. Holland, however, had built a string of submersibles over the previous decade and won the contract on the strength of experience.

After emigrating to America, he worked as a teacher in Patterson, New Jersey, while associating with an Irish resistance group called the Fenians. At the time, the Irish rebellion against England was supported very heavily by the many Irish immigrants in the United States. As a means of breaking the British blockades of Irish seaports, Holland demonstrated to the Fenians at Coney Island, New York, a 30" model submarine propelled by clockwork and spring power that was complete with a miniature hull, dive planes, rudder and propellers. The Fenians funded Holland to build a full scale version for $5000.

Called *Holland No.1*, construction began in 1876. The steel sub was 14'-6" long with a 3' beam and a height of 30", not including the turret. It was launched in the summer of 1878. Most systems worked well, but the mid-ship dive planes were useless and stern planes were indicated. Once this experimental sub had proven the design to be correct, the hull was deliberately scuttled, to be recovered 50 years later and put in the Patterson Museum.

With an additional $13,000, the Fenians or "Irish Associates" then sponsored a bigger sub, which Holland designed in 1878. Constructed in 1879, it was named *The Fenian Ram*. The *Ram* weighed about 19 tons and was 31' long by 6' in diameter by 7'-4" in height, tapered to a point at both ends. The hull was 11/16" steel and its gas engine was rated at 15-17 H.P. which drove the sub at 8 kts. The crew consisted of a pilot, an engineer and a gunner. Two joysticks controlled the rudder and dive planes. The engineer adjusted ballast and buoyancy by the operation of flood valves or air valves from a compressor.

Both the pointed bow and stern were used for compressed air reservoirs, water ballast tanks were located adjacent to and inboard from the air reservoirs. The crew compartment was centrally located amidships. The gunner operated a 9" diameter pneumatic launching tube from the crew quarters for launching the newly designed Ericson torpedoes.

The first test dive series was made in June, 1881. These tests indicated a need for a lantern for deep operations, that submerged operations with the engine running were practical, that torpedoes could be fired while submerged, and that diver lock-outs could be made through a bottom hatch by pressurizing the crew compartment first.

The next dive series consisted of two dives a day consisting mainly of repetitional practice runs. Between tests, Holland spent his spare time supervising the construction of a 16 ton, 16' smaller replica of the *Fenian Ram* which incorporated improvements learned from the on-going trials.

Trouble developed between the Fenian organization and Holland when they failed to pay him what was promised. The Fenians stole the 16' replica and the *Fenian Ram* itself, but were incapable of running them, accidentally sinking the replica in the

New Haven Connecticut Harbor. Noting their inexperience, the Harbor Master enjoined them from operating the remaining *Fenian Ram* in the harbor. Holland himself was unable to claim part ownership or back pay.

Without his submarines, Holland found employment as a draftsman in New York City. While there he met a Captain Zalinski of the Army who was an instructor of Military Science at Michigan Institute of Technology (MIT). Zalinski offered to be partners and financial promoter with Holland on his next sub if it included one of Zalinski's new dynamite guns. The partnership began.

During 1884, the Zalinski sub was assembled near Brooklyn New York. It was 50' long with an 8' beam and constructed of iron frames covered by a wood shell. The Zalinski dynamite gun was designed to fire a 200-500 lb. explosive projectile almost half a mile. The sub hull was configured as "a spindle of revolution" that was bigger and slimmer than the *Fenian Ram*, with a conning tower from where the pilot could

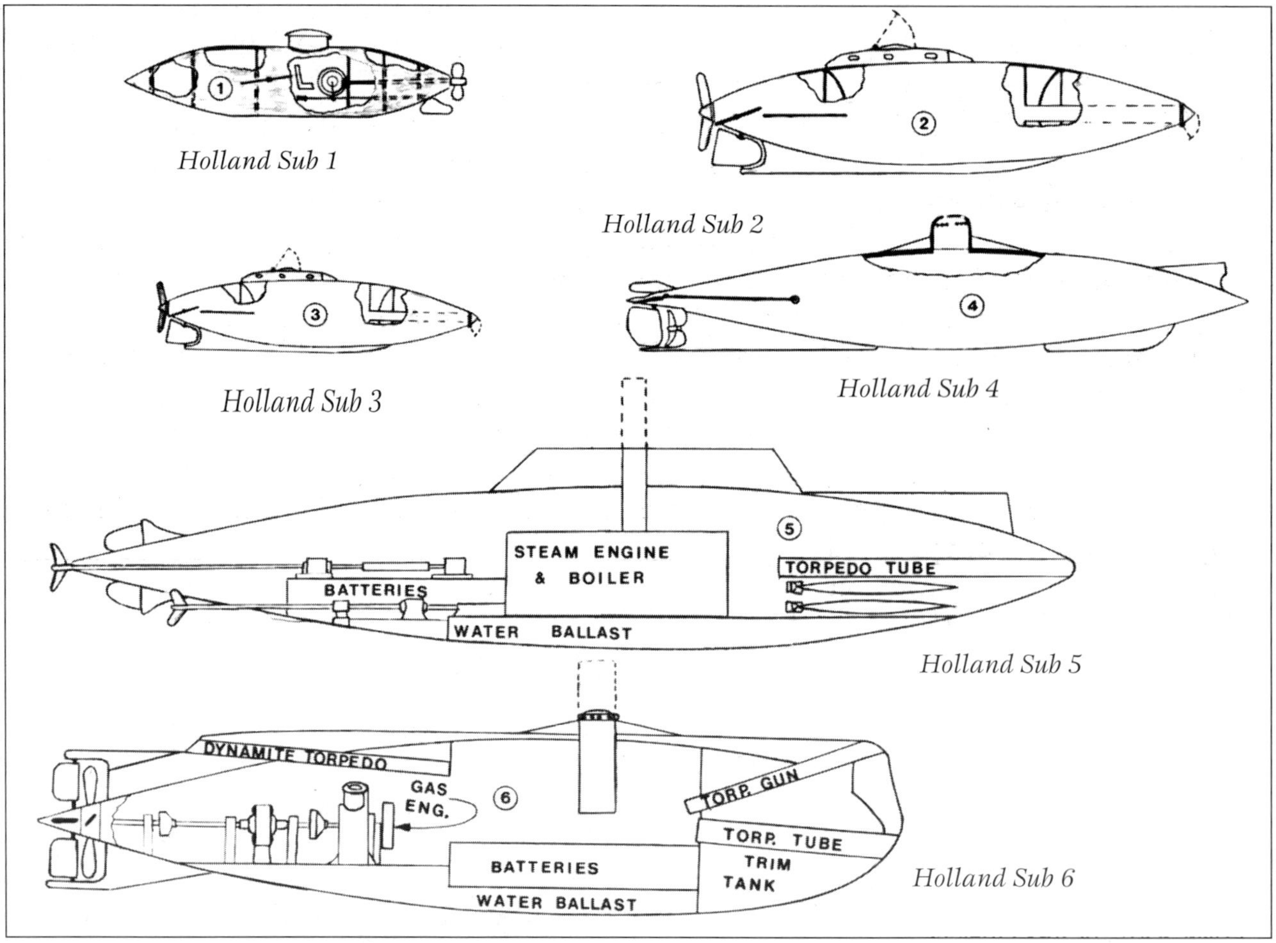


***Name:*** *Holland Subs 1-6* ***Builder:*** *J. Holland*

| ***Name*** | ***Length*** | ***Capacity*** | ***Maximun Depth Rating*** | ***Year Launched*** |
|---|---|---|---|---|
| *Sub 1-Holland #1* | *14'6"* | *1 Man* | *200'* | *1878* |
| *Sub 2-Fenian Ram* | *31'* | *3 Man* | *150'* | *1881* |
| *Sub 3-Fenian #2* | *16'* | *1 Man* | *200'* | *1883* |
| *Sub 4-Zalinsky Boat* | *50'* | *4 Man* | *150'* | *1885* |
| *Sub 5-Plunger* | *85'* | *5 Man* | *125'* | *1897* |
| *Sub 6-Holland VI* | *53'4"* | *4 Man* | *150'* | *1898* |

navigate and observe targets as well as handle all the controls. At the launching in September 1885, the sub flew off of the ways into some pilings causing severe damage. It was disassembled and sold for parts to reimburse the investors.

Although John Holland won the 1888 Navy Submarine design contest, the money for construction of the sub was diverted to finish a new cruiser. Congress passed an appropriation in 1892 to reopen the contest which Holland won again. On 3 March 1895, the John P. Holland Torpedo Boat Co. received $200,000 for construction of the Navy submarine *Plunger*.

The original plans for *Plunger* called for it to be 85' long by 11'-6" in diameter and to have a submerged displacement of 168 tons. Holland was secretary/treasurer of the Holland company, but had been forced to sell a majority of the stock to finance the project. In 1896, he realized that the *Plunger* construction was not going at all right. Changes in construction that he wanted were not being approved while changes by Navy brass, who knew nothing about submarine designs and operations, were approved and incorporated until there was no chance that the sub would work. Two unnecessary propellers had been added and the boiler selected was an oversized, overheating monstrosity. In general, the *Plunger* was the product of a committee of amateurs. The submarine was all wrong.

By early December even the Holland Company's majority owners could see that. Now, they granted his request to be released from work on the *Plunger* to start construction of his 6th sub to be known as the *Holland VI*. The company would honor the *Plunger* contract, but felt that any new contracts could only come as a result of an all-Holland design rather than repeat the *Plunger* fiasco. *Plunger* was launched on 7 August 1897 but proved even at dockside trials to be useless, so it was abandoned. The funds received for *Plunger* were returned to the Navy.

Holland continued work on the *Holland VI*. Managerial problems were eliminated and he was able to get the best people that he knew of, to do the most meticulous work possible on the culmination of all his dreams. Upon completion, the *Holland VI* was 53' long by 10'-3" in diameter. The internal frames were circular at "a little more than a foot apart." A flat deck, used as a walking platform, sat on top of the pressure hull which concealed exhaust pipes and external hardware, such as anchors and lines. A 24" diameter turret extended above the pressure hull and contained the only entry hatch and the only place from which the sub could be operated. The sub was propelled, at the surface, by a 50 H.P. gas motor and lead acid batteries submerged.

*Holland VI* Courtesy of U.S. Naval Historical Center, Washington, D.C.

Two dive planes aft, a single rudder and a 3 blade ships propeller provided the maneuvering and propulsion. There was no periscope since good ones had not been invented yet. The armament consisted of one 18" torpedo tube with 3 torpedoes and two 8" dynamite guns.

The sub had a surface speed of 7 kts., a submerged speed of 5 kts., a range of 1500 miles surfaced or 65 miles submerged. *Holland VI* made her first dive on 15 February 1898. A series of Navy acceptance tests were completed on 6 November 1898. The Navy officially purchased the *Holland VI* on 11 April 1900, for $150,000 and commissioned it as the *U.S.S. Holland* (SS-1). Almost immediately, Japan bought three of the *Holland's* and England bought five in 1902 – the first submarines owned by the Royal Navy. The vast majority of all U.S. submarine types built until WW I were based on Holland's design. The Holland submarines rapidly gained a reputation for excellence. Admiral Dewey, the hero of the Spanish-American War of 1898 said that, if the Spanish had possessed two submarines in Manila, I never would have held it with the squadron I had."

John Holland's early submarine patents were released to the Electric Boat Company in the process of selling the Holland subs. After his business agreement with Electric Boat expired, he was able to obtain new patents for improvement in submarine design. However, Electric Boat managed to tie him up in court, preventing him from getting new financial backing and hence competing again. Holland claimed foul play to the Navy, but got nowhere. Broken in body and spirit, he died in 1914 while others reaped the profits of his genius.

***Name:*** *Argonaut* ***Builder:*** *S. Lake* ***Length:*** *36'*
***Capacity:*** *2-10 Man* ***Maximun Depth Rating:*** *< 200'*
***Year Launched:*** *1897* Courtesy of Maryland Historical Society

## Simon Lake

Born in 1868, Simon Lake began dreaming of and designing submarines when he was 11 years old after reading *Twenty Thousand Leagues Under the Sea.* His favorite experiments as a teenager were to see how long he could stay in the trapped air in an overturned canoe. He later claimed that most of the "submarine elements" of his successful designs were the same as those he made at age 14. At age 19, he read of the Navy contract that Holland won. Lacking funds to construct a full size submarine, he decided to build a miniature one, using money borrowed from his Aunt and Uncle.

Called *Argonaut Jr.*, it had all of the features Lake thought important, yet was designed so that he and his cousin could build it themselves. It was shaped like an old-fashioned flat iron. From the flat broad fantail the sidewalls bowed slightly to a point at the prow. The deck was flat with a small square box for a conning tower on top. Port holes fore and aft in the conning tower, on each side of the prow and a third of the way forward from the stern, provided light and external viewing. The sub was 14' long by 5' high, plus the height of the conning tower. It was equipped with 24" diameter wheels located near the bow on both sides. In addition, a set of double wheels were located at the stern, placed close together and made turnable like a tricycle wheel for steering while on the sea bottom. The hull was made of two layers of yellow pine planking with waterproof canvas between them. The outer planks were caulked to complete the waterproofing. The vessel displaced about 7 tons submerged.

Propulsion while on the bottom was achieved by foot-powered bicycle pedals which turned the big wheels at the bow. A pressurized air cylinder was used to build the pressure up in the interior of the sub to equal the outside sea pressure so that a bottom hatch could be opened and bottom objects such as clams, oysters and rocks could be recovered.

Lake's demonstrations at Sandy Hook, New York, in 1894, enabled him to sell $2,500 in stock to finance his next submersible *Argonaut* which, after two years of design and construction, was launched the same day and month as *Plunger*: 7 August, 1897. Shaped like an iron dirigible, *Argonaut* had a diameter of 9' and a length of 36'-9". Like *Fenian Ram*, it was equipped with an internal combustion engine. The 30 H.P. gas engine powered a propeller when in mid-water or, when on the bottom, turned two 7' diameter wheels. These were located about a third of the way aft from the prow giving *Argonaut* the appearance of a large wheeled, farm tractor. Two long, hollow masts served as air inlet and engine exhaust. Batteries provided power for lights, and a lock-out trunk with a hatch permitted viewing or visiting the bottom and recovering objects.

Although youthful, Lake appears to have been an aggressive promoter. The fiery, freckled red-head would take reporters, scientists, Naval personnel and social or political figures on dives to look at, photograph or bring back samples from the bottom. Still he could not sell *Argonaut* or obtain a new construction contract. When the Navy mined Hampton Roads during the Spanish American War, Lake submerged and made a map of the mines to show how he could have disarmed them or could clear the enemies mines in Santiago de Cuba Harbor. The Navy turned him down and the Army ordered him away from all U.S. mine fields.

In disgust, he cruised *Argonaut* from Norfolk, Virginia to New York in 300 miles of open ocean without escort in the worst storm in years.

***Name:*** *Protector* ***Builder:*** *S. Lake* ***Length:*** *65'* ***Capacity:*** *12-15 Man*
***Maximun Depth Rating:*** *< 200'* ***Year Launched:*** *1902*
Courtesy Navy Photographic Center, Washington, D.C.

Over 200 ships were lost in the area in that one storm, but Lake lashed himself to the helm and made it safely, soaked to the skin and covered with ice. As a result, he emerged a hero to the press. His childhood hero, Jules Verne, congratulated him from France.

Lake then redesigned *Argonaut* as a "seagoing submarine" by adding 20' to the pressure hull and a lightweight superstructure. The features were soon copied by German, Italian, Russian, and French submarine designers. He moved his operations to Bridgeport, Connecticut where he invited thirty Bridgeport notables, including bankers, the mayor, and other prominent men of the city, to an "undersea picnic." He submerged the *Argonaut 2* to the bottom with the vent masts sticking out of the water and proceeded to let his guests act like overgrown boys, picking shellfish from the bottom, making chowder, passing the jug and pretending they were diving with Captain Nemo. Two hours overdue, he made for port – but not until he had obtained pledges for all the money he needed for his next sub, *Protector*.

*Protector* was completed in 1902, also with the intent to sell it to the U.S. Navy. It ran on a gasoline engine at the surface and batteries submerged. It was 65' long and displaced 130 tons submerged. It dived while maintaining a level keel by using both bow and stern planes simultaneously –

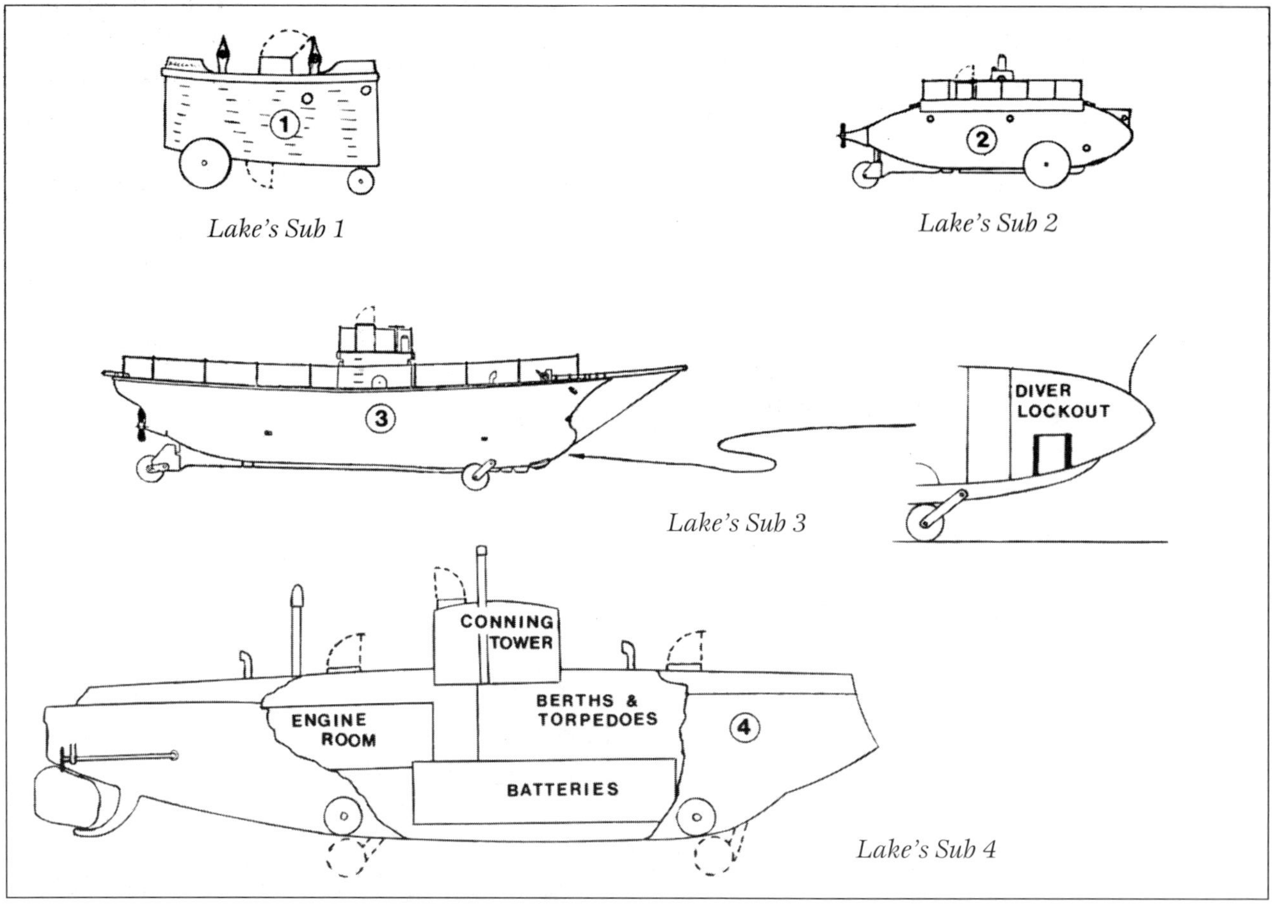


**Name:** *Lake's Subs 1-4* **Builder:** *S. Lake*

| ***Name*** | ***Length*** | ***Capacity*** | ***Maximun Depth Rating*** | ***Year Launched*** |
|---|---|---|---|---|
| *Sub 1-Argonaut Jr.* | *14'* | *1-2 Man* | *< 50'* | *1894* |
| *Sub 2-Argonaut I* | *36'* | *2-10 Man* | *< 200'* | *1897* |
| *Sub 3-ArgonautII* | *56'* | *1-31 Man* | *< 200'* | *1899* |
| *Sub 4-Protector* | *65'* | *12-15 Man* | *< 200'* | *1902* |
| *Sub 5-Seal* | *161'* | *24 Man* | *< 250'* | *1912* |

unlike the Holland subs where the bow and stern plane operated opposite one another so that during a dive the bow would pivot downward as the stern pivoted upwards. Submarines today can do both but prefer the Holland techniques as it allows the most rapid means to dive or surface.

The *Protector* was designed to fill needs expressed by the Secretary of War, which it easily met. Nevertheless, political sculldugery involving special interest groups and hindrance lawsuits by Electric Boat had it turned down on two occasions.

Four or more foreign governments were lined up at the door of Lake's office so he sold the *Protector* for a great price to the Russians with a guarantee to build them five more in their country. The batteries and sub had to be shipped separate by dead of night to avoid restraining orders lobbied into law by the competition. Lake and his crew made their way to Kronstadt, Russia by separate passage and stayed seven years to build subs and train crews. Lake built Austria's first two subs and took out German patents in preparations for building subs for Germany. The Germans used the patents as blue prints and built their own without him. He then tried to sell his tenth sub design to the U.S., but was again turned down, so he sold it to the Russians.

Lake next advertised his latest sub publicly, targeting the U.S. Congress and showing that the specifications for his submarine Seal exceeded those of subs currently being purchased. He stated that, if the specs were not met, the sub was free. The politicians, influence peddlers and special interest groups could no longer prevail and a Navy contact was let to Lake.

On 28 October 1912, the Navy accepted the submarine *Seal*. It had a surface displacement of 400 tons, a length of 161' and a beam of 13', carried six torpedoes, a crew of 24 and tested to a depth of 256'. The Lake Torpedo Boat Co. sold a few submarines to the Navy during WW I and a total of nine between WW I and 1925 when the Navy started designing their own subs. Simon Lake died at age 78 in 1945, but not before he had talked Louis Reynolds of Reynolds Aluminum into someday building aluminum submarines.

## Beebe/Barton

On 3 June 1930, two Americans in a very small steel sphere dangled in the ocean at the end of a cable, below a barge near Bermuda, British West Indies. The men, William Beebe and Otis Barton, initiated manned deep-sea exploration on that day during that dive. It was the beginning of the

**Name:** *Beebe's Bathysphere* **Builders:** *W. Beebe and O. Barton*
**Length:** *4'6"* **Capacity:** *2 Man*
**Maximun Depth Rating:** *3,028'* **Year Launched:** *1930*
Courtesy of *National Geographic Magazine*

modern era in oceanographic research by manned vehicle.

Once the two tall men and all of their equipment were inside the 4'-6" diameter sphere, named the *Bathysphere*, a crane hoisted the 400 lb. hatch onto 10 bolts sticking out from the hull around the entry hole. Next, the deck crew tightened nuts onto the bolts, first finger tight and then tighter by hitting a long wrench on its handle with a huge sledge hammer, which nearly deafened the occupants. After a dive, the process was reversed to get the people out.

Beebe, at 53, had written 17 scientific books, been curator of Ornithology for the New York Zoological Society (NYZS) for 31 years and was at that time director of the Department of Tropical Research for NYZS. Barton was half of Beebe's age, an engineering graduate of Columbia University in New York, and had just finished a "hop around the world playing at being explorer"on a sizable inheritance from his grandfathers' estate that also funded the diving sphere they were using.

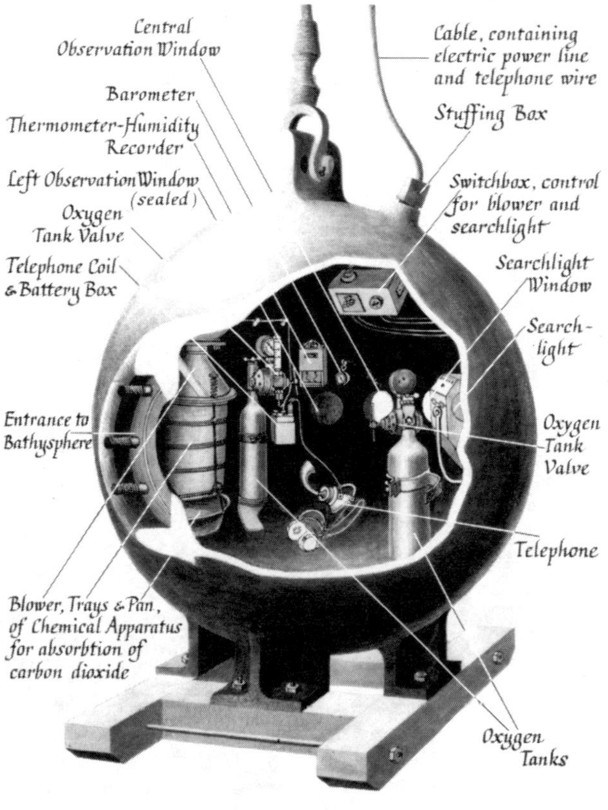


*Beebe's Bathysphere, Detailed Illustration*
Courtesy *National Geographic Magazine*

Three viewport housings were located at the front of the sphere opposite to the hatch on the equatorial line. Inside of each housing was an 8" diameter, 3" thick viewport of fused quartz made watertight by paper gaskets. The axis of the three viewports converged a few feet in front of the sphere so that cameras, lights and viewing were all aimed at the same area of view.

A special 7/8" cable terminated into a pad eye on the top of the sphere. The electrical and communication wires, which attached to the hoisting cable, entered the hull through stuffing glands. The glands were fitted with a threaded tube and tightening nut to secure the wires from being squeezed into the cabin by sea pressure. The 110v electrical cable powered a 250 watt flood light. Communication wires served for the telephone to the surface.

Oxygen was metered from two storage tanks with another tank as standby. Each tank contained enough oxygen for two men for two hours. Carbon dioxide was scrubbed by trays of soda lime and moisture was scrubbed by calcium chloride. A Japanese hand-held palm leaf that had to be continually fanned manually "kept air in circulation."

The 5,000 lb. *Bathysphere* was tethered by a cable 3,500' long, made by Roebling and good for 29 tons in tension. Barton estimated the crush depth of the sphere at 4,500 feet.

Unmanned tests of the *Bathysphere* were conducted which proved the sphere watertight, but resulted in the conductor cable getting tangled around the hoisting cable.

Manned tests rewarded the researchers with views of live marine life as opposed to the dead samples available to them previously by trawl. The concentration and excitement was so much that Beebe sat on a monkey wrench for over an hour without realizing it, leaving him with a sore tail long afterwards. Of the two men, Beebe seems to have been the calmer. Barton relates: "We were descending. Pieces of seaweed and small fish moved upward before our window. It was at this time I had my first fright. A stream of water was coming in under the door, wetting my shorts. 'Dr. Beebe,' I called, 'There's a leak! Shall I tell John to pull us up?' Dr. Beebe glanced at the door and studied it for a moment. 'No, I think not,' he said. 'Don't frighten them at the surface.'"

The deepest dive of the first season was to 1,428'. Most of the many other dives were in shallower water where contour dives were practiced. These were accomplished by putting the barge into a desired location and letting it drift with the sphere below. If obstruction appeared or the bottom came up, the researchers would call up to the winch to be raised out of danger.

***Name:*** *Barton's Benthoscope*
***Builder:*** *O. Barton*
***Length:*** *4'6"*
***Capacity:*** *2 Man*
***Maximun Depth Rating:*** *5,000'*
***Year Launched:*** *1949*
Courtesy Naval Ocean System Center

In 1932, dives were resumed with an unmanned test of the *Bathysphere* to 3,000'. Upon recovery, the sphere seemed too heavy. When a 4" diameter plug in the hatch was removed, water pressure that had built up inside the sphere from leakage at depth shot across the deck and dented a steel winch. A second 3,000' test gave the same result, but the sphere returned dry on the third try and manned dives commenced. The last dive of the season was broadcast over NBC from Beebe and Barton's new record depth of 2,200 feet.

Thanks to the publicity received from the press and radio, several U.S. companies furnished some newer equipment for the project. The Air Reduction Co. provided new oxygen tanks and an electric blower to replace the palm leaf fan. General Electric provided new quartz windows while Bell Telephone provided an improved telephone system.

The 1934 dives began in the Bahamas in August. A three hour dive to 2,510' was the longest in the series while the biggest fish seen occurred on dive No. 32 when an 20' long unidentified specimen swam by their viewports. The deepest dive, to 3,028', was made until only a few turns of the hoisting cable were left on the winch drum and rough surface weather required the dive to be terminated. The *Bathysphere* was retired to the aquarium at Coney Island, New York.

Following his adventures with William Beebe, Otis Barton tried various endeavors. Some involved filming undersea or jungle life, another a new diving suit he invented and, when WW II began, he joined the Navy. After his Navy discharge, he had a new diving sphere fabricated which was quite similar to the *Bathysphere* which Beebe had designed. Made of a stronger chrome-nickel steel, the inside diameter was the same as that of the *Bathysphere* (54"), but a half inch thicker (1 3/4"). Barton calculated the crush depth at 10,000'. Named the *Benthoscope*, it weighed 7,000 lbs. in air and had only two viewports. One was centered amidships on the equatorial line of the sphere and the other, smaller one, was angled downward and located below the first. Like those on the *Bathysphere*, the quartz viewports were 3" thick and sealed with paper gaskets. The new hatch was concave in shape which reduced its weight to 150 lbs. External lights consisted of a 1,500 watt spotlight and a small high intensity photography light. Six thousand feet of electrical, phone line and hoisting cable connected the sphere and the topside winches and operators. The *Benthoscope* cost $16,000 compared with the $12,000 cost of the earlier *Bathysphere*.

After an unmanned test of the sphere at 6,000', near Santa Cruz Island off Southern California, Barton made the first manned dive in August, 1949. On his fourth dive, he went to 4,500'. In October, contour dives were made from Catalina Island to the south of Santa Cruz Island until winter weather caused the operations to be ended.

There followed a five year sabbatical from the oceans in Africa, after which Barton headed in 1954 for the San Clemente Basin, off the coast of Southern California with the *Benthoscope*. Still photos were taken for *Life Magazin*e at 3,100 feet. After the lights malfunctioned, the dive series was terminated. The *Benthoscope* was last seen on exhibit in the old Marineland of the Pacific in Los Angeles.

William Beebe continued his work in the natural sciences, writing many books and papers. His last article in *National Geographic* in 1958 at age 81 was as sharp, clear, and instructional as his several contributions to that magazine were in the 1930s. He died a few years later.

Almost two centuries of development occurred before electrical equipment replaced candlelight and human power in submarines. Beebe and Barton's undersea research by direct observation was the first of its kind in the U.S. Their work also represented some of the earliest submarine research to be sponsored by non-military funding and a forerunner of many more oceanographic programs to follow.

Technically, none of the projects described in the preceding chapter could be described as deep submersibles. Their inclusion is necessary for historical background as well as for the appreciation of the work of the giants that lead the way for others.

# *Trieste*

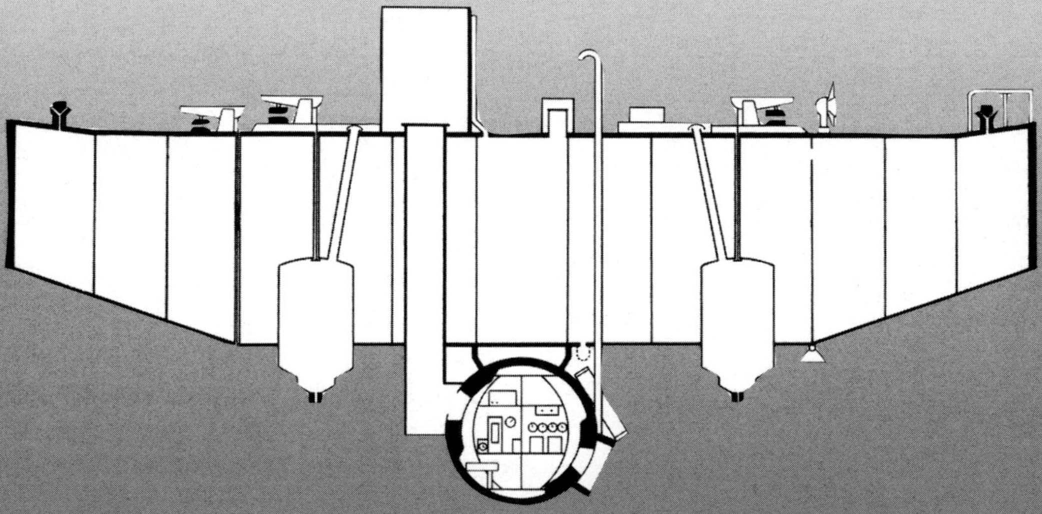

## *1958–1963*

# Trieste

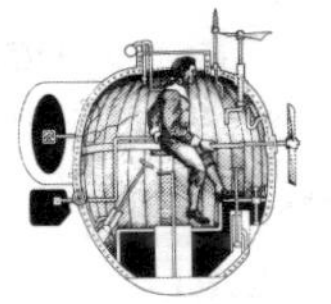

The Swiss-designed, but mostly Italian-funded *Trieste*, was an important bridge between the early U.S. submersibles and the American-built deep submersibles that followed its purchase and operation by the U.S. Navy. The catalytic affect of the Navy *Trieste* program was a major factor in the initiation of deep submergence technology in the U.S.

The designer and builder of the *Trieste* was Auguste Piccard. His extensive knowledge as a physicist contributed to practical engineering and established many of the basic design standards still in use today over sixty years after he introduced them. As a university student, Piccard dreamed of oceanographic expeditions in submersibles. Before he ever went to sea, the same concepts led to his stratosphere ballooning. About a year after the Beebe/Barton dives began, Piccard set a balloon altitude record over Switzerland.

After WW II, Piccard requested and received funding for a bathyscaphe from the same Belgian group (F.N.R.S.) that funded his balloon explorations. Named the *FNRS-2*, since the balloon project was the first FNRS, it operated with a system similar to those used on balloons, with gasoline for buoyancy and iron shot for ballast. Piccard chose an oil heavier than water in which to submerge the motors. The rotating shaft stuck up through the oil into the surrounding sea water vertically and, consequently, did not require a shaft seal. Gravity kept the oil in place, in or out of the water.

The spherical steel hull had an internal diameter of two meters (6'-6 3/4"), with a nominal wall thickness of 9 centimeters (3.54"). The steel castings of the hull were made into two hemispheres clamped together at the equator with a hatch containing a viewport at one pole and a viewport only in the opposite pole. The sphere was positioned so that the viewport opposite the hatch was facing forward and downward.

After extensive testing, Piccard chose a 90 degree conical segment shaped viewport with a 10 cm (3.94") internal diameter, and a 15 cm (5.91") thickness. Piccard's pressure tests indicated that the viewports would not fail at less than 7.4 miles deep.

The tests of the *FNRS-2* during the winter of 1948 off Dakar, Senegal, Africa were pioneering in nature since they tested the validity and watertight integrity of Piccard's designs, both manned in shallow water and unmanned at 4,620'. The French Navy witnessed the *FNRS-2* tests and were greatly impressed. In response, the funding group at FNRS sold the *FNRS-2* to them so they could cannibalize it and build the *FNRS-3* to their specifications. Piccard was kept on as an advisor, but the amateurs in the French Navy thought they knew more than he did so he was pretty much ignored and lost interest.

In early 1952, the Italian city of *Trieste* offered Auguste Piccard funding and industrial support to build his own bathyscaphe. He accepted. With some Swiss donations and the help of his son, Jacques, he began building the bathyscaphe *Trieste*. Most of the basic design was already completed and tested based on the experience with the FNRSs, but some minor yet significant improvements were needed. The new sphere was identical to the one on *FNRS-2* except that, instead of being cast, it was forged and machined at the Terni plant in Italy. This made it much stronger. The volume of the new float was increased to 3,742 cu. ft., but the battery, buoyancy and ballast control systems would be the same as in the *FNRS-2*. Where *FNRS-2* had to be entered prior to launching, the *Trieste* could be towed to the dive site and entered in the open sea by means of a watertight entry tube extended from the top of the float to the sphere below.

Compartmentalization of the float into 12 segments provided integrity if one or more of the compartments became damaged. The fore and aft sections of the float tapered toward the ends for streamlining purposes. Corrugations in the bulkheads between sections provided for expansion/contraction due to temperature and pressure changes. Two ballast tubs containing nine tons of releasable iron shot were provided for variable ballasting. One ballast tub was forward of the cabin and one was just aft.

Two motors with propellers were located on top and at the forward part of the float for propulsion and maneuvering. During diving, the entrance tunnel, running vertically from on top of the float down to the sphere, was flooded with seawater. While at the surface, it was blown dry to permit access to and from the sphere. The polar axis of the spherical cabin, which intersects the two viewports, was tilted 18 degrees with the horizontal to provide a downward and forward viewing angle. Overall length of the submersible was 49'-6".

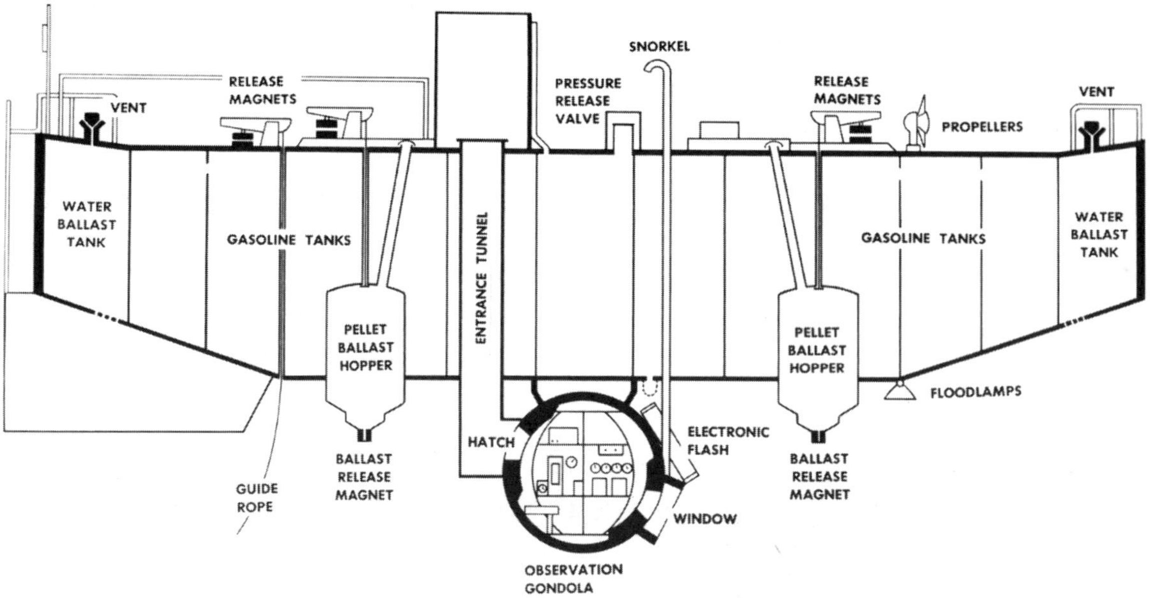


***Name:*** *Trieste* ***Builder:*** *A. Piccard* ***Length:*** *49'6"* ***Maximum Depth Rating:*** *12,000'*
***Capacity:*** *3 Man* ***Year Launched:*** *1953* ***Description:*** *Cross Section*

In 1953, the first dives with *Trieste* began with Auguste Piccard and his son Jacques Piccard, making most of the dives together. The maximum dive depth for the year was made to 10,390'. It was Auguste's last dive a few months before his 70th birthday. A few shallow dives were made in 1954, but none were made in 1955 due to a lack of funding. A few dives were made in 1956 with some modest funding from Swiss and Italian sources.

Many people associated with oceanography, science and the Office of Naval Research (ONR) were aware of the *Trieste* and its capabilities. The London office of ONR brought the *Trieste* to the special attention of ONR in Washington D.C. Interested parties in the National Academy of Science got the Secretary of the Navy to lean on the head of ONR to "do something about the *Trieste*." Thus, in 1957, almost all of the 26 *Trieste* dives in Europe were made with U.S. scientists or ONR representatives evaluating the bathyscaphe operations. As a result, *Trieste* was purchased by ONR from the Piccards for $200,000, and shipped to the Naval Electronic Laboratory (NEL) San Diego, California in August 1958.

Early in 1959, Lt. Don Walsh USN volunteered and became Officer in Charge with Larry Shumaker as Executive Officer, and Andy Rechnitzer as Scientific Advisor. Giuseppe Buono came with Jacques Piccard, as an all-round chief of the boat and master of all its idiosyncrasies. Soon after the purchase, plans for Project *Nekton* started. *Nekton* was the name given to all of the preparations and modifications necessary to reach the bottom of the Challenger Trench at over 35,000' of depth in the Pacific. A heavier and stronger sphere was needed, as well as a larger float to buoy up the heavier sphere. The Krupp Werkes in Germany committed to a delivery time of six months at a price of $65,000 for a three-piece sphere with 3.5" thick walls in the non-reinforced areas. The three-piece design was comprised of a midsection and two end caps to form a sphere. The three pieces were machined to a near perfect fit. A thin coat of epoxy glue, acting in compression, bonded the pieces together.

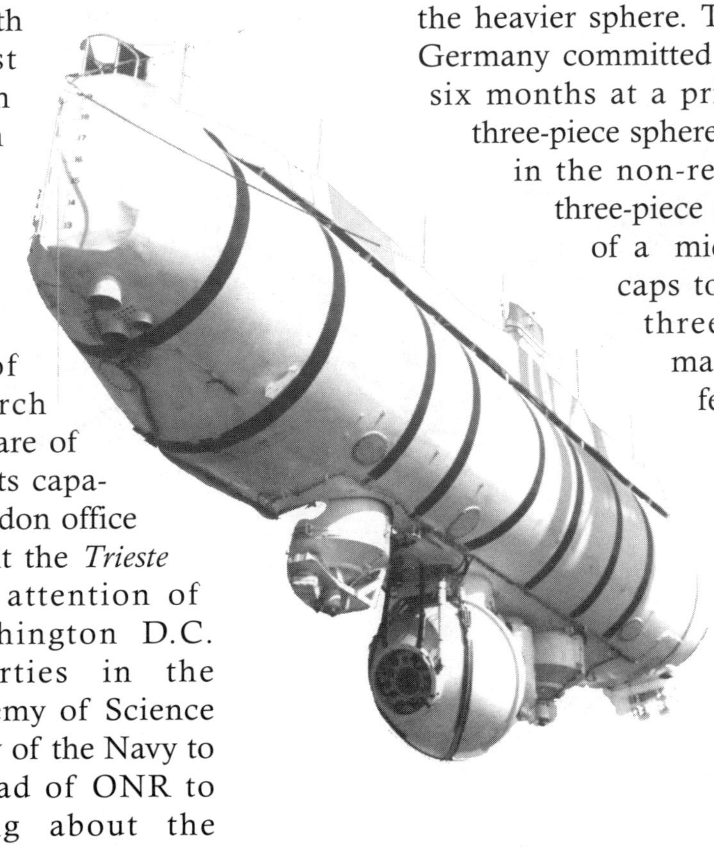

*Trieste*
Courtesy of U.S. Naval Photographic Center, Washington D.C.

After the Krupp sphere was installed and the float lengthened an extra 8 feet, the vessel was shipped to Guam, reassembled and dive tested to 18,150' in November 1959. While on the way back to the surface, the epoxy joint popped with a loud noise. Even with the glue cracked, the external sea pressure sustained a tight fit in the joints both at depth and at the surface. The best explanation for the epoxy bond breaking was that the larger central hull had undergone a change in dimension – due to pressure and/or temperature affect – that was different from the smaller end pieces.

Large screw jacks were obtained from a Navy yard in Japan, and the parts were moved back into near perfect alignment. The joint, after reseating, was then given an additional external seal, and an extra retainer ring was made and placed around the sphere to prevent reoccurrence. Without calling official attention to the field-repairs, Don Walsh test-dived the *Trieste* to the bottom of the Nero Trench at 23,000' on 8 January 1960 – fortunately, with no further leaks or noises.

On 23 January, the *Trieste* was towed 220 miles from Guam to a dive site directly over the deepest spot in the Marianas Trench for the 65th dive of its career.

*Trieste with Krupp Sphere at Marianas Dive*
Courtesy of Naval Electronics Lab, San Diego

The *Trieste* flooded down and ejected gasoline until it submerged. A brief stop on the first thermocline at 300' was made while descending, as they hit the denser, colder water. More gasoline was ejected to make the boat heavier, and the dive continued. Additional thermoclines stopped them momentarily at 425' and 530'. After 600' the craft attained a constant rate of descent at about 3'/sec. They observed luminescence at 2,200' and at 20,000'. At 26,000' the rate of descent was reduced to 2'/sec, and at 30,000' to 1'/sec. Walsh, in the *Trieste*, and Shumaker, at the surface, made phone calls to one another at 5,600', 10,000' and 13,000 feet. In the sphere at 26,000', Walsh and Jacques Piccard could hear the conversations between the tugboat and the Navy destroyer at the surface.

At 32,500' they heard a dull crack-like sound coming from somewhere in the bathyscaphe. They stopped and looked about them to see what the problem was, but finding nothing wrong they continued downwards. At 1256 hours, the bottom sounder indicated that the bottom was 300' away. Ten minutes later, at 1306 hours, they landed on the bottom near a fish of the halibut family. A few minutes later, a shrimp swam by. Walsh called to the surface, "This is *Trieste* on the bottom, Challenger Deep. Six three zero zero fathoms, over."

They heard him topside, weak but clear. He repeated the message adding, "We will surface at 1700 hours." Later, Walsh looked through the aft viewport located in the hatch and noticed that the viewing window in the access trunk was cracked which explained the noise they had heard on the way down. The thin acrylic window was not normally subject to stress, since the access trunk was filled with water prior to leaving the surface and was not subject to pressurization. The window would be needed in place at the surface, so that the bathyscaphe crew could blow the trunk free of water and go up the trunk to the surface.

After 20 minutes of observations and note-taking, they decided to leave the bottom and head for the surface. The ascent was without incidence. Their vertical velocity increased as they ascended, due to the gasoline expanding as the water pressure decreased. The less buoyant, warmer waters at the thermoclines were barely discernible. Bioluminescence was noted at 3,300' and 2,000'. They surfaced at

*In spite of time and the approaching fall weather being against them, they found a tangled mass of torn and twisted metal and debris of all kinds.*

---

1656 hours and blew the access trunk very slowly and carefully. The acrylic window held. Fifteen minutes after surfacing, they exited into daylight after a dive that lasted a total of eight hours and thirty-five minutes.

The crew were each given a special award and handshake by President Eisenhower for their part in taking a man-made vehicle to the deepest part of the ocean at 35,800' and safely returning for the first time in history.

After a brief home stay, the American crew returned to Guam planning to revisit the Challenger Deep two more times, but they were enjoined by the Navy and limited to depths of 20,000' or less. On the seventh dive of their return, the last mercury vapor light burned out ending the Marianas dives.

On 10 April 1963, the *U.S.S. Thresher* was lost at sea with 129 men aboard during a routine sea test after a shipyard overhaul. The sinking of *U.S.S. 593* occurred 260 miles off the coast of New England in 8,400' of water. Results of the official inquiry would indicate that the cause of the disaster was due to a rupture in the seawater piping; jetting water wiped away power and control cables in its path as if they were cobwebs.

The only submersible in the U.S. capable of exploring the ocean at 8400', at the time, was the bathyscaphe *Trieste*, in San Diego, one small canal and one large ocean away.

*Trieste* was immediately ordered to Boston by the Navy. It had been over three years since the Challenger dive. Don Walsh had been relieved by Don Keach, who was now the officer in charge of *Trieste*. It was a frustrating time for Keach when he got the call. The old *Trieste* float was full of holes, plugged with wooden pegs to prevent the gasoline from leaking out. A recent overhaul had not included a badly needed new float. Under tow, the old float could actually stretch as much as a foot in length. On the positive side, the field-repaired Krupp sphere had been replaced with the original Terni sphere which was still good for thousands of dives as deep as 12,000'.

Keach got word to deploy to Boston on 11 April 1963. *Trieste* was immediately hoisted aboard the Landing Ship Dock (LSD) *U.S.S. Preserver*. After a long slow trip to Boston followed by a long tow to the dive site, the first dive was made on 24 June 1963. After months of back-breaking effort and observing endless pieces of nondescript debris in dive after dive, the wreckage itself was not discovered. However, a plastic shoe cover used for radiation protection with part of the *Thresher's* hull number visible on it was found. The *Trieste*, lacking a manipulator, could photograph it, but not recover it.

Needing repairs and additional batteries, the *Trieste* was returned to Boston, where among other things, a manipulator was added. Returning to the dive area on 18 August, they began their second dive series. In spite of time and the approaching fall weather being against them, they found a tangled mass of torn and twisted metal and debris of all kinds. Positive identification was made when a piece of pipe was recovered with the inscription "593 Boat" on it. The trick was to maneuver the underpowered 60' long blimp-like structure while looking out of one small viewport in a steel sphere hanging below it. Twisting and dodging the bathyscaphe among the wreckage they photographed everything in sight with the automatic cameras. After dive No. 10, the badly worn and damaged battery compartment finally failed, and operations were halted. Upon return to San Diego, the faithful old *Trieste* float, along with the Krupp sphere, was retired to the Naval Historic Museum at the Washington D.C. Navy Yard. The *Thresher* search was later to be continued by *Trieste 2*.

As a result of the *Thresher* loss, a Deep Submergence Review Group (DSRG) was established by the Secretary of the Navy. Just about every civilian and Navy expert on oceanography, undersea exploration and ocean salvage was called upon to assist or join the group to determine the best approach for the U.S. to take based on the lessons learned from the *Thresher* accident.

It was decided that DSRG should:

- Review the Navy's plans for locating and rescuing in the deep ocean
- Recommend changes in such plans
- Recommend plans for future systems
- Develop a five-year underwater search, rescue and salvage program
- Recommend organizational responsibilities for the program parts

The report by DSRG to the Secretary of the Navy resulted in the forming of the Deep Submergence Systems Project (DSSP), with the following goals set by the Chief of Naval Operations:

- A submarine location system with escape from 600' and rescue from collapse depth
- A large object salvage system, within continental shelf depths
- A deep-ocean search system with investigation and small object recovery down to 20,000' depths
- A man-in-sea program with 600' operations for extended periods

The above goals led to the deep submergence rescue vehicles operating at 5,000', the research vehicle *Sea Cliff* operating at 20,000', and the construction and operation of Sea Labs 1, 2, and 3.

The *Thresher* incident was a pivotal event which initiated the Navy's interest and funding in deep submergence and led to many of its underwater projects in the years that followed.

# *Deep Jeep: Development*

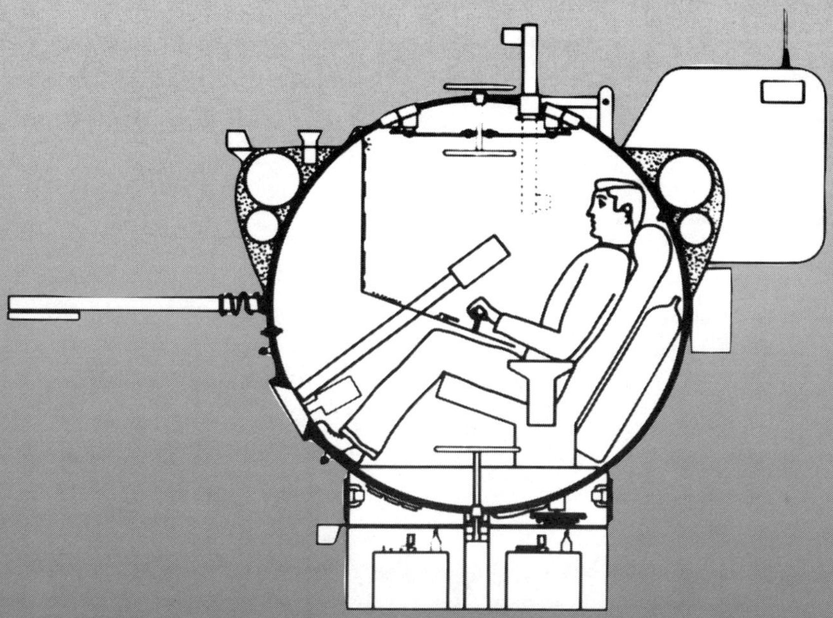

## *1960–1963*

# Deep Jeep: Development

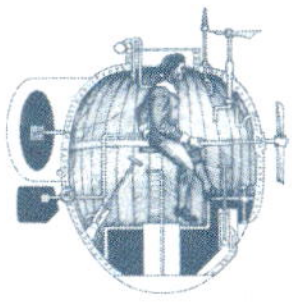

In the late summer of 1960, six months after I joined the Naval Ordinance Test Station (NOTS) at China Lake, California, as an engineer, word came down from the Director's Office at NOTS that he would welcome and consider proposals for a two man deep submersible, capable of operating at a depth of 2000'. The Test Department at NOTS had recently purchased a remotely operated vehicle (ROV) from Vare Industries for use on the test range at San Clemente Island, for the recovery of malfunctioning torpedoes from the sea floor. The purchase price of $100,000 led Director William B. McLean to believe that a manned vehicle could do a better job at a lower cost than the ROV, and I agreed. To accomplish the range requirements, the submersible would need a 2 kt. velocity capability, be highly maneuverable and, above all else, safe and effective in operation.

Since the depth, speed, and complement had been specified, with economy as another parameter, several design approaches could initially be established. For maximum volume with the least stress in the shell, a spherical pressure hull was selected. Maneuverability, sufficient to capture and recover a torpedo lying on the sea floor, indicated almost helicopter like control and propulsion management. Darkness at 2000' indicated the need for artificial light. Safety and the practicality of operations required a topside hatch for both shipboard access, as well as for emergency in-water egress. Droppable weights for emergency surfacing would be another required feature. A normal operating time of four hours seemed realistic, based on aircraft pilot's effectiveness over extended periods of high stress and concentrated effort. The existing state of the art in underwater propulsion indicated battery powered d.c. motors with propellers as the first choice for the thrusters. The French Soucoupe system of two fixed position rotatable thrusters was chosen, using electrical driven propellers.

Stability during normal operations was easily attained by placing the batteries below the spherical pressure hull. Once the additional safety feature of dropping the batteries in an emergency was added, it was necessary to add a fixed flotation ring around the top of the pressure hull for stability without the batteries. Pressure resistant syntactic foam, made with very small hollow glass spheres, embedded in epoxy casting compound, provided the fixed buoyancy desired. Two hollow torii (donut shapes) were encased inside the flotation ring. The torii were filled with colored water during normal operations. They could be blown out with compressed air to obtain

*On 31 October, I received a memo providing \$50,000 in funding for "Mobile Underwater Test Vehicle and Associated Instrumentation." That was the best Halloween ever.*

---

emergency buoyancy under water or, when at the surface, to signal the submersible's location to the surface ship with the 150 ft. vertical jets of colored water blown out.

I selected reclining seats for long term operations, with individual optical viewing scopes used by both crewmembers looking through the single viewport. The large diameter lenses of the system were located at the viewing end with small optics on the opposite end near the viewport. The optical viewing was corrected to provide true dimensions and distances between underwater objects and compensated for the enlargement usually associated with looking through face plates and viewports.

A periscope was included in the design for use at sea while surfaced, to locate ships from the low freeboard of the submersible. Later, this proved to be especially effective, since the submersible could sometimes spot the large support ship when the ship could not see the submersible. High quality, boiler steel, hemispherical heads were selected for cost effectiveness, as well as for quality control. Machining to accurate dimensions and curvatures was done at NOTS.

The welding on the first hull was done at NOTS after machining, but the original hull cracked so badly after the hemispherical weld was done that a second hull was fabricated and sent this time to a boiler fabrication plant in Los Angeles for welding. The welding contract included very strict specifications regarding welding preheating and post-welding stress relieving. But for the loss of the first hull after all the heavy machining had been done, *Deep Jeep* could have been launched at least four months sooner, costing some \$20,000 less.

The design of the viewports in the hull and the periscope used conical acrylic units based on Auguste Piccard's designs. Articles about Beebe and Barton from the 1930s somewhat influenced the initial plans for the environmental system. The Medical facility at China Lake provided information on the chemicals required and the data needed for humidity and $CO_2$ scrubbing as well as oxygen supply requirements.

I submitted a nine page compilation of design requirements and solutions, including calculations and a page of references on 1 October 1960 to the Engineering Department, for forwarding to the Director's office. On 31 October, I received a memo providing \$50,000 in funding for "Mobile Underwater Test Vehicle and Associated Instrumentation." That was the best Halloween ever.

After the funding was authorized, the long term delay items were ordered, such as:

- Two steel hemispheres which would be hot spun by Lukens steel
- Two 3/4 H.P. 24 v.d.c. motors for the propulsion units and one 1/4 H.P. 24 v.d.c. motor for angling the propulsion motors
- The batteries and charger
- The environmental equipment, instrumentation, and chemicals

A mock-up out of plywood of the instrument and control panels was begun at once, to figure out the locations of equipment and instruments, where the controls would go, how the seating would fit in, etc. About a hundred post-cards requesting manufacturer's catalogs and brochures were sent out during the first week or two after the project was funded, using the Thomas Register for addresses of manufacturers and suppliers.

A cross section drawing was made as soon as the instrument and control locations were determined and a detailed fabrication drawing was made for the shop's use in machining and welding the steel hull. My shop and field engineering experience at the Interior Department in Oregon made it possible for me to do all the wiring diagrams and fabrication drawings of the submersible used for both the in-house and contract fabrication. The electrical wiring assemblies were let out on contract in Los Angeles with the specifications for length, size, and end connectors for the various assembled bundles of the wiring. Upon arrival at China Lake, assembly for continuity and circuitry was checked.

The electrical switching system for selecting the power to the propulsion motors was simple. Six volt deep cycle golf cart batteries were connected in combinations to provide 12 v, 18 v, and 24 v power to the propulsion motors in either the forward or reverse mode. Finding no suitable hermetically sealed circuit breakers that would hold up against the salt spray, high humidity environment, we used fuses instead. The propulsion motors could be pivoted simultaneously from the horizontal to vertical by actuating the 1/4 H.P. angling motor. This technique permitted the mixing of the horizontal thrust with the verticle.

The China Lake hospital had several pamphlets which were useful during the preliminary environmental work. The U.S. Navy Bureau of Medicine and Surgery for Submarine Medicine Practice specified physiological requirements with both minimum and maximum acceptable oxygen and carbon dioxide levels, as well as oxygen consumption and carbon dioxide production rates for various conditions.

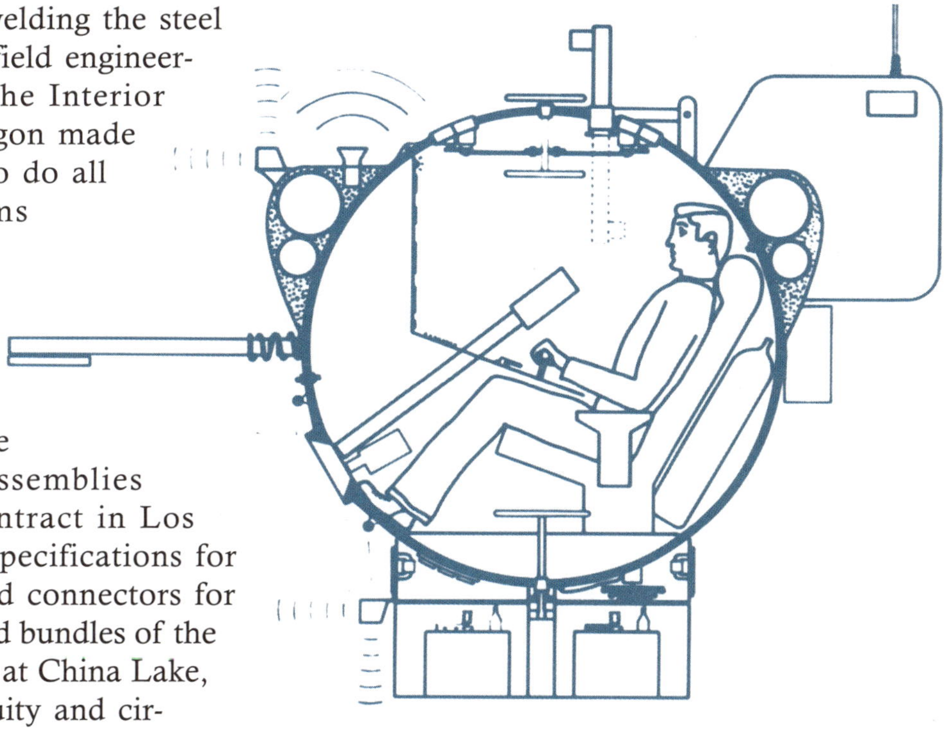

***Name:*** *Deep Jeep*
***Builder:*** *Naval Ordinance Test Station* ***Length:*** *9.5'*
***Maximum Depth Rating:*** *2,000'* ***Capacity:*** *2 Man*
***Year Launched:*** *1964* ***Description:*** *Cross Section*

Books on propeller performance, fluid dynamics, human engineering, panel display techniques, etc, available at the Michelson Laboratory library, were very useful.

Once the design was fairly well established in principle, and a good part of it represented on drawings, it was desirable to get some outside critique and comments.

The *Trieste* people expressed a willingness to help out, so a trip to San Diego was arranged. Just a few months earlier, Don Walsh, Larry Shumaker, and Andy Rechnitzer had been part of the Marianas deep dives, so their experience was of extreme value. One very strong suggestion was that the lights on the front of the vehicle should be placed as far out in front as possible, to minimize backscattering of the light off of the various detrital and snow-like biological material in the water. For a full day, they shared their experience and answered numerous questions. The data were recorded for future use, mentally as well as literally. Ultimately, this meeting saved a lot of lost effort, besides contributing toward safer operations.

For several hours a day, over a two month period, I conducted extensive testing, by trial and error, of electrical motors in oil. The goal was to get the *Deep Jeep* d.c. motors to run in a non-conductive oil efficiently, under pressure. The motors were immersed and operated in a number of different oils for comparative purposes. Silicone oils worked the best. The thickest (10 centistoke) had about the viscosity of kerosene. The next thinner (1 centistoke) had about the viscosity of water and the thinnest (1/2 centistoke), which was the best and most expensive, had about the viscosity of gasoline. After 3000 hours of running the gearmotors in the 1/2 centistoke fluid, no detectable wear could be found on the armature or the gears of the gearmotor, in spite of the fact that silicone oils have a low lubricity. Amazingly, a 3% increase in efficiency, at all power settings, was attained in the oil-filled motors, compared with the as-received condition of the motors running in air.

At China Lake, it was necessary to give status reports and demonstrations to small groups of V.I.P.'s about once every 60 days. My co-presentors at these events were the Moray project people. Moray was a manned torpedo project that was well funded by operational money from Washington D.C. Their budget supported about 38 people and some very sophisticated technology. My skunk works project had bare bones funding for just myself and a machinist. It was the scrub team on the same program with the varsity. The presentations were made in the small conference rooms, so elaborate demonstrations were not feasible. Lacking visual aids, such as movies, still photos, charts, and graphs, I would demonstrate my motor-in-oil development system by dipping a small slot car motor, running off of a hand held battery, in and out of a water glass full of kerosene while explaining the test data and efficiencies obtained.

Since there were no commercial underwater lights available for deep submersibles some inovation was required. In another presentation, a homemade 450 watt light, made up of three Sylvania home movie lamps, sealed in a pressure resistant Pyrex tube, would be lighted using a hand held lantern battery, momentarily blinding the audience. About four brief, bright light-ups were all you could get before the little, lantern battery ran down, so no more than three were ever done. The point was made, though.

Just about the time the project was well underway, the two man, 1,000' depth, French designed *Soucoupe* was brought over from France by NOTS and I was asked to act as liaison between the Navy, Westinghouse that would provide the surface support and the Cousteau group. The latter was to operate the submersible at the Navy facility at San Clemente Island for a $1,000 contract which included the round trip military air transportation of the submersible and crew. It was agreed that NOTS would get two weeks of demonstration dives in the vehicle at the island. After additional operations in Scripps Canyon, the submersible and crew would be returned by military air to France.

Andre Laban was in charge of the Cousteau group and he invited me to go back to France with them. We had shared some good times and made some dives together and were still friends. More to the point, I needed to see how the French operated with their own equipment and support people.

The French operation and support activities were vastly different from what the *Soucoupe* group had been using in the U.S. The submersible and the people were the same, but everything else looked like it was half scale. The buildings, trucks, cranes and even the unloading gear at the airport were small, by comparison. Our equipment had been very big; theirs at first looked a little too small, but was not. Their operations were simply more efficient, even though on a smaller scale.

On the first launch of *Soucoupe*, just outside of the harbor at Marseille, the submersible – still ballasted for the Pacific – stuck out of the water so far after launch, that the pilot, Falco, stuck his head out of the hatch and yelled "kees ke seh?" I was standing on the second deck of the ship, *Calypso*, with Cousteau overlooking the work deck. Everyone was looking at each other with quizzical expressions as the sub continued to roll in the swells, so far out of the water it was funny. I quietly said to Cousteau, "Commandant, do you suppose that the Mediterranean is different than the Pacific?" A light bulb went on and he began to lecture the crew below, on the difference in the salinity and density of the two bodies of water, and how the Mediterranean was one of the saltiest seas in the world. The matter was soon corrected.

When the *Deep Jeep* hull was returned to China Lake after welding, a lot of very fine finish machining was left to be done. At my request, the shop assigned John W. Ball to the project on a full time, permanent basis. He had done all of the pre-welding machining and his work was excellent. In addition, he was the kind of guy you wouldn't mind going to sea with. John had machined 16" diameter gun barrels at the Navy facility in Idaho and built gyroscopes at China Lake. I asked him if he'd ever machined a submersible before. He indicated that he had not, but said he was sure he could do it O.K. Later, when he saw my name on the fabrication drawings, he asked if I'd ever designed a submarine before. I told him I had not, but that I was sure I could – and since I was to be the test pilot, we'd better both do it right!

*True* magazine ran an article about a Navy Doctor, CDR George Bond, making practice escapes from a submarine from 300' under water in their November 1960 issue. Drawing on the details in the article, I installed a 1" Diameter ball valve with which the sub could be flooded so that the crew could escape (theoretically) from as deep as 600' in an emergency. A procedure was established from graph curves and calculations as to how to exit and make an ascent. As luck would have it, our Safety Officer for the *Moray* and *Deep Jeep* projects was CDR, Linc Lippincott, who had been the executive officer of the submarine that Bond had escaped from. Lippincott invited Bond to review the environmental systems on our two submersibles.

Bond arrived on 1 December 1961 to give a technical presentation to the laboratory and then, spent the next two days reviewing the environmental systems of our projects in every detail. CDR Bond liked the flood valve and the escape procedures. Once during the review, as he was told that the entire environmental system cost around $200, he looked at Lippincott and there was the slightest twitch in his face. He had just come from a review of the *Moray* project and learned that their continuous read out, oxygen analyzer alone cost $50,000.

The *Deep Jeep* environmental system consisted of a scrubber assembly made up of four round lab sieves, normally used for sifting and grading sand, all fitted together like Beebe and Barton used in the 30's. The bottom two trays held silica gel for absorbing moisture from respiration and perspiration. The top two trays, for absorbing carbon dioxide, contained soda lime from the hospital's stores for basil metabolism measuring equipment. On top of the assembled trays was a lid with a 12 volt d.c. motor and blower, borrowed from the family station wagon, where it was formerly used as a defroster. For carbon dioxide measurement, I had obtained a WW II surplus Navy submarine unit made by and called a Dwyer, that was obtained for $39. For an oxygen indicator, we used an inexpensive Mine Safety Apparatus. A thermometer in the soda lime scrubber trays indicated if the scrubber was working to reduce the carbon dioxide level, a system learned from Auguste Piccard by the *Trieste* team and passed on to us. An aircraft altimeter, zeroed before diving, indicated the pressure inside the hull for a second indirect measurement of the oxygen content inside the hull. Excess pressure, indicated by negative altitude, was the sign of excess oxygen and vice-versa. Data from numerous manned tests inside the hull, both in the cold altitude chamber and in the hot shade, were made available to Bond.

Bond took notes as we described the systems, sometimes scowling as he wrote, occasionally looking at Lippincott with a smile, whatever that meant. After a thorough inspection, he left to consider the data. The next morning as he came down to the submersible work area, he was all smiles, saying he would give his official O.K. to our environmental system.

Four years later, and in charge of Sea Lab, the now Captain Bond stowed one of the Dwyers in among all the other spare gear that went down in the sea with Sea Lab, which was an underwater home for divers at 200' for two weeks. When the helium environment caused the $50,000 carbon dioxide indicator to give incorrect readings, he pulled out the Dwyer. Operated manually, it was labor intensive, but it let the team stay down the two weeks instead of aborting the first day. Bond often told the story of how he first saw one of the old Dwyers at China Lake.

Bond thought that all Navy submersible pilots should go through the submarine escape tower at Pearl Harbor for practicing escapes. I was all for it, since the *Deep Jeep* flood valve made emergency escape a real possibility. The first to go were the three military pilots from *Moray*. Two of them were badly injured while making a 100' free ascent in the tower. The base Doctor at China Lake, after much deliberation, decided that the remaining pilots should go through the tower, but acclimate in Hawaii for ten days, away from the dry desert, before going through escape training. Al Berryman, a civilian from the *Moray* project, and I were the only submersible pilots left that hadn't gone through the tower. We took our wives and scuba gear and spent ten days at a friend's house at the beach on the far outskirts of Hana, Maui, on per diem and travel pay, diving by day and eating pig and poi by night, acclimating like crazy – just like the doctor ordered!

A tenth scale working model was made to practice operating the full size sub. Slot car electric motors with propellers served as thrusters. The actual control box from the submersible was used as a controller. Very fine enamel coated wires ran from the model to the controller, which was attached to a battery pack. In a large glass-walled tank in the Research Department Laboratory, the model was operated in every conceivable mode of operation, including dropping the simulated battery pod. Many questions were answered and the practice with the model as a simulator was extremely useful.

A test of the pressure hull was required and Dr. Bob De Hart of the Southwest Research Institute (SwRI) offered their facilities and services for $10,000, which was reasonable. We shipped the hull to San Antonio and after its arrival there, I followed. I potted the electrical pass throughs with epoxy in my room in the Menger Hotel the night before the pressure tests began. This close timing was necessary because the metal pass throughs had not been finished by the shop until the day I left for Texas. The pressure hull was the first of many deep submersibles to be tested at SwRI. Strain gages in 255 locations recorded the stress in the metal hull at every conceivable area of importance. The results indicated that in the very highest stress areas, a collapse depth of 5,000' could be anticipated. A proof pressure equal to a depth of 2,700' was held for 15 hours, followed by cyclic pressure testing.

After the hull was returned to China Lake, the complete vehicle was assembled and operated, manned in the welding shop quench tank to test the motors and underwater phones. The amount of syntactic foam buoyancy was proven adequate and the hollow torii that could be filled with air from internal scuba tanks worked fine. The syntactic foam formula, developed by Ray Hinton of the *Moray* project, worked out great. John Ball and I mixed it in a mason's hod trough by hand and poured it in place inside a form which fitted around the greased hull.

Dr. Dave Potter came by one day and made an offer for the use of the General Motors Research ship Swan for $2000/day which included food, lodging, and the *Swan* support crew for each 24 hour day. G.M intended to build the *DOWB* submersible and wanted to first profit from handling our sub. It was the last piece in the puzzle involved in getting the submersible manned at sea, and the price was right.

Actual cost at the date of the first manned dives, after all testing was done, was $110,000. Cousteau had stated in a presentation about the *Soucoupe* made at NOTS China Lake in 1960, that his submersible had cost a little more than double that amount.

# *Deep Jeep: Operations*

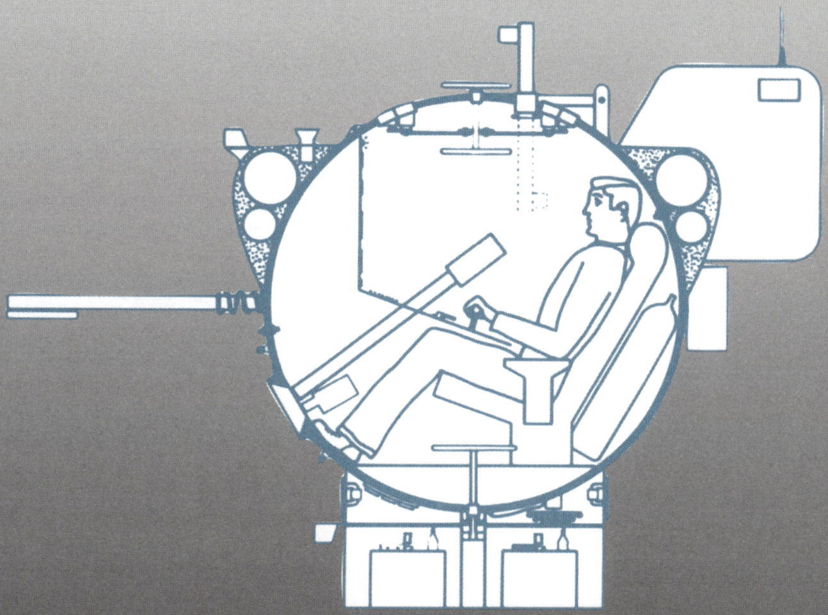

## *1964–1966*

# Deep Jeep: Operations

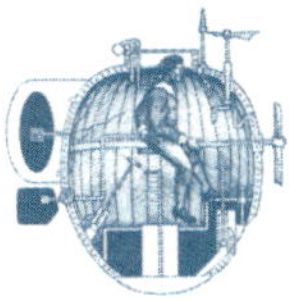

On 21 January 1964, the *Deep Jeep* was christened and launched near Santa Barbara, California. Not knowing any other way to prepare and write up the operation plan, I mocked up the entire operations through each step with my son's Tonka trucks and boats, and used the tenth scale model of the sub as well. This method of preparation worked exceptionally well and we had very few surprises during the actual operations.

At China Lake, an ad hoc committee was set up for the review of all *Deep Jeep* operation plans. The committee, consisting of all the Department Heads on the base, made it necessary for me to see each individual Department Head separately and pitch my written plan for signature. It was a good way to become well known, but not necessarily popular.

The General Motors (GM) Defense Research Laboratory let us use their shop facilities in preparation for the ocean dives. One night, working late on the sub Scott Dobbins, the designer of the G.M. submersible *DOWB*, introduced me to Otis Barton, whom I'd been trying to contact for two years. He very kindly said he thought we knew what we were doing.

We next began a long series of operations aboard the *Swan*. We'd spend a week at sea and then return to China Lake to fix or modify things and then do it all over again. The full-size sub behaved so much like the model that it was spooky. Three submersibles later, I wrote a booklet titled "Remote-Controled Models as Submersible Simulators," still in awe of the benefits of dynamic modeling.

During one of the next test series, two of us took the *Jeep* down to the bottom, in the open sea, in about 100' of water and unscrewed the battery pod, which weighed 1020 lbs., to see if the electrical connectors would disconnect and if the sub without batteries would somersault or come up straight. After unscrewing the batteries and bracing ourselves for the ride of a lifetime, nothing much happened: we came up like a bat out of hell but without swerving. We came half-way out of the water and remained upright and stable at the surface, just exactly as the model had done.

On one of the dives in the Anacapa Island area, the observer was an oceanographer who had warned me before the dive that he was a hyperthyroid, a condition which caused him to constantly nibble food to keep his energy up. I noticed during the dive that he had enough cheese and crackers about his slim body to fill a banquet table. A storm came up after the dive, just as we were getting

*The skipper headed for the lee of Santa Cruz island into Smuggler's cove. An observer later reported that the Jeep looked like it was in a spider's cocoon.*

---

ready to be brought back on board *Swan*. As luck would have it, it was the only time the overhead monorail recovery crane on *Swan* ever broke down.

We were being pulled up to the fantail of the ship by hand and were about to go under the stern and into the spinning three foot diameter propeller, when I blew the torii to gain buoyancy at the surface. It had been a deep dive with more gear being carried than the sub was designed for, so I had filled the torii with diesel fuel to get a little extra buoyancy, while submerged. When the torii blew, the diesel fuel blasted everyone on the *Swan* and made a big slippery mess of the deck. Slipping about, the deck crew tied the *Jeep* to the fantail, still in the water, as tight as they could in a bouncing seaway.

The skipper headed for the lee of Santa Cruz island into Smuggler's cove. An observer later reported that the *Jeep* looked like it was in a spider's cocoon; it had been tied to the stern with all the line available. Inside the sub, the viewport was in and out of the water and looked like the window on a vertical washing machine. We were being thrown about and bumped against the hull with a lot of thumps. The observer was hastily pumping crackers and cheese into his mouth with both hands, at about 90% efficiency with the remaining crumbs flying all over me, him, and the sub. An hour later, we were hoisted on board with only two casualties. The port motor was bashed up a bit and we had one half-starved oceanographer to feed.

That dive series ended the good times with the GMC crew aboard *Swan*. We next flew everything by C-124 from China Lake to San Clemente Island. Our first significant dive was to our operating depth of 2010' in the San Clemente escarpment. It was then, that we learned that the charts of the ocean bottom don't indicate how rugged the terrain can be in many places. In fact, the walls were as vertical as the Grand Canyon in places with the outcrops and stepped formations making descent awkward, to say the least. We came up simply by dropping weights and noted the first hint of natural light at 750' on a bright, early summer afternoon.

When working at San Clemente Island, we usually had as our surface support ship the *YFU 53*, a converted landing ship. We often performed our operations underneath the *YFU*, on or near a lighted platform that could be dropped to over 1000' beneath the ship. The platform, when at the surface and out of the water, filled a large hole in the center of the ship. It could be lowered to the sea floor by four cables and winches at each corner of the platform. Under water, we practiced landing on mock up submarine hatches to get data for the design of the Deep Submersible Rescue Submersibles. With special brackets attached to our submersible, we demonstrated the sub's ability to capture and recover lost torpedoes.

On one occasion, while operating under the *YFU*, a sudden surge caused the *Jeep* to rub against one of the lifting cables on the lowered platform which shorted out one motor and all thirty, 4 lb. ballast plates held by the electromagnets, causing an immediate ascent. We pointed the good motor down and ran it full power forward to slow the ascent. We removed the internal periscope parts and replaced them with a plug made for just such an emergency. Happily, we reached the surface without the crash we expected from hitting the bottom of the ship. When we replaced the periscope optics, nothing could be seen but Navy gray paint. The edge of our sub had cleared the ship's bottom by 6" in an out of control 300' vertical ascent.

At the end of this series, we finished all testing and operational qualifications and prepared the submersible for an indefinite inter-lab loan to Scripps Institute of Oceanography. As part of the inter-lab loan, it was decided in June 1965, to give demonstration dives to the potential *Deep Jeep* pilots and project people at Scripps. A series of dives were made using the B Street pier near downtown San Diego, employing a mobile crane for launch and recovery. In spite of the low visibility in San Diego Bay, each of the four or five Scripps people selected had a chance to dive briefly in the submersible and/or observe the operations from topside.

After the demonstration, dives the project was returned to China Lake where all the spare parts, miscellaneous handling and operating equipment and expendable provisions were being assembled for shipment to Scripps. The project was in a semi-dormant status as several of the temporary personnel were transferred off-station and the official paper work was in process for the transfer of the sub.

On 17 January 1966, during a 24 hour Strategic Air Command (SAC) flight out of North Carolina, a B-52 had finished its first refueling over Spain, made a loop around the periphery of the Mediterranean Sea and was preparing to refuel for the second and last time, by air tanker, over Spain before heading back to North Carolina. The B-52 was a little fast as it approached the KC 135 tanker. In a standard procedure, the pilot ducked down 50' below the tanker until the speeds of the two planes became equal, so the refueling process could be initiated. On board the B-52, there was first a rumbling noise then a bang as the plane broke apart and disintegrated. When pilot Messinger felt flame on his face, he pulled the ejection handle and was shot out of the plane with minor injury. Second Pilot Wendorf stayed aboard while the plane tumbled forward in a somersault-like motion, until he too ejected.

Later, it was concluded that the right longeron, which acts as one of two structural beams lengthwise in the fuselage, failed back around the bomb bay and wing root causing the rear of the plane to fly off and crash into the tanker. Aboard the tanker were a pilot, co-pilot, navigator and a boom operator. The tanker went into a descent without transmitting any radio distress signals and disintegrated at about 2000' above the ground. Wendorf witnessed the tanker blowing up from his parachute shortly after ejecting. Co-pilot Rooney, Messinger, and Wendorf all landed at sea and survived to retire much later as Lieutenant Colonels. Navigator Buchanan landed on land and was the most severely injured. He was unable to fly again. Neither the tanker personnel, nor the three other B-52 crew members survived.

Four H-bombs were dispersed by the explosion from the plane. None were found in any of the aircraft wreckage on the ground. A search located three of the bombs over a wide area of ground, but the fourth was missing. Each survivor was interviewed individually. Afterwards, Messinger and Wendorf compared notes and agreed that the bomb, which had a large, white, ribbon-type equipment parachute, landed in the water near where they had landed, about 12-15 miles from shore. The personnel chutes were smaller and orange and white, so the larger bomb chute stood out among the other airborne chutes like a sore thumb. Both pilots indicated to the investigators on a map, by cross reference points on land, where they saw the bomb go in the water.

At China Lake on the 24 of January 1966, while working late on a new project *(Deep View)*, I received a request to attend a meeting in the Assistant Director's office. Since it was 5:30 at night and an hour after quitting time, some kind of "Big Medicine" was suspected. Upon entering the meeting room, I was surprised to see so much Navy brass. It looked like a varsity high school letterman's rally with all the striped sleeves in the room. The subject was the lost bomb at Palomares. The group asked if the *Deep Jeep* could search for the lost H-bomb off Spain until *Alvin* and *Aluminaut* were to arrive in 3 weeks. I told the group we could leave in thirty hours, but it would have to be on a limited basis since no spares existed for our motors or underwater electrical cables and that we had a limited period of reliable operations due to the developmental nature of the project. The limitations were already known to the group and acceptable, so it was indicated that I'd soon hear from the duty officer who was in charge of the Salvage Operations in Washington D.C. The duty officer called me at my office about thirty minutes later, other calls continuing well into the night.

One man, besides myself, was all that was left of our last *Deep Jeep* crew due to its demobilization. A call to a Navy carrier squadron got my submersible co-pilot, CDR. Phil Johnson, back (he was a flier by his Navy designator, but had previously volunteered half his time with our group while at China Lake). Another call to Westinghouse got a valuable technician, Glenn Minard, an ex-Navy warrant officer back. The only civilian submersible pilot on *Moray*, Al Berryman, joined up for the duration of the emergency. Thirty or forty phone calls to Washington D.C. to or from different Navy Captains and Air Force Colonels got aircraft reserved and the paperwork started. I got very little sleep, but with the excitement and a lot of sandwiches brought in, passport arrangements, material ordering, and travel order arrangements, somehow everything went humming along fairly smoothly. My lone machinist and

*Deep Jeep crew at Palomares 1966. Left to right: Berryman, Johnson, Forman, Ball, Minard*
Courtesy of Phil Johnson

only employee, John Ball, spent several days packing all the spares, tools and extra chemicals into crates with very little help from me.

About 28 hours after the initial meeting we were loading our project on a C-130 at the Naval Air Station, China Lake. Passports were obtained by those that needed them and the out-of-town crew were on hand. We took off for Charleston S.C. on the morning of 26 January. We had a presidential priority, of some kind, that cleared the field of all other traffic at Charleston while the pilot made a straight in approach. On the ground, you would have thought we had found a cure for cancer or had five medal of honor winners on board.

The Air Force ground staff had our gear unloaded, reloaded on special pallets, stowed aboard a new C-141 and all lashed down in less than two hours. A doctor arrived with trays of shots, which he gave us even while we were signing papers, making phone calls, or directing the lifting of delicate equipment, like our submersible. It seemed like only a few minutes until we were out of there, still working on unfinished preparations for diving. About an hour out, we heard on the radio that the Navy had en route a bathyscaphe (us) and that "the bomb had been located off the coast of Spain in 1900' of water." After that, the shots and the delayed fatigue hit us and we woke up in the landing pattern at Madrid, Spain.

The Safety Officer and I were briefed by the Atomic Energy Commission people and the Air Force. They had, in loose leaf form, a manual full of all kinds of airborne nuclear bombs with pictures and descriptions just like a *Betty Crocker* cookbook. The presentation could have been scripted in Hollywood. The only person under the rank of colonel in the room was a warrant officer. He flipped through the cookbook, describing the bombs as he went: "This is your Mk 32, 1.6 Meg (equivalent to 1.6 million tons of conventional explosive); this is your Mk 38 at 4 meg." and so forth. The smallest one, as I remember, was somewhere around 50 times more powerful than the one at Hiroshima and 280 times bigger than the total number of bombs dropped in a 1000 plane raid on Germany in WW II. After the briefing, a driver took the Safety Officer and me out about eight miles from the main part of the base to a large old armory building with 2' thick concrete walls and big heavy steel doors. While everyone else waited in the car, I was escorted to an inner room where one of the recovered bombs was located. I measured everything about it, for use later, if needed, for fabricating a capture and lift device. The bomb which I inspected now rests in the Atomic Museum in Albuquerque. For all the power inside the weapon, it was only about 15' long and 30" in diameter.

While we were looking at the cook books and bombs, the Air Force had transferred our project to a propeller driven plane for a flight into a field with a short runway at Torrejone, Spain, where the Spanish Air Force Academy is located. From there, U.S. Navy trucks hauled everyone and everything to the seaside fuel pier at Cartagena, Spain to await the arrival of our Navy support ship.

The next day, 29 January, *Deep Jeep* was loaded onto the 205' long *U.S.S. Kiowa* (ATF 22), a fleet tug. Just before *Kiowa* was to pull out from the pier, a man from Scripps Institute of Oceanography introduced himself as a sonar specialist. He had just been running sonar side scanning operations with a towed "fish" or torpedo shaped device off of *Kiowa*. He had found an anomaly during one of his many sonar transits across the sea floor. His sonar display paper looked like a sheet of very coarse sandpaper to me, with a few grains of sand out of place. He was firm in his declaration that while he didn't claim the disrupted sand was a parachute, he did think it was man-made. He admitted that it might be a fishing net except "they don't normally float." I noted the cross reference points on land that he cited and kept the sonar display sheet. The anomaly was far from shore in deep water.

*Deep Jeep in Action*
Courtesy of NWC, China Lake

The next day, 30 January, we made the first manned dive assigned by the Task Force Commander close to shore in 200' of water. For over two hours, we saw nothing but bumps in the sand. The beach was lined with Navy divers who could readily go to 200', so we became suspicious for the first, but not the last time, that the leadership might be flawed. Why assign a 2,000' deep submersible to shallow water search?. The next day, 31 January, we were assigned an area even closer to shore and dived for three and a quarter hours, again to see nothing but sand banks at 175 foot depths.

By this time, I was mad as hell. On 1 February, *Kiowa* was assigned the job of picking up a large aircraft section close in shore. We couldn't dive, so I got a whale boat and an interpreter from the Army general in charge of the bomb search and disposal group on the beach. We took the fisherman that had rescued one of the Air Force pilots and had reported seeing the bomb hit the water, out in a whale boat. The fisherman's name was Francisco Simon Orts. He said he knew the area like the back of his hand and indicated several hill prominences ashore to confirm the location he directed us to. I marked a map and, upon comparison to the Scripps sonar anomaly, found the two locations to be very close. Indeed, Francisco's location was in 2000' of water which was close to the 1900' of water where the radio had reported the bomb to be. I was sure that "we had broken the code" and relayed all of the information to the Task Force Commander by dispatch, requesting permission to dive on the spot the next day.

We received no reply and the next day, instead of the proposed area, were assigned an area far from the "Fisherman's Spot," as it was called thereafter. We spent four hours at 890' without seeing anything man-made, only a rock cliff. Another request was made to dive at the "Fisherman's Spot," but spent four hours the next day at 200' depth looking at more rock outcrops. After a tiring day of useless effort and with a bad toothache, I wrote a dispatch that would either get us a dive at the "Fisherman's Spot" or get us fired and sent home (since there was no other trained pilot for *Deep Jeep*). The Safety Officer didn't send it, and it was just as well, since he could have been hurt in the fallout, which was not intended.

The useless dives went on in hazardous seas in very deep and very shallow water far from the fisherman's spot with dispatches sent every night asking to dive at the "Fisherman's Spot." We were often miles from shore in an experimental submersible in rough seas with no other submersible for rescue in the area — and the toothache continued. On 7 February, as we entered the submersible for the second long deep dive of the day, my co-pilot said, "this is the first time my ass has even hurt getting in the sub." Posterior pain was common after a dive, but previously had always subsided overnight between dives.

On 8 February, *Alvin* arrived aboard a landing ship dock (LSD). *Deep Jeep* had been operating for a day without reverse on one motor but could still maintain excellent maneuvering and control. The visibility was as good as could be attained in the murky water, even with our wide field of view of 140 degrees. However, the maneuverability slowly deteriorated as the electrical cables began to flood out. Since *Alvin* was on hand and our submersible was unrepairable at sea, a dispatch was sent to the Task Force Commander requesting detachment.

While awaiting orders home, our crew went aboard the LSD to brief the crew of *Alvin*. I knew Bill Rainnee, the chief pilot, quite well and we met Dr. Hays, the chief Scientist in charge of the expedition for the Woods Hole Oceanographic Institution (WHOI). Both men were held in pretty high respect at WHOI in their respective positions. I relayed to them what a screwed up mess I thought there was and told them the whole story about the Scripps sonar, Francisco's trip in the skiff, and the 1900' radio message. I told them that my secret plan had been to take the *Deep Jeep* to 2500', if necessary, since it had been tested to 2700' and had a calculated crush depth of 5000'. I strongly emphasized that we were not allowed to even try looking in the Fisherman's spot. We also pointed out that it was now up to *Alvin*, with help from the *Aluminaut*, and that they had a good chance if only they could dive on the fisherman's spot during a test dive or – by some terrible error in navigation – stray from an assigned area. We left all of our equipment and spares that they might use and were allowed to depart the area a few days later.

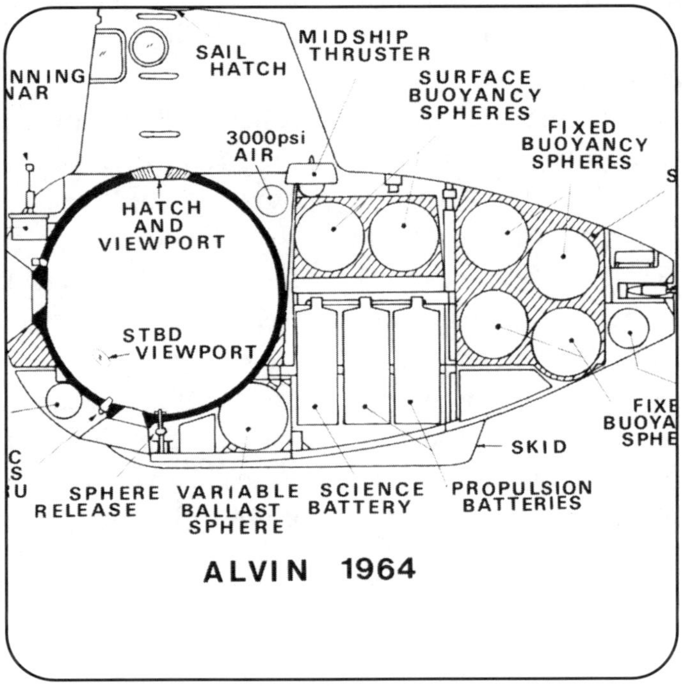


***Name:*** *Alvin* ***Builder:*** *General Mills/Litton*
***Length:*** *22'* ***Maximum Depth Rating:*** *6,500'*
***Year Launched:*** *1964*
Courtesy of Naval Photographic Center, Washington, D.C.

As a historical footnote, when the bomb was finally located some six weeks later, after several severe storms, the location was at the end of a groove and down slope from the point where Francisco had indicated that the impact had occurred in 2000' of water.

We had no priorities going home so it took awhile. On meeting the Director and Captain of the base upon arrival at China Lake, they seemed quite complimentary on the job we did in Spain. I was a bit ashamed of not having done better, but to blame someone else is a tactic that never works. A year later, I received a Navy Meritorious Civilian Service Award from the Assistant Secretary of the Navy for participation in OPERATION AIRCRAFT SALVOPS MED. and made a Fellow in Ordnance Science from the Naval Ordnance Test Station at China Lake. Ideally, it would have been appropriate to tear those awards into a lot of pieces and present them to the many people that helped with the work.

The bomb was finally located and retrieved months later near where it had gone in at the "Fisherman's Spot." Inside the military the word gets around, and, almost always, justice prevails in a slow, frustrating, but inevitable way.

*Deep Jeep* was transferred to Scripps Institution of Oceanography (SIO), after it returned from Spain. Fred Spies, of Scripps, thought that the one million dollar initial certification cost, plus the periodic certification and constant maintenance costs would be more expensive than other research that they could do of equal value. *Deep Jeep* then went into storage at Scripps, but is nowhere to be found at SIO today.

Like so many developmental deep submersibles, *Deep Jeep* lacked enough funding to get through the first 18 months of sea trials necessary before reliable operational status could be attained and, thus, be eligible for major program funding and long term usage. Nevertheless, it was the first of its kind and, consequently, had an impact on much that followed.

# *Alvin*

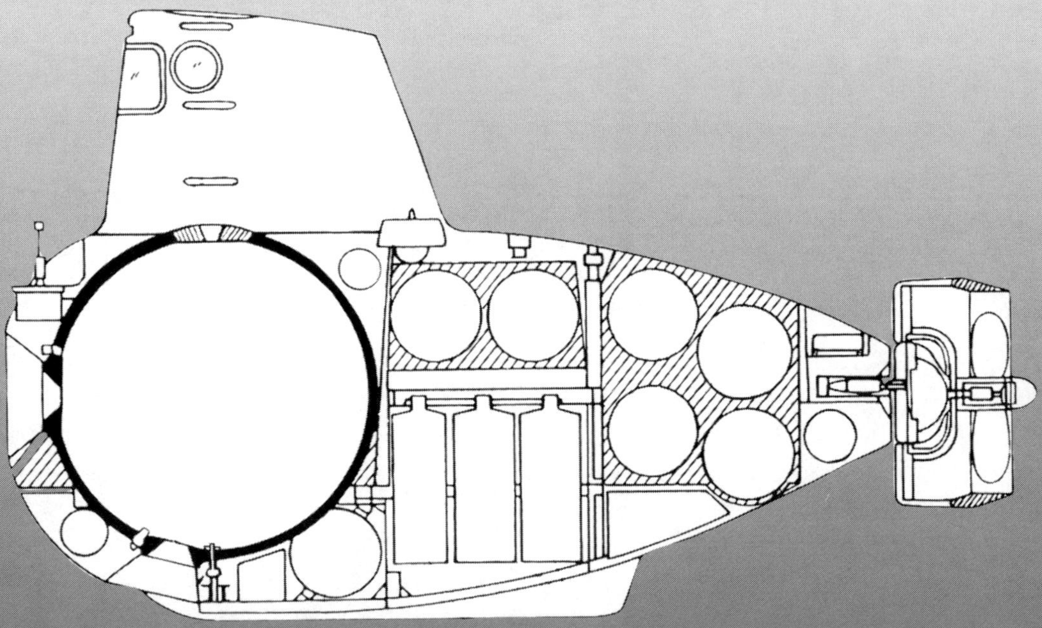

**1964–(        )**

# Alvin

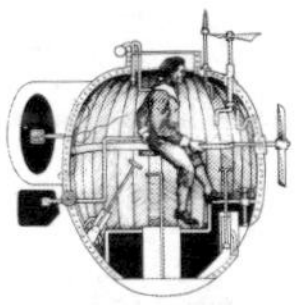

Harold E. (Bud) Froehlich worked in the Nuclear Equipment Department at General Mills, developing new markets for nuclear research manipulator arms which had been originally designed for laboratory use with radioactive materials. After visiting a customer at Scripps Institute of Oceanography (SIO), Bud called on the *Trieste* group at nearby Point Loma Naval Electronics Laboratory (NEL). Having just finished the deep dive in the Marianas, the *Trieste* Group was looking ahead to the scientific use of the submersible, for which a manipulator would be necessary. They modified a manipulator design for *Trieste* purposes which General Mills ultimately delivered, followed by operational training.

Over several sessions between Froelich and the *Trieste* crew during the purchase and training process, it became clear that the *Trieste* crew was of one mind about the need for a much shallower diving, cheaper to operate, "Volkswagen" type of research submersible. It had taken them seven days to prepare the *Trieste*, whether for a 4,000' or a 38,500' dive plus the towing time back and forth to the dive site, and another seven days to return *Trieste* to its shore cradle. They agreed, in concept, upon a highly maneuverable submersible with a 6,000' depth capability, small enough to be launched and recovered off of a 200' long fleet tug or similar vessel. The Volkswagen concept should permit many dives per trip to sea, economically, aboard a "mother ship." Bud, however, knew about the *Deep Jeep* being built without fanfare at China Lake. He wondered if an outside contractor had a chance against the regular Navy bureaucracy to sell the sort of equipment he was sure many Navy-sponsored laboratory scientists would want.

On his next trip to Washington D.C., Don Walsh, the Officer-in-charge of the *Trieste*, reported to his Office of Naval Research (ONR) boss, Captain "Swede" Momsen, on the proposed new submersible idea, now nick-named *Seapup*. Momsen was very receptive to the idea, more to solve one of his own dilemmas than to give NEL a small sub. Woods Hole Oceanographic Institute (WHOI) had been urging ONR to find them a research sub for several years. The closest Momsen had come was an offer from Reynolds Aluminum. Reynolds wanted ONR to pay for most of the construction of a large aluminum submersible called the *Aluminaut*, and then for WHOI to lease the sub from Reynolds with ONR money – an offer that Momson rated as less than generous. Meanwhile, to

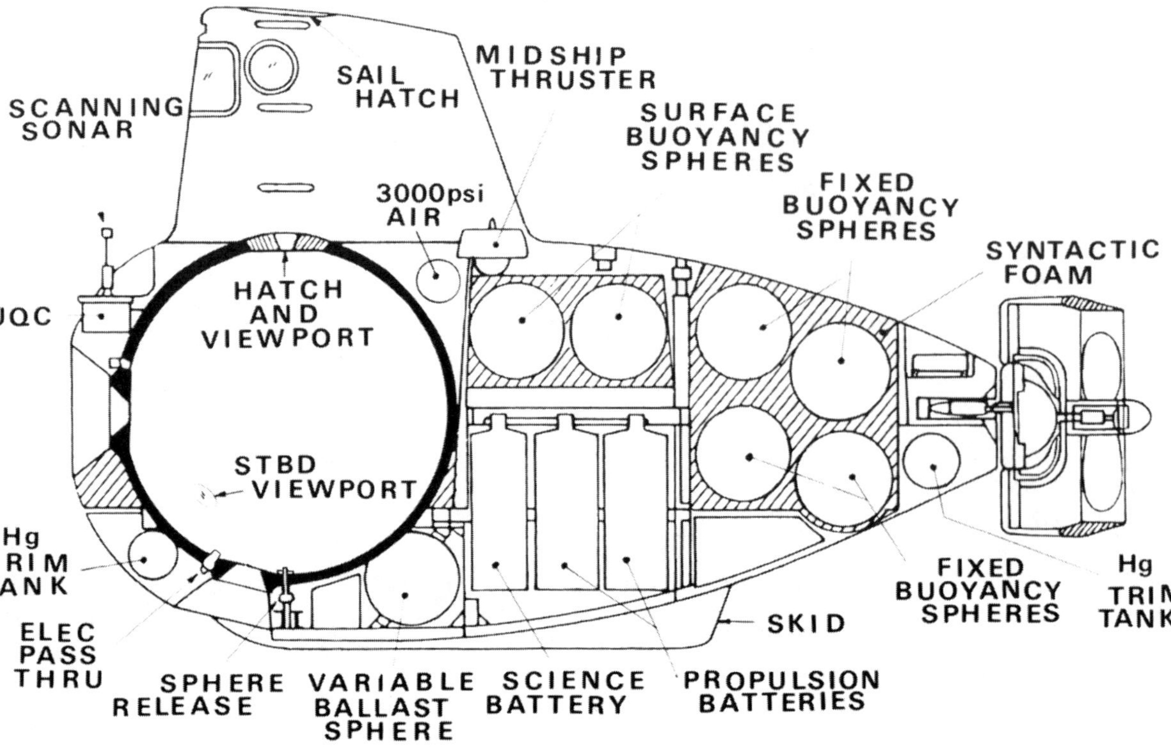


***Name:*** *Alvin* ***Builder:*** *General Mills/Litton* ***Length:*** *28'* ***Maximum Depth Rating:*** *6500'–13,124'* ***Capacity:*** *3 Man* ***Year Launched:*** *1964* ***Description:*** *Detailed Cross Section*

compound his problem, the Bureau of Ships was trying to get authority for complete control of the design and administration of anything funded or used by the Navy on or under the water. Momsen, with a limited annual budget, but wanting to get a safe submersible before the bureaucrats tied everything up in red tape, counter offered to lease the *Aluminaut* for three years, after which time the Navy would own it.

Several times, when Momson thought he had an arrangement ready for signature, someone at Reynolds changed his mind. *Seapup* looked like a way out. With money set asidc for leasing *Aluminaut*, he advertised for competitive bids for a Seapup-type submersible. General Mills gave strong backing to Froehlich. Their lowest bid of $500,000 won them the contract, which included a guaranteed performance clause. Now, the submersible was to be called *Alvin* in honor of Al Vine who had spearheaded the drive for an undersea research vehicle.

The original *Alvin* design, so simple and straightforward, has remained almost unchanged twenty-five years after its first dive. While almost all of the original equipment has been replaced or improved, the basic design remains with few alterations. In a cross section view, the largest single component is the personnel sphere. After the original contract between ONR and General Mills was let, an amendment was added to allow a total of three personnel spheres to be fabricated of HY-100 steel (100,000 psi). Wherever imperfections were located in the spheres, hatches, and viewport areas were cut out of the imperfect areas and replaced with inserts of thicker material in the case of viewports and better metal in the case of the hatch.

Southwest Research Institute (Sw.R.I.) tested spheres #1 and #2 to approximately

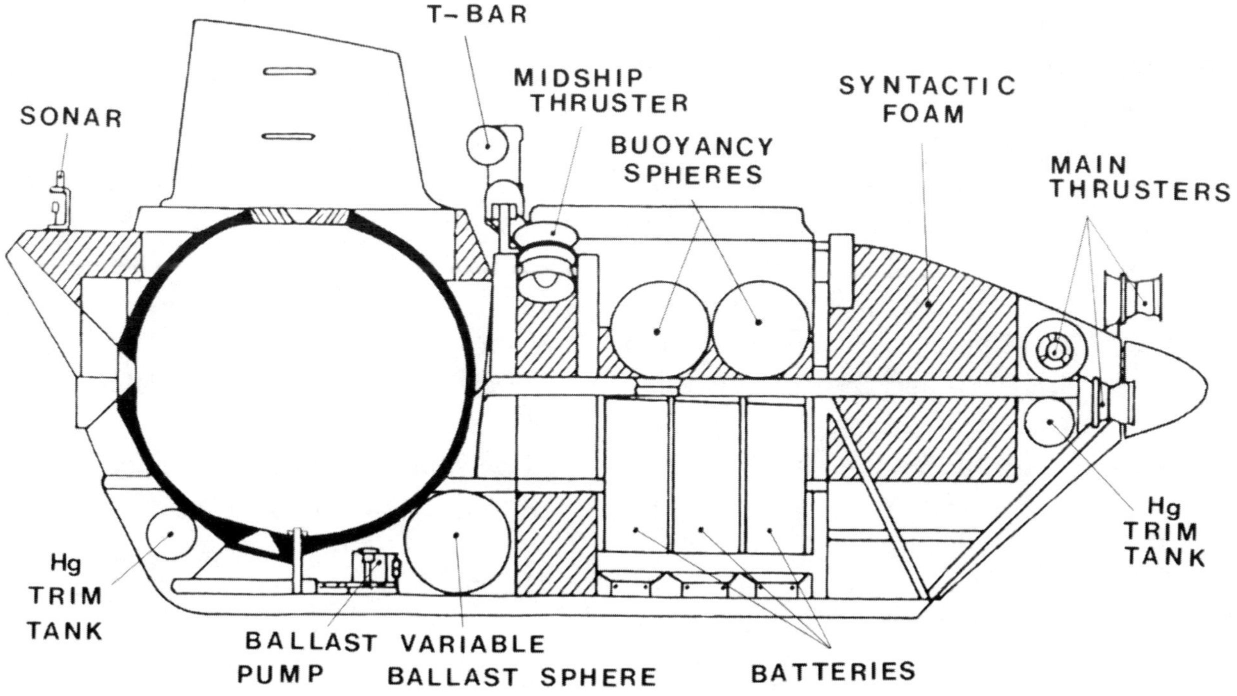


*Alvin, 1994, Detailed Cross Section*

1.25 times their operating depth of 6,000' in a pressure test tank. Instrumented tests with strain gages indicated that both hulls could stand much greater pressures before failure. The #2 hull was sent to Minneapolis for assembly, while the #1 was prepared for proof testing to failure in the SwRI pressure tank. At just short of the equivalent pressure of 10,000' of sea water pressure, the tank blew up, but the submersible hull was virtually undamaged.

The construction and assembly of the #2 hull for *Alvin* continued after the pressure tests. The finished sphere weighed 8,000 lbs., had a nominal wall thickness of 1.33" with a safety factor of 2.7, and an outside diameter of 6'-10". Operationally, it was detachable from the rest of the submersible along with the forebody and sail, retaining a buoyancy of 400 lbs. when released. The original submersible had a 1,200 lb. payload, a 13 ton displacement, an overall length of 22' with a width of 8'. Cruise speed was 2 kts. with a maximum horizontal speed of 6 kts. for brief periods.

The five viewports in the personnel sphere each had a safety factor of 2.8. One, located forward on the "equator," looked straight ahead. Another, near the bottom of the sphere, looked slightly forward from straight down. Two side-viewing ports, one on each side and slightly forward of the beam of the sphere, looked outward, downward, and forward. A smaller viewport was located in the center of the hatch.

Twelve thru-hull electrical penetrators around the bottom and forward viewports provided up to 216 electrical leads of 10 amps each. Directly above the personnel sphere was the five foot high sail. To give the submersible adequate freeboard and to project the sail completely out of the water when surfaced, 3,000 psi air tanks were used to blow the water from the two main ballast/buoyancy tanks, which were flooded when diving. Vent and flood valves drained or flooded the sail near the surface. A large window was located in the upper forward section of the sail, with smaller viewports on each side, while the entry/exit hatch was located at the top of the sail.

When submerged, the buoyancy/ballast trim was adjusted by pumping up to 600 lbs. of oil between six pressure resistant aluminum spheres and two compartments with flexible membranes for expansion and contraction. Up to 350 lbs. of buoyancy could be provided from the displacement of the seawater by the lighter oil. Nine aluminum, air-filled spheres, two feet in diameter with a total of 4,000 lbs. of lift, provided fixed buoyancy. Syntactic foam, shown in cross hatch in the cross section, surrounded the spheres for additional lift. Pitch was controlled by transferring up to 600 lbs. of mercury between the forward and aft trim tanks for a maximum of 25 degrees of pitch, up or down.

Three shrouded propellers powered by hydraulic motors provided thrust for fore or aft, as well as up or down and turning motions. The reversible 10 H.P. motor and 50" diameter, five-bladed propeller aft pivoted 50 degrees left or right for turning and horizontal propulsion. The two pivoting 7 1/2 H.P. motors with 14" diameter propellers outboard and amidships, primarily for vertical thrust to aid in ascent or descent, could be used to aid in turning or for greater horizontal speed. Three droppable containers housed the oil-filled, lead-acid golf cart batteries–two for propulsion, one for powering scientific equipment. The forward looking sonar and underwater telephone (UQC) transducers were located on the brow of the submersible, just below the sail. A manipulator arm, on the port side of the forward window, was droppable in an emergency.

After the commissioning ceremonies on 5 June 1964, *Alvin* was put through its weight, balance, and stability tests. After installing additional syntactic foam buoyancy in the sail, *Alvin* made its first manned, untethered dive on 4 August 1964. Making all its dives in Woods Hole Harbor or nearby Buzzards Bay under 70 feet, three pilots were trained to operate *Alvin* and 64 other men dived as observers.

Extensive overhaul of *Alvin* began in early October. Construction of *Lulu*, the surface-support craft named after Al Vine's mother, began the same year. *Lulu* was a catamaran composed of two, surplus, 96' long, Navy pontoons, which in the final configuration were 9' high, connected together by two 50' wide archways, resembling oriental arched bridges. A local builder assembled it for just under $43,000. The ship's control bridge, workshop buildings, and galley, atop the arches, provided surface support while at sea. The port pontoon held the power plant, while the starboard pontoon provided quarters for up to 25 people. A single, large outboard motor, located centrally and forward on the ship, made towing required for all but local operations.

After the overhaul in March of 1965, *Lulu*, *Alvin,* and crew headed for Cape Canaveral for shallow sea trial dives, followed by shallow dives in the Bahamas and an unmanned tethered submergence to 7,500'. The next series of dives led to a manned dive to 6,000' on 2 August 1965, earning the sub its certification by the Navy and completing the ONR purchase contract for $950,529. The Artemis underwater listening arrays in the Tongue of the Ocean provided ONR sufficient justification alone for *Alvin*'s construction as a range tool. The project group next sailed for Bermuda for five dives before hurricane Betsy hit in late September. Two shallow dives in Woods Hole wound up the 1965 season.

In late January 1966, with *Alvin* in the middle of its annual overhaul preparing for a February inspection of the Artemis range in the Bahamas, the ONR notified project leader Earl Hays of the B-52 accident and a lost H-bomb in Spain. With very little information to go on, the project members worked night and day to finish the overhaul and prepare for travel. On 1 February, after lengthy, intensive preparations, *Alvin*, the shipping vans, and the crew were loaded into two C-133s to fly to the U.S. Navy base at Rota, Spain, near Gibraltar. A harbor test, run soon after arrival, revealed a battery leak. After three days of preparation and repair and a second test dive, *Alvin* was ready. The project vans, submersible, and personnel were loaded onto the Landing Ship Dock (LSD) Plymouth Rock along with the submersible, *Aluminaut* for transport to Palomares, Spain. At Palomares, during final transfer of the project to their surface support ship, the LSD Fort Snelling, a 70 Kt. gale blew up. Both *Aluminaut* and *Alvin*, with their crews, had to be tied off on buoys for twenty hours until the storm blew out.

Finally, on 14 February, *Alvin* could take its first dive at the search site. Three pilots, Bill Rainnee, Marv McCamis and Val Wilson, took turns as Pilot or Surface Controller. The underwater theater of operations generally consisted of a gentle slope from the shore seaward, gradually increasing in steepness to about 24 degrees at around 2,400', then to about 35 degrees at 3,000', where some leveling took place. Various gulleys, valleys, and ridges appeared randomly in a shore to seaward pattern, McCamis remarking that the bottom looked like "wrinkled old skin." *Alvin*'s mercury vapor lights provided 20-25' visibility, despite the suspended material in the water.

The Navy assigned half-mile square areas, the first "square" was about five miles offshore in 1,800' of water. The search work, dull and monotonous, strained the eyes. Eight more dives were made without seeing anything of interest. By this time, the "fisherman's area" where the Spanish fisherman had seen the bomb impact near his boat was well-known and plotted on Fort Snelling's charts. The fisherman had been brought out to the area many times and always selected the same spot. Having fished there all his life, it was second nature for him to triangulate with the mountains and other landmarks. Moreover, he had perfectly described the odd pattern and colors of the cargo chute on the bomb. But, despite repeated requests, the Navy absolutely refused permission for *Alvin* as they had for *Deep Jeep* to dive in the "fisherman's area."

Since the Navy "square" on 1 March was fairly close to the "fisherman's area," McCamis, the surface topside controller, and Dr. Earl Hays, the project leader, contrived to "accidentally" search off-course, and, with the support ship captain's knowledge, directed *Alvin* toward the presumed bomb impact area. The *Alvin* pilots, in spite of many passes, saw nothing. Later, while viewing videos during the post-dive debriefing, one project member noticed a track in the mud resembling a bomb skid groove. McCamis asked Earl Hays to get permission to take another look. Meeting strong resistance, "Earl Hays bluntly told Rear Admiral William S. Guest, leader of the Task Force, that if the Navy did not cooperate, he would take *Alvin* home. The Navy cooperated." This was reported in the

*In the excitement, Wilson forgot the prearranged code words and started shouting over the underwater phone, "We found it, we found it!"*

---

Woodshole publication Oceanus Vol. 31 #4. On March 2, 3, 4, and 7 dives, no groove could be found. On March 8 and 9, they dove at Navy assigned areas. On March 12, they returned to the "fisherman's area." Four hours into the dive, they found the track on the 35 degree slope, but it was so late they had to surface. The next day, they returned to the area, but the larger, less maneuverable *Aluminaut* had been there and torn up the bottom. After a fruitless — and, needless to say, frustrating — day, they surfaced.

On 15 March, the Navy wanted the *Alvin* group to shift to another area, but the pilots decided on one last attempt. While traversing the slope, they found the track and followed it down slope. Watching through the side windows, Val Wilson and Art Bartlet, an electrical technician, both shouted out. In the excitement, Wilson forgot the prearranged code words and started shouting over the underwater phone, "We found it, we found it!" Told to stand-by until *Aluminaut* could relieve them, they waited eight hours with their lights off to save the batteries. *Aluminaut* rendezvoused successfully and remained on station 22 hours until *Alvin* could recharge its batteries and install a mechanical arm.

Returning to the site, the *Alvin* pilots discovered that *Aluminaut's* lights in the background allowed them to see 150 feet, instead of the usual 20' when looking through the backscatter from their own lights. This close-working relationship resulted on one near-incident. While hovering near and beneath the horizontal stern of the *Aluminaut* which was resting its bow against the steep slope, the subs almost collided. *Aluminaut's* stern suddenly began to drop, forcing *Alvin* to take – as aviators like to say – evasive action. What happened was that a heavy crew member in *Aluminaut* had walked suddenly to the toilet at the back of the sub. Soon afterwards, *Aluminaut* left and the *Alvin* pilots firmly anchored a pinger in the chute fabric with fish hooks. The bomb, visible among the shrouds and chute fabric, was still attached to the bomb racks from the plane. Wilson and McCamis knew this was the bomb, as weeks earlier an Air Force Sergeant had illegally shown them a picture of the bomb and chute and given them all the pertinent dimensions of each.

The very information vital to recovery had been denied to the *Alvin* crew under some misguided, ignorant pretense of security. McCamis reported that, upon showing their underwater pictures to Admiral Guest, he asked sharply, "How do you know that it isn't a bag of mud?" McCamis wrote, "In all my life, I'd never had my intelligence so insulted. We returned to our mother ship tired, discouraged and disgusted" (also in Oceanus vol. 31 #4).

On 18 March, after a day of bad weather, the *Alvin* pilots carried a 3/8 " diameter line, paid out from the surface down to the bomb and plunged a harpoon into the seabed. A float was attached to the surface end of the line as a marker. On 23 March, after four days of bad weather, *Alvin* tried to put a clamp on the bomb, but were prevented from doing so by the mud around it. On 24 March, the *Alvin* pilots were able to pull the chute clear of the bomb, shroud by shroud.

On 25 March, a heavy-framed apparatus called a "poodle," with various hooks and grapples attached, was lowered into the area near the bomb. The *Alvin*-arm attached a grapple from the poodle to the top of the bomb chute. A cable tethered the poodle to the surface for lifting. However, instead of lifting the bomb vertically to the surface, they attempted to drag it up the steep slope to shallow water, finally dropping the bomb because the lines could not withstand the frictional forces. On 26 March, *Alvin* found it again after hours of search.

On 27 March, *Alvin* again painstakingly pulled the chute down slope, clear of the bomb and left a grapple with a line to it in the chute. CURV, an unmanned, tethered, remote operated vehicle with a television camera viewed from topside, was attempting to clamp onto the chute when someone topside gave the order to yank CURV's umbilical to get an unobstructed view of the bomb. While almost snaring *Alvin* in the chute, the bomb was lost again. After four days of exhaustive search, *Alvin* again found the bomb, but a sudden billowing of the chute almost ensnared the submersible. They were ordered to surface.

On April 7, while the *Alvin* crew observed, the CURV controllers tried to grapple the bomb, totally ensnaring CURV in the chute and shrouds. In a gamble, CURV's controllers tried lifting the chute and bomb by simply hauling CURV up by its cable. It worked. When 200' from the surface, divers attached additional lines to the bomb for the airlift onto the ship's deck.

During the operation, *Alvin* had made 34 dives as deep as 3,200', averaging 6 1/2 hours per dive. After returning to Wood's Hole for a change of batteries and minor repairs, *Lulu* left for Bermuda with *Alvin* on deck in late June. Inspection of the Artemis listening arrays, near Argus Island, began on 1 July and took up most of the nine dives made there. At the next stop, the Tongue of the Ocean in the Bahamas, the Naval Oceanographic Office had *Alvin* make fifteen performance evaluation dives. *Alvin* and *Lulu* returned to Woods Hole in mid-October 1966 to finish the overhaul of both vessels that had been aborted to accommodate Palomares.

During the overhaul, which lasted until April, 1967, fifteen hollow 24" diameter titanium spheres were substituted for nine of the aluminum spheres in the syntactic foam and the six used in the oil ballast system. Two of the aluminum spheres, formerly used for fixed buoyancy in the syntactic foam, were substituted for the scuba tanks previously used to blow the water out of the sail at the surface.

*Alvin* made the first dives of 1967 in the Bahamas and the Tongue of the Ocean for the Naval Oceanographic Office. On 6 July, at 2,000 feet, during the next dive series for biological and geological research in the Blake Plateau area off Miami, a 200 lb., 8' long swordfish attacked *Alvin*, lodging its 18" sword into the external hull above the forward window, just missing a set of electrical cables. Positive identification as to the sub-species was impossible, as the crew ate it.

*Alvin* next dove offshore and south of New England for geological, biological, thermal, and sound measurement studies. On 13 August, 1967, an F6F WW II fighter plane was spotted at 5,214 feet. It had been there since 1944 and was in amazingly good shape. On 24 September, the mechanical arm was bumped loose in a rough sea impact with *Lulu* and sank to 4,400'. On 15 October, the arm was recovered using a special probe and repaired inexpensively, saving the $50,000 that a replacement would have cost.

The 1968 season began in mid-April with whale listening and watching until diving for re-certification on 5 June. Through early September, *Alvin* inspected the Azores and Bear Seamounts, potential listening-post sights for the Navy. On 16 October, while launching during a biological research dive series 88 miles east of Nantucket Island, both of the forward cables on the *Alvin* launch cradle snapped, pitching the submersible forward into the sea, causing it to submerge with Bland, the pilot, and two observers, Weaver and Stimpson, on board. *Alvin* popped back to the surface briefly, while flooding through its sail and open pressure-hull hatch. Bland, in the sail, jumped out first, followed by the two observers who had been in the sphere. Within several seconds, the sub sank a second time, disappearing in mile deep water.

The only casualties were bruises and Bland's sprained ankle. *DOWB*, the General Motors Research Lab submersible, was airlifted to Woods Hole, making one search dive for *Alvin* before winter weather made submarine operations impractical. On 4 June, 1969, the U.S. Naval Ship (USNS) *Mizar* located and photographed *Alvin* on the bottom. On 1 September, with *Mizar* at the surface, *Aluminaut's* crew spent 17 hours of painstaking labor rigging a toggle bar inside the open *Alvin* hatch. After tying a 7,500 lb. nylon lift-line to the toggle bar, *Mizar* reeled *Alvin* up to 85', where Navy divers slung a net around the submarine and attached three pontoons to it. *Alvin* was towed, submerged at 40', to Martha's Vineyard where a crane hoisted it onto a barge for transfer to Woods Hole. The depth gage still read 5,052'.

Damages to the propeller and sail were repairable; the hull and batteries were in good shape, as were most of the other parts. The cause of the accident was determined to be the continual flexing of the lift cables over a sheave that was not accessible for inspection or maintenance. The cable was replaced by heavy chain, which does not fatigue in bending. In all operations thereafter, the submersible would be lowered into the water unmanned with the pressure hull hatch shut.

Due to repairs and refurbishment required after the unmanned dive that lasted 10 1/2 months, *Alvin* did not dive again till dive #309 on 17 April, 1971. In 1971, *Alvin* completed 82 dives in spite of a late start, the first 14 in or near Woods Hole Harbor, mostly to check on the rebuilt portions of the sub. A June series of dives was made off Martha's Vineyard, very near to the accidental submergence area of 1968. Deep Ocean Station #1 (DOS#1), intended for various deep ocean time evaluated experiments, was dedicated and marked off on the bottom at 6,000'. In July and August, during 23 geological and physical oceanography dives in the Gulf of Maine, the first returns and data recorded at DOS#1 were obtained.

On 10 August 1971, a large marlin attacked *Alvin* at 1,500'. It circled three times before smashing into the lights, breaking two of them and injuring itself. During debriefing, the pilots decided to turn off the interior lights when in the presence of marlin or swordfish since the fish were apparently attacking the lights themselves and not some other part of the sub. Mechanical tests run on some spare 3 1/2" thick viewports determined them safe from marlin or swordfish attack.

*Alvin* completed a second series of dives in the Gulf of Maine in September and, in October, made 14 dives in the Florida straits. In November and December, *Alvin* dove 25 times in the Tongue of the Ocean, some for Defense Advanced Research Project Agency (DARPA), many to inspect the Atlantic Underwater Test and Evaluation (AUTEC) range equipment. Beginning on 5 May, *Alvin* made 76 dives. In 1972, Val Wilson and Ed Bland piloted all of the dives until August when Jack Donnelly became a qualified pilot. After harbor tests, the sub visited DOS #1 three times, once to establish recertification. A new ceramic pump replaced the old oil-displacement system, simply pumping sea water in or out of a pressure resistant sphere changing the ballast/buoyancy status. After testing a new rock drill in deep water in the Gulf of Maine, *Alvin* dove from 1,000' to 6,000' in Hudson canyon to survey the draining and/or dumping grounds for New York and other cities on the Hudson. With normal visibility as much as 50' or 100', less than 2' could be seen in any direction due to the terrible pollution. *Alvin* returned to the Gulf of Maine, testing the new navigation system on 16 dives and using a new hydraulic chipping-hammer for geological research. The sub made return trips to DOS #1 and the Gulf of Maine in August, September, and October, prior to overhaul in late October.

Bill Rainnee had convinced WHIO Director, Paul Fye, of the need for a titanium pressure-hull for *Alvin* back in 1967. ONR had rejected funding the proposal but, in 1968, the Naval Applied Science Lab agreed to have two hulls made as part of their ongoing research into titanium. Luken's Steel rolled and shaped four hemispheres for Mare Island Naval Shipyard to weld into two spheres of the same outside diameter as the original *Alvin* hull, but .60" thicker, for a total wall thickness of 4.9 cm (1.93"). The safety factor to be used was greatly debated, but all equipment, such as windows, light, and TV were designed for 12,000' dives. After assembly in 1973, the sub was tested twice, unmanned, to the equivalent of 12,000 feet in the test tank at Annapolis. Then, on 10 July, pilots Val Wilson and Ed Bland, with a Mr. L. James, made a manned, simulated dive in the chamber for 10 hours at 12,000' equivalent depth, which allowed Navy certification for operations to that depth. Due for extensive overhaul, *Alvin* made only sixteen dives in 1973, all of which were in or near Woods Hole Harbor.

Operations in 1974 began on 4 January, continuing the Tongue in the Ocean work started earlier. *Alvin* conducted biological research on the AUTEC Range in the Bahamas, including establishing a new Bahamian bottom station. Early in 1973, with funding at an all time low, project FAMOUS (French American Mid-Ocean Undersea Study) received funding by the two governments to study the Mid-Atlantic Ridge (MAR), using the French bathyscaphe, Archimedes, their deep submersible, Cyana, and the U.S. furnished *Alvin*. From February to May, *Alvin* conducted training dives while the Naval Research Lab's *Mizar* photographed the area to provide pre-dive data for the submersibles.

*Alvin*'s expedition series of 18 dives began on 30 June. On 17 July, with Jack Donnelly piloting, bottom currents pulled *Alvin* backwards at 8,868' of depth, wedging her in a crack on the bottom where no visibility was possible. On an unused radio channel, according to the prearranged emergency plan, *Alvin*'s sub-boss,

Shumaker, called Lieutenant Commander de Froberville, commander of the French subs, and described his plight. Immediately, *Archimedes* began its preparation for diving. However, while *Archimedes*, like *Trieste*, could dive extremely deep, she would take as much as 12 hours to get ready. Meanwhile, Donnelly was wiggling, rocking, and rolling and, after 90 minutes, freed *Alvin*.

On project completion, the three subs had covered much of the 20 mile long by 1/2 to 2 mile wide rift between two undersea volcanoes. Taking several rock samples and observing many highly unusual, underwater volcanic formations made by still erupting lava, the researchers concluded that the rift was formed not by volcanic activity, as supposed, but by tectonic plate movement. Over a 12 day period on the return leg to WHIO in late August, *Alvin* visited seven different seamounts in the New England Seamount Range, including the deepest dive of the year to 9977' on the Mytilu Seamount. After its 59th dive in 1974, *Alvin* went into overhaul on 5 October.

In April and May of 1975, *Alvin* conducted 25 biological and geological dives for the Naval Underwater System Center (NUSC) on the AUTEC Range in the Tongue of the Ocean. On 19 May, the sub reached 12,000' for the first time since the Annapolis pressure tank tests. From June through September, *Alvin* inspected the NOAA radioactive dump site and the garbage dump site in Hudson Canyon. Several dives were made on the Deep Ocean Station #1 (DOS #1 ) at 6,000' and a new DOS #2 was established at 12,000' south of Cape Cod. To obtain desperately needed funding, *Alvin* was declared a National Facility in November, giving the selection and control of the dive projects to a committee. WHOI, however, retained control of the *Alvin* operations and was assured of funding for three more years.

Dives in 1976 began with pilot training at Guantanimo Bay, Cuba, on 13 January. Sponsored by the National Science Foundation (NSF) for geological research, the Naval Facility Engineering Command (NAVFAC), and the Naval Air Development Command (NADC), *Alvin* spent the next two months at the interface of the Caribbean and the American Tectonic Plate, the torn up area of the ocean floor which includes the Puerto Rican Trench and the Cayman Trough. Most of the plate interface is of the "sliding" type, but the dogleg made by the Cayman Trough resembles the spreading MAR, which separates and lets magma exude.

*Alvin* reached or approached its new certification depth of 4,000 meters (13,124') nearly a dozen times during its exploration of the 4,921' trench, which is four times deeper than the Grand Canyon. *Alvin* conducted fifteen more dives for inspection and salvage in the Tongue of the Ocean, and then monitored DOS#1 at 6000' & DOS#2 at 12,000', recording the data in June. In July and August, the sub did biological research on the Eastern U.S. Continental Slope near Veach Canyon and the Atlantis Canyon, including five dives in the radioactive waste dump south of Cape Cod, retrieving drums for detailed surface inspection. The diving year ended with inspection and survey of the AUTEC range, the last and 90th dive of the year in December.

In early January 1977, dives continued in the AUTEC Range. After making deep dives of around 12,000' near Puerto Rico and St Croix, *Alvin* then transited the Panama Canal aboard *Lulu*, accompanied by the research vessel, *Knorr*. Twenty-five scientists and 26 technicians were eager to explore the northern interface between the NAZCA and the COCOS tectonic plates a few miles north of the Galapagos Islands. Called the Galapagos Rift, the area represents the separation zone between the two oceanic plates.

Their eagerness was rewarded by an extraordinary discovery. Thermistors, attached to a towed camera system called Angus, detected hot water, 63°F in one area of an otherwise near-freezing ocean. Photos of the warm water showed hundreds of giant clams among vast fields of lava beds. Diving with alternate pairs of scientists, *Alvin* took water samples indicating the strong presence of hydrogen sulfide. Some microbes seem to thrive on the sulfides, and clams and other organisms thrive on the microbes, creating a food chain without sunlight or photosynthesis. Animals around the warm water vents, immune to the toxicity of the water, reproduced rapidly and in larger sizes than normal. Some of the warm water vents were dubbed "oases" because they were isolated and tidepool-like, even at depths of 9,000', and contained 10" long clams, numerous crabs, lobsters, new types of fish, and 18" tube worms.

In the 24 dives made in 31 days, the scientific observers acted like small children on Christmas morning. Pilot Shumaker said that the volcanic caverns, gaping pits and glistening lava formations made an "Alice in Wonderland" scene of scintillating splendor in some areas. In late March, *Alvin* returned through the canal for six more geological and biological research dives in the Cayman Trough, then to the Bahamas and the Tongue of the Ocean, returning to WHOI in June. *Alvin* made 30 dives on the continental slope south of Cape Cod in Oceanographer's Canyon, Atlantis Canyon, as well as DOS #1 and #2 for geological and biological data retrieval, making its last of 101 dives for the year in October.

In an extensive overhaul which lasted through April 1978, a new titanium, external frame and a second manipulator were installed, increasing *Alvin*'s weight and length. *Lulu*, completely renovated for better habitability and speed, had its submarine cradle system improved. May and early June, spent in recertification and biological investigation off of Capes Hatteras and Charles, included the Environmental Protection Agency (EPA) dives on a radioactive dump site. Fifteen dives on the MAR, continued the investigation of tectonic plate spreading. *Alvin* rounded out its 81 dives for the year in late September in Oceanographer's Canyon, while devoting November and December to dives in the Tongue of the Ocean, Bahamas, St. Croix, and a new Deep Ocean Sight #3 (DOS #3) near Puerto Rico.

After transiting the Panama Canal, *Alvin* made another 29 dives on the Galapagos Rift in 1979. Researchers again observed tube worms similar to the 18" ones seen on the earlier expedition, but after one dive, they brought up on deck one worm 8' 6" long. ONR, NSF, and NOAA sponsored the Galapagos expedition as well as the East Pacific Ridge (EPR) work at 21 degrees north, which followed. In April, the Scripps ship *Melville* rendezvoused with *Lulu* and *Alvin*. Two dozen geologists and geophysicists from Scripps and WHOI of the USA, as well as many from other countries, were aboard *Melville*. The rendezvous point – and dive area – was at 21 degrees north Latitude, southeast of Cabo San

Lucus, Baja California, near the intersection of the Middle America Trench, the EPR and the San Andreas Fault. The East Pacific Ridge, estimated to be 5 million years old, spreads as much as 8" a year. For nearly 30 years, the Scripps scientists had been searching the area with surface deployed instruments for the temperature anomalies without success. Now, deep towed instruments quickly picked them up.

Soon after the first dive, the first "Black Smoker" thermal vent was discovered and observed up-close. A thermistor failed to work and the plastic pipe that it was on melted. Later, a temperature of 662°F was measured, more than hot enough to melt acrylic viewports. White Smokers were also found. Water samples were taken and animals collected and preserved in formaldehyde. The thermal vents were later found to account for the quantities of the elements and compounds in sea water, not river run off or weathering. The chemicals and compounds, from very deep in the earth, were considered to be from volcanic origins.

The scientists estimated the concentrations of suspended food available around the vents to be 300-500 times that of the adjacent ocean areas and four times that of the rich surface waters. The vent biological communities were as strange as if from another time or planet. Once-molten lava beds gave evidence of frequent volcanic action in volumes too great to immediately solidify in sea water. One scientist commented, "The undersea ore forming system of the EPR produces materials is similar to the enormous deposits on Cyprus Island in the Mediterranean Sea, which also may have been under the seas in ancient times in a crustal spreading zone."

After 20 dives in the EPR area, *Alvin* made 44 more in the San Clemente basin and San Diego area. After a brief overhaul in San Diego, the expedition moved to the tip of Baja California for eight dives in the Tamayo Fracture Zone. After five more dives in the EPR area, followed by 11 dives in the Galapagos, *Alvin* finished 1979 with total of 117 dives.

The first dives in 1980 took place at the Galapagos site in the Panama Basin, the sixth trip to the area for *Alvin*, and on 15 January Ralph Hollis piloted *Alvin*'s 1,000th dive. After a quick trip home to WHOI through the canal for a three month overhaul, *Alvin* made nine dives in the Kane Fracture Zone of the Mid-Atlantic Ridge near 23 degrees north, followed by ten dives in the Oceanographer Fracture Zone of the MAR near 35 degrees North, and six dives on DOS #2. Various dive sites were then selected on the continental slope of the east coast from off Cape Cod Southward, which included Oceanographer, Lydonia and Wilmington Canyons and Blake Spur off Miami. The year ended with dives in the Tongue of the Ocean and the Bahamas, with the final dive at DOS #3 near San Juan, Puerto Rico in December, for the 92 dive of the year.

*Lulu* and *Alvin* moved from Puerto Rico to an area near St Croix to begin the 1981 season, the ONR/NSF/NOAA triad sponsoring 18 biology/geology dives. After overhaul and recertification at WHOI, *Alvin* conducted 16 geological dives on the continental slope in Lydonia Canyon and the Delaware Capes. Following a quick return to WHOI in July, the team entered the Pacific for ten dives near the Galapagos, 14 near Panama, and four near Guatemala, before returning to 21 degrees north latitude on the EPR for 20 more dives in the area of the "Black Smokers." New instrumentation permitted more accurate measurements of the temperatures in the "Smokers."

The first dive series in 1982 examined a whole system of plate spreading zones in the Gulf of California near Guaymas, Mexico. An international research team discovered large mounds exuding tars, waxes, and oils, as well as the largest bacteria on record. Ten dives followed on the EPR, then 23 more in the Catalina, San Clemente, and East Cortez basins off the California Coast.

In April and May, the *Melville* and the *New Horizon* from Scripps, with 40 scientists from around the world, joined *Lulu* to study the ecology and physiology of the deep-sea, hydrothermal vent communities. On 19 dives on most of the two dozen known vent communities, *Alvin* used its new collector to gather samples of hydrothermal vent fluid, *in situ*, undiluted by sea water. After ten dives in the Panama Basin, *Alvin* returned to the MAR, 1800 miles east of Miami, for eight dives in July. One scientist from NOAA reported seeing a massive mineral deposit at 10,000' that looked "like a world of manganese." *Alvin* dove four times on the Continental slope including DOS #2, four times near Cape Cod, then eight times in Oceanographer's Canyon. After maintenance at WHOI in mid-August, *Alvin* and crew made four dives in the Florida Straights and the Caribbean, and 26 near the Providence Channels, St Croix, Puerto Rico, and DOS #3, returning to Woods Hole in early December.

Due to major modifications, the sub and its new support ship, *Atlantis 2*, remained idle the first five months of 1983. A single lift-point on the sub required a complete alteration of the titanium exterior frame that had to fully support the sphere and all ancillary equipment, during launch and recovery. The *Atlantis 2* modifications included a huge pivoting A-frame on the fantail, a high ceiling-hanger on deck near the A-frame, better crew and passenger quarters, a large galley and mess space, and an enclosed hanger to work on the sub in foul weather. The modifications dramatically up-graded the dive operations. After recertifying the sub to 4,000 meters on 29 June, the crew made 12 dives on the Continental slope, including dives to DOS #1, DOS #2, and one on the *Thresher* crash sight. The dive year ended in September after 28 additional dives were made for new pilot training.

The first expeditionary use of the *Atlantis 2/Alvin* combination began with four dives in February, 1984, in the Blake Plateau area on the eastern continental slope, about the same latitude as Georgia. Rough sea launch and recoveries demonstrated the capabilities of the new ship with the A-frame and a single point lift in the sub. Passengers appreciated boarding *Alvin* from a dry deck instead of the rubber boat, which had often pitched about in a rough seaway. The operations included five dives in the AUTEC Range and seven in the Gulf of Mexico, west of Florida, where cold water vents, with the same kind of animal life and fluid emissions as the hot vents found in the Pacific, were observed for the first time. Passing through the canal, *Alvin* continued work in the Panama Basin, diving 28 times to reexamine the EPR parallel with Guatemala, and ten times parallel with Nicaragua for geological and biological research. The sub then made dive #1391 on 5 June, the 20th anniversary of her christening.

After a brief stop-over in San Diego, the expedition headed north to the Gorda Ridge for five dives at a latitude parallel to Medford, Oregon, investigating hydrothermal vents in the spreading plate zones to evaluate possible mining prospects. In the Juan de Fuca Plate boundary zone, near Washington state, six National Science Foundation (NSF) scientists researched the hydrothermal vents. In late July, *Atlantis 2* picked up researchers from Oregon State College at Newport, to investigate the extremely large, black smokers in the Juan de Fuca plate boundary. In early August, University of Washington scientists made nineteen dives in both subduction and spreading zones at the Juan de Fuca plate boundary, and a USGS exposition made nine dives on the plate boundary parallel with Eugene, Oregon. *Alvin* went into port at San Diego after 39 dives in the San Diego, San Clemente, and Catalina basins brought its 1984 dive total to 174.

After brief visits to San Clemente, East Cortez, Valero, Long, West Cortez, and Catalina basins in January and February of 1985, *Alvin* dove 19 times on the Galapagos Rift near the Equator, including the original "Rose Garden" vent areas, visited years before. Geophysicists diving in the Galapagos area, some 600 miles west of the first vent areas, saw for the first time, what appeared to be the birth of a new section of sea floor spreading zone. "It was the most tectonically disrupted piece of seafloor we have ever seen," remarked one geophysicist. None of the pillows were intact; all were completely broken up. New fissures and faulting dramatically demonstrated how seafloor spreading is born.

During the last half of April, *Alvin* made 15 dives at 10 degrees north on the East Pacific Ridge, investigating the tops and sides of a series of seamounts 500 miles off the coast of Central America. This was soon followed by 20 dives near the Clipperton Isles. The sub then made 40 dives in the Guaymas Basin of the Gulf of California to investigate the spreading zones which lie in a line south of the San Andreas Fault, followed by eight more dives on the EPR at 21 degrees north, near the Revillagigedo Islands parallel with Mazatlan. After five dives on the Larson Seamount and 18 dives on the Galapagos Rift for geological/chemical and physical sampling by NOAA, *Alvin* departed in early November for Massachusetts via the Panama Canal. The last dive brought the year's total to 163. Among other significant developments, the new deep ocean sampling and photographic equipment, designed for control from within *Alvin*, tested satisfactorily.

During the November 1985 to February 1986 overhaul, *Alvin* received several significant upgrades. Because carbon from the brushes contaminated the oil that the old, oil-flooded D.C. motors ran in, they were replaced with six reversible, brushless, D.C. electric thrusters. The two mid-ship thrusters were fixed in place in the vertical position. Three of the new thruster units replaced the large propeller aft. The unit farthest aft, located amidships, was fixed in its position for fore/aft thrust, but the two outboard units were rotatable 90 degrees to the horizontal, for vertical thrust and some pitch movement at the expense of forward or aft propulsion. A lateral thruster was installed at the fantail for turning as an active rudder. New, more efficient lead-acid batteries replaced the old ones. A joy stick controller made "flying" the submersible possible, an improvement over the switches in a control box. The unused sail windows were eliminated. A 120 volt electrical system replaced the 60 volt system. A new and more reliable system of emergency release of equipment using explosive bolts was installed. The personnel sphere was upgraded for comfort and efficiency. The payload capacity was increased, and the *Alvin* certification period was extended from one to five years.

After pier side checkouts, *Alvin/Atlantis 2* began combined certification and working dives in Bermuda in May, followed by MAR dives and a *Titanic* search in the North Atlantic in August and September. Scientists from NOAA, M.I.T., and Duke sweated out the Bermuda dives for fear of losing their dives on the MAR. On 23 dives in the MAR spreading zones, scientists discovered "Black Smokers" in the Atlantic for the first time. One NOAA scientist said, "It's very much like Old Faithful going off under 12,000' of water." The find suggested that vents exist all over the world. New species of shrimp and fish were also found. After seven dives on DOS#2, *Alvin* returned to WHOI to pick up the *Titanic* observers.

The late 1970s development of suitable towed side-scanning sonar made feasible the discovery of the *R.M.S. Titanic's* crash site. But unfortunately, the vast area in which it was presumed lost, made even sidescan search a horrendous task in the brief 60 day period of suitable weather in the North Atlantic. Several serious attempts to locate the wreck had been made in the 1980s using towed sonar and magnetometers. Bob Ballard, aboard the WHOI ship *Knorr*, towing a sonar sled named Argo/Angus, spotted the wreckage on 27 August, 1985. On 1 September 1985, a picture of one of the *Titanic* boilers confirmed the location at 41-46 W and 50-14 N.

On 21 July 1986, the *Alvin/Atlantis 2* left WHOI with a load of excited scientists, photographers, and observers. Photographs from towed cameras and DOTs (deep ocean transponders) were used to navigate toward the ship and find places in the ship's hull to land on. The ROV (remote operating vehicle) Jason Junior was deployed from the equipment rack on *Alvin*, making its way into the various passages of the ship. During the twelve dives, *Titanic* was mapped and photographed in detail. Salvage of massive parts of the hull was obviously impossible. Pilot Ralph Hollis remarked, "If you told an artist to draw the most spooky graveyard scene he could imagine, that would be what we saw."

Following eight dives in the Laurentian Fan, south of New Foundland, *Alvin* dove 12 times in the Thresher crash-sight area on the Continental Slope. NOAA investigators then made 15 geological dives in the Delaware canyon, the oceanic continuation of the Delaware River. Nineteen biological dives followed in the West Florida Escarpment of the Gulf of Mexico. After a brief stop at Tampa, a canal and ocean transit lead to the December dives in the San Clemente Basin west of San Diego, with the last of 116 dives for 1986 on 18 December.

After spending the holidays in port, the *Atlantis 2/Alvin* group made 12 dives in nearby Catalina Basin, before sailing to the Hawaiian Islands and the Loihi seamount just south of the big island. Probably one of the most seismic instrumented underwater areas in the world, it is monitored ashore by the University of Hawaii's Institute of Geophysics. Towed instrumentation had mapped hundreds of chimneys six feet high on the seamount that rose 13,000' above the seafloor at 3180' below the surface of the sea. Eleven dives, made with observers from UCSD, Scripps, and the University of Hawaii, found warm (86°F) but not hot water, as well as very high levels of $CO_2$, helium, and methane. The volcano crater contained vast numbers of terraces, spotted with white-rimmed holes exuding warm water, and researchers observed many near extinct angler fish.

En route across the South Pacific, the expedition stopped and made ten dives on the Horizon Guyot, about 250 miles south of Johnson Atoll, followed by seven dives on the Magellan Rise, about 800 miles north of Howland Island (USA). From 7 April to 7 August, *Alvin* made 65 dives around the Northern part of the Marianas Islands, on both eastern and western slopes. Many seamounts were found, most of them exuding cold water from the vents. Hot springs, with strange deep-sea life near the hot vents, were later found in many of the areas west of the island chain. In some of the areas, anemones predominated; in others, snails were the most numerous. Missing were the giant clams and ridged tube worms found in hot vents in the EPR.

In the Marianas Trench, the deepest place on earth, the Pacific crust descends under the volcanically formed island chain of the Marianas and Bonins, as well as islands such as Japan, further to the north. The area is a prime example of subduction on a grand scale. Many of the vent mineral souvenirs and animal specimens eaten or collected were accidentally found to be radioactive and capable of creating radon if confined to closed spaces. Although differences in biological samplings occurred, almost identical vent creatures were found in far distant areas of oceans. All vent creatures differed appreciatively from nearby ocean life. Whether an area was a spreading or subduction zone did not seem to alter the similarity between animal life in different areas. Some vent areas did not have equivalent animals in the other vent areas examined. Eleven dives were made in the Bonin Arc, a continuation of the Marianas Trench northward, about 600 miles South of Yokohama.

Following the Bonin dives, the *Alvin/Atlantis 2* group made a brief stop in Tokyo holding an open house for the Japanese Royalty. The *Atlantis 2* then crossed the Pacific to Astoria, Oregon for 36 dives, made from August to mid-October, examining vents off the Oregon and Washington Coasts, including the Oregon Slope, Juan de Fuca, and Endeavor Ridges. *Alvin* then made 16 dives in the San Diego Trough and Catalina Basin, on one dive observing a 60 ft. skeleton of a whale, the first ever seen in deep water, leading to speculation that animal bones might contribute to the food chain of vent areas. After the 169th dive of the year on 11 November, the ship returned to San Diego for hull maintenance in dry dock.

Diving in 1988 began with nine dives in the Catalina Basin, 24 in the Guaymas Basin of the Gulf of California for geological and ridge process studies, and 20 off Guatemala for geological studies on the ridges of the EPR. On 3 March, *Alvin* made dive 2,000, followed by 28 dives in the Galapagos Rift before a stop-over in San Diego. Ten dives followed, to depths near 11,000' in the Encaba Trough near Eureka, California, before 13 dives examining the Oregon Continental Margin. Thirty-nine dives on the San Juan de Fuca Ridge area, studied heat-flow rates from hot vents and evaluated geological and geochemical processes. Ambient faint light was photographed emanating from a 50' high castle-like structure in the ridge zone. Twenty-two dives followed on the Endeavor Ridge, about the same latitude as Seattle. Returning southward, *Alvin* dove ten times for algae and geological studies in Monterey Canyon, seven times in the

Catalina Basin and 12 times on an underwater volcano off Acapulco, Mexico. *Alvin* passed through the canal on the long trip home, returning to WHOI in early December for a complete overhaul.

*Alvin* spent the first seven months of 1989 in overhaul, having, among other things, its thru-hull penetrators replaced. *Alvin*'s 25th anniversary occurred in June during the overhaul process. During the 25 years, 4,500 observers made 2,150 dives averaging 6.41 hours per dive at an average depth of 6,115 feet. After certification dives in Bermuda in mid-August, *Alvin* examined the Continental Rise near New Jersey and the New York ocean dump site, followed by geological sampling on the rise south of New England and in Carteret and Tom's Canyons.

*Alvin* began its three year voyage in early January 1990, beginning in the MAR area, then to the south of the Azores, investigating the geochemical make-up of hydrothermal vents on 15 dives. In a 13 dive series in May, visiting scientists made geological structural measurements in the Ness Deep Rift Valley west of the Galapagos Islands. Beginning in July, *Alvin* made 37 dives for geophysical crustal studies in the Juan de Fuca ridge area off the Washington coast. On 20 August, Cindy Van Dover became the first woman command pilot on dive #2,260. Beginning in early December, during the 13 dive series on Fieberling Seamount, the first all-female pilot and crew made dive #2,320.

*Alvin* conducted whale bone biological community studies at the Catalina basin on a six dive series in February 1991, followed by 11 dives for hydrothermal vent system research in the Guaymas basin in the Sea of Cortez, and 27 in March and April in the EPR for hydrothermal, volcanology, and geochemical process studies. During a six dive series on Fieberling Seamount, Van Dover and an all female crew made dive # 2400. On 1 July, after four dives, *Alvin* recovered CURV 3, the Navy's Remote Operated Vehicle (ROV) lost off Southern California in 7,400' of water. A 63 dive series in the San Juan de Fuca area continued the multiple-discipline study of hydrothermal flange evolution, as well as biological examinations of the mollusk population. A NOAA sponsored program followed, for chemical and geological investigations of the southern Juan de Fuca Ridge. Thirteen dives made in the EPR studied plankton at the hydrothermal vent plumes.

The 1992 season began with a five dive series in the San Pedro Channel for engineering tests and to evaluate cameras, videos, lighting and the submersible's graphic displays. Fifty dives in the EPR area completed earlier experiments of hydrothermal and geothermal processes. *Alvin* returned to Wood's Hole in June after completing a 2 1/2 year expedition. In early August, *Alvin* conducted a ten dive study of the biological communities at the deep water dump site off New York, before undergoing major overhaul in mid-August after 75 dives in 1992.

The first cruise of 1993 began in March, off Bermuda for the collection and photography of benthic vertebrae. Members of Duke University then conducted investigations to correlate geologic and seismic structures in the MAR, near 22-39 degrees north latitude. Twenty dives at 26-08 degrees north of the MAR included geothermal and geological investigations by sampling and measurement in the hydrothermal active sites, followed by preliminary investigation dives at newly discovered hydrothermal vent sites at

32-17 degrees N in the MAR. Scientist/pilot Cindy Van Dover led investigations into the ecology and microbiological transformation in deep hydrothermal vents, which included extensive observation of the deep sea shrimp. In July, objective, long term observation studies of the ecosystem were made at Deep Water dump site 106 off New York in 8,000' of water. In August, off the Oregon coast in the EPR, *Alvin* investigated fluid forces and movements in the Cascadia Margin and the geomorhphology (forms) of submarine canyons. In September, *Alvin* retrieved data from an instrumented deep water drill-hole and coring, and then investigated recent eruptions in the North slope of an axial seamount off the coast of Northern Oregon in the EPR.

In the last dive series of the year, which extended into 1994, the sub returned to a dive site previously visited in 1991 on the EPR. Called location "Nine North", the area lies 500 miles off the coast of Mexico, between latitude 9 and 10 degrees north. The dives were video graphed and described on PBS television by Bill Kurtis Productions Ltd. in 1994. *Alvin* revisited smokers not observed for 18 months, which had been formed after a lava flow just prior to the April 1991 dives. The vents had grown from 4' high to 21' in the interim, and the former biological desert had become extremely rich with life.

Dives in January 1994, continued those started in December 1993 in the EPR at 9-10 degrees north. These were the first dives to study the origin of the abyssal hills about halfway between the Galapagos Islands and the southern tip of Baja California. In February, *Alvin* examined the distribution of the fluid vents just off the coast of Costa Rica in the Costa Rica Accretionary (geologically growing) Wedge zone. March dives continued earlier work studying the evolution of the hydrothermal vents, and the volcanic and geological properties of the EPR following the 1991 eruptions there. April dives conducted near-bottom gravity measurements and microbiology changes in the EPR. In July and August, dives in the Juan De Fuca Ridge zone at the northern extension of the EPR made a time study of the vent fluids which included temperature, chemical, physical, and radio isotope analysis and variation. In October and November, *Alvin* again visited the EPR at 9-10 degrees north for continued studies of the changes of the biological community structures and associated geological changes in newly formed hydrothermal vents. On December 16, *Alvin* returned to Woods Hole for a well deserved stand down after 180 dives for the year and a submersible total of 2,875.

In early March of 1995, *Alvin* conducted a series of dives to investigate hydrothermal vents on the MAR. By early April, *Atlantis 2* and *Alvin* were back in the Pacific at the "Nine North site," the same site *Alvin* scientists had discovered years before, just after a new lava flow. It was revisited later to observe the extensive growth and biological population increase. This third visit measured and observed the even greater increase in growth and biological make up. From May through September, *Alvin* made a series of dives on hydrothermal vents in the San Juan de Fuca area, making dive #3,000 on 20 September 1995. In October, *Atlantis 2* and *Alvin* made port in San Diego and, after a quick turn around, returned to the Nine North area to investigate sea mounts with University of Hawaii scientists, followed by dives on the hydrothermal vents. Dives in the 9 to 12 degrees north area continued into 1996.

At the end of 1995, some 31 years after its first launch, *Alvin* had made a little over 3,000 dives in all parts of the world. About 85% of all dives were for scientific investigation, involving observers some 6,000 times. Some of the scientific achievements in that time were:

- The first underwater expedition to the spreading Mid-Atlantic Ridge in 1974
- The discovery of hydrothermal vents and later, their existence world-wide, and that they emitted a very low light level glow which could be seen by nearby animals using special receptors
- The discovery of life near the hydrovents which uses chemosynthesis for food which is created at a greater rate and larger size than would have been produced by photosynthesis on the earth's surface

Some of the engineering achievements for the same period included:

- Installation of the first titanium hull for a submersible
- Development of very deep operating ceramic pumps
- Replacement of all variable angle propulsion units with interchangeable fixed position thruster pods for all functions

The *Alvin* project is an American success story that spans more than three decades. Following initial design and birth pains, years of fruitful operations for scientific observations followed. The Woods Hole library estimates that there were "over a thousand articles and publications written as a direct result of the *Alvin* dives."

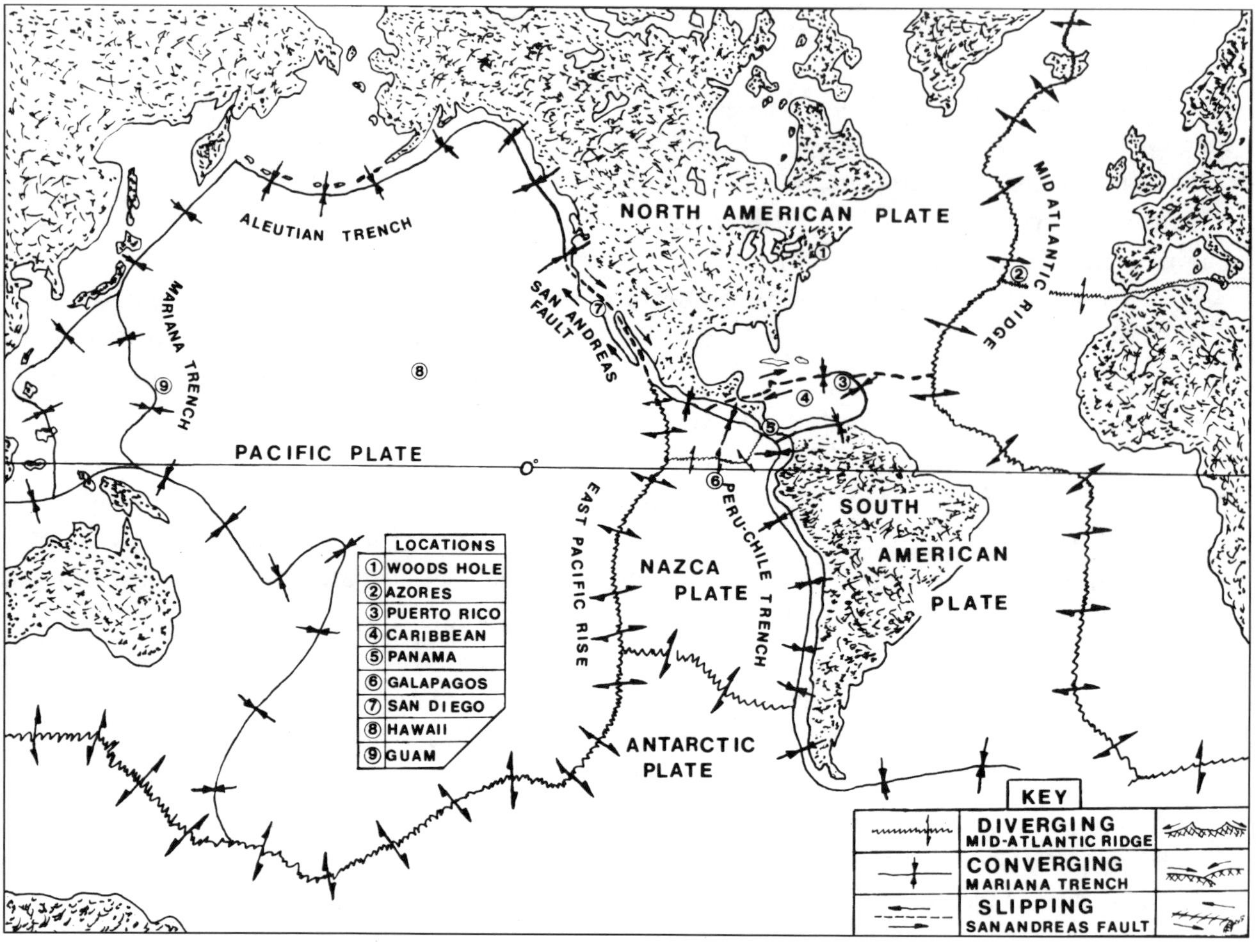


Fracture Zones of Historical or Potential Research

# *Aluminaut*

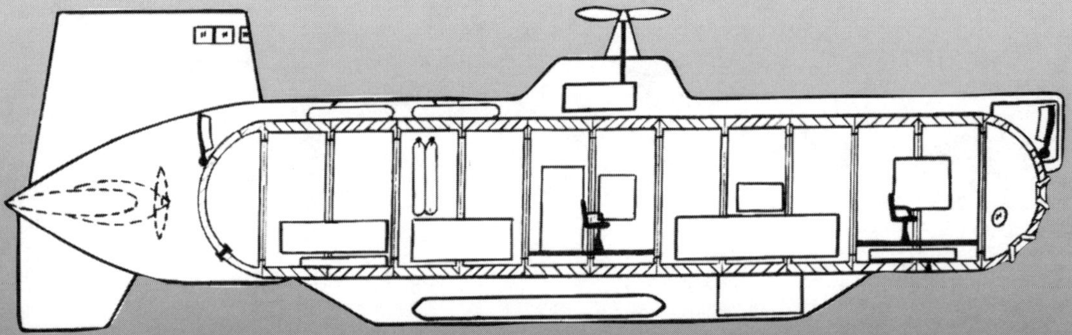

**1964–1972**

# Aluminaut

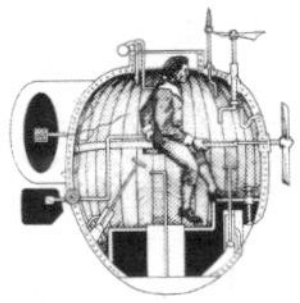

J. Louis Reynolds, of the Reynolds Aluminum Company, was the driving force behind the submersible *Aluminaut*. Simon Lake (1866-1945), with his *Argonaut* subs, and Jules Verne, with his 1869 book *20,000 Leagues under the Sea* were Louis Reynold's inspirations. Simon Lake had also been inspired by Verne's book and, in turn, encouraged Reynolds with his dream to "submerge in a ship." During WW II, Simon Lake visited the Reynolds plant when he was about 77 years old, relating to Louis his sea stories from over fifty years of building and diving submersibles. At the time of their meeting, Reynolds was a recent college graduate and a newly appointed Vice-President of the Aluminum Company. They discussed an aluminum model of a submersible cargo ship being considered by the company.

*S. Lake (far left) and L. Reynolds (third from right) in 1945*
Courtesy Jeffrey Lake

In the 1940s and 50s the metal company, in a continuous search for new markets for aluminum products, had been studying its use for surface ships. In the study process, it was discovered that aluminum would have even greater application for submarine hulls. It took some arguing but, by the late 1950's, Louis had convinced his brothers and the rest of the company heads to begin feasibility studies on an aluminum submersible.

As a result, the design and construction of submersible models began in early 1959. The goal was a large volume submersible with long underwater endurance, capable of operating at 15,000'. The high strength aluminum alloy (7079-T6) was selected for optimum strength and its relative ease of fabrication for the large and thick pieces required. The alloy was virtually unweldable, so it required a bolted-flange technique for assembly. To prevent corrosion, six coats of specialized paint had to be applied to the pressure hull after fabrication. Three years of designing and pressure testing many models to destruction at Southwest Research Institute, Texas, followed. Another two years of finalizing the designs and specifications led to the start of a three year fabrication program in 1962.

Reynolds cast the initial ingots that weighed over 17,000 lbs. for each of the eleven cylindrical sections fabricated as well as the two castings weighing 14,000 lbs, each made for two end hemispheres. The twelve interfaces between the thirteen parts were carefully hand-lapped to one another for an 85% contact, making the bolted connecting "cracks" almost invisible to the eye. Over 400 bolts were shrunk in liquid nitrogen before assembly in the flanges.

All viewports, hatches, and pass-throughs were located in the end hemispheres which, due to their geometry, were less stressed than the cylindrical sections. The Electric Boat Division of General Dynamics made the final assemble, by installing corrosion resistant aluminum superstructure, keel, and ballast tanks at Groton, Connecticut.

Initially, the Office of Naval Research (ONR) had indicated an interest in purchasing or leasing the *Aluminaut*, but the design depth of 15,000' was deeper than Woods Hole and other Navy sponsored users would need. The expense of the construction of *Aluminaut*, due to size and the precision fabrication required, put the price way over the ONR budget. Unable to negotiate an acceptable lease/purchase agreement, ONR went out on open bid to purchase what was basically a smaller submersible of less capabilities and lower cost.

*Aluminaut* had an overall length of 51'-3" with a 44'-9" long pressure hull. The pressure hull consisted of eleven cylinders having an 8'-1" outside diameter and a 40" length, plus the hemispherical end closures. The 6 1/2" thick cylinder walls were reinforced by the connecting flanges which, when joined together, resulted in a 5" deep by 3" wide flange extending inside the nominal wall of the pressure hull. The pressure hull's displacement provided the primary buoyancy for the submersible.

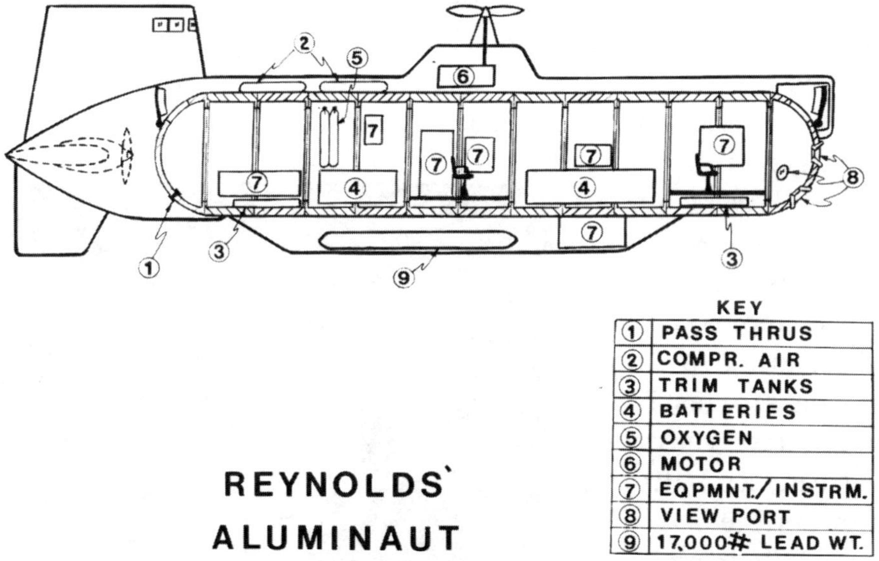


**Name:** *Aluminaut* **Builder:** *L. Reynolds* **Length:** *51'3"*
**Maximum Depth Rating:** *8,000'* **Capacity:** *6 Man*
**Year Launched:** *1964* **Description:** *Detailed Cross Section*

*Suddenly, the submersible became quite heavy in the less buoyant fresh water and began to sink out of control to the bottom.*

---

Since the submersible was less compressible than seawater, it became more buoyant with increasing depth. The seawater ballast tanks, normally flooded when diving, provided 2000 lbs. of buoyancy when blown free of water at the surface. The variable ballast was provided by iron shot which could be released by electromagnetic valves. A large, droppable, lead keel provided additional ballast for release in an emergency. Pitch trim could be accomplished by pumping water between the fore and aft internal tanks. Internal weights could be shifted manually for additional trimming. The combination of water and lead weight shifting made possible the angling of the bow of up to thirty degrees above or below horizontal. The submerged displacement of 81 tons required a marine railroad or a 100 ton crane for removal or transfer of the submersible from the water to shore or shipboard or visa versa. Towing at 6 1/2 kts. from major ports was the usual means of transport to the dive sites.

Propulsion was provided by two reversible 5 H.P. thrusters, mounted on the front of the dive planes at the stern. Vertical control was attained by a 5' diameter horizontal propeller, driven by a reversible motor. A rudder aided in steering. Four viewports, 4" in diameter were located in the forward hemisphere, which provided 160 degrees of horizontal viewing and a vertical field from 25 degrees above the horizontal to 80 degrees below. The sub's power was obtained from internal silver zinc batteries. The environmental system was designed for 32 hours of normal operations for the usual crew of three, with emergency provisions for 72 hours. The interior instrumentation included those for submersible status and control read outs as well as those for oceanographic data. Continuous transmitted frequency modulated (CTFM) and side scanning sonars were provided. The first two bays of the cylindrical hull were for scientific instrumentation. The next three for batteries, the next two for the pilot's station with its submersible status read outs and controls. The next two bays were for batteries and the last two for auxiliary equipment. The forward observer from his position in the hemispherical head, could also steer and control the submersible.

*Aluminaut* was launched on 2 September 1964, by Electric Boat, at Groton, Connecticut. During the first sea trials in Long Island Sound, the stability, maneuverability, and handling characteristics were checked out. These took place from December 1964 to November 1965. During the early trials in Long Island Sound, a casualty was narrowly averted when the submersible, trimmed for salt water density, crossed the freshwater mouth of the Connecticut River. Suddenly, the submersible became quite heavy in the less buoyant freshwater and began to sink out of control to the bottom. It was estimated that 3,500 lbs. of buoyancy was lost. The alert pilot quickly blew the water out of the ballast tanks, dropped shot, and put the horizontal propeller in full power setting upward. His fast action prevented a dangerous collision with the sea floor.

During the next sea trials, a lengthy submergence was conducted using a crew of six instead of the usual three, while transiting the 70 mile-wide Straits of Florida from the Bahamas to Miami over a 33 hour period at an average depth of 1,250'. At the time of surfacing, 24 hours of life support remained, as well as 12% of the battery capacity. Strain gage data, monitored and recorded during the dive indicated no unusually high stress areas.

The 1965 sea trials began in mid-July when the *Aluminaut* made a controlled drift in the Gulf Stream for a 25 mile distance, at depths of 1000' or more, in currents of up to 3 1/2 kts. The surface escort ship was the R.V. Privateer, later to be used by the Ben Franklin on its 30 day Gulf Stream drift mission. In August, several sunken vessels were detected off of Ft. Lauderdale, Florida. A large WW 1 concrete barge section was observed about 80% buried in the accumulated sediment. Admiral Rickover, Vice-President Hubert Humphrey, Glennie Reynolds (the sponsor's daughter), and Claude Kirk, governor of Florida were among the notable personalities that dove in or visited *Aluminaut* during the sea trials near Miami.

In November, while demonstration dives were being made for the University of Miami and Department of Interior scientists, a cold, counter-flowing current was encountered on the bottom of the Straits of Florida at a depth of 2,450'. It had been first discovered in 1957 by an American-British oceanographic surface expedition. During the demonstration dive, the observers saw what looked like odd shaped rocks scattered over a large area of the bottom. On later dives, when the manipulators had been installed, samples were found to be fossilized dugong bones that were dated at 25 million years of age. In late November, a deep diving trial to 6,250' was conducted twenty miles west of the Great Abaco Island in the Bahamas. During the deep dive, systematic strain gage data from both inside and outside the hull showed there were no exceptionally high stress areas. The successful conclusion of the sea trials made the $3.0 million submersible *Aluminaut* ready for charter by the defense, scientific, and industrial communities.

*Aluminaut at Sea* Courtesy of Naval Ordinance Test Station, China Lake

In January 1966, *Aluminaut* ran tests at depths of 2,500' for the Deep Submergence Systems Group's Doppler navigation equipment, which was later installed on *Trieste 2.* Targets of 100' long aluminum pipe were laid on the bottom for calibration tests of the Doppler system. Later in the month, *Aluminaut* was shipped to Spain aboard a landing ship dock (LSD) to help search for the lost hydrogen bomb near Palomares, Spain.

During the H-bomb search, *Aluminaut* and *Alvin* made round-the-clock operations, stopping only for recharging batteries and foul weather. In the early dives, in shallow water, at 120' a 15th Century galleon was sighted by *Aluminaut*. It was lying on a pile of rocks with a light sediment covering it.

"Tattered sails, broken masts and cannons" were supposedly seen in near mint condition, but the bomb search could not be interrupted.

Approximately, 30 dives were made during the H-bomb search program, each averaging seven hours at 2,300'. The deepest submergence was 3,000' in the area where the bomb lay after it had been dropped during the first recovery attempt. During the initial discovery of the bomb's location, on 15 March 1966, *Aluminaut* rendezvoused with *Alvin* to stay on station while *Alvin*'s batteries were recharged and the surface fleet repositioned itself over the bomb sight – the first time *Aluminaut* stayed on station for 24 hours. On a second occasion, after the bomb was dropped during recovery, *Aluminaut* waited on station 22 hours so as not to lose sight of the weapon prior to the recovery.

On one occasion during the search, *Aluminaut* was skimming over the bottom in a deep flat area and scooped up mud in the keel. The keel had a fairing that was perforated to keep most things out, but the stirred up clay bottom was almost liquid until it settled in the keel and became solid and trapped. The problem was noticed when the submersible couldn't ascend under power. It was necessary to drop the iron shot and blow the water ballast. The submersible was then flown off the bottom, using both forward and vertical propulsion. Later, aboard the *LSD*, it was estimated that 5,000 lbs. of mud were washed out during a five hour hose-down.

In July, after returning from Spain, two large manipulators were installed, each capable of lifting 200 lbs. An external storage compartment was also added, which could hold retrieved material and many different tools for the manipulators.

During August and September, *Aluminaut* was used for a ten day in-house sponsored exploration of the Blake Plateau. The 3,000' deep plateau is a 150 x 100 mile flat, underwater terrace off Northern Florida, Georgia, and North Carolina. Removable, free flooding aircraft wheels were mounted on the submersible, to permit easy transit of the flat black, almost macadam like manganese oxide, pavement-like surface. The manganese oxide pavement was cluttered with nodules containing phosphates, manganese, nickel, cobalt, copper, and platinum. The compounds in the manganese oxide conglomerate had been experimented with, in an effort to find catalysts for removing sulfur and nitrous oxide from the exhausts of fossil burning fuels. The pavement of manganese oxide appeared to be of as high a commercial quality as the nodules usually found at depths seven to ten times deeper.

In October and November, *Aluminaut* made an extensive ocean bottom survey for the U.S. Naval Oceanographic Office. The first phase was to make *in situ* oceanographic readings and take samples to compare with data gathered from the surface. The second was to recover a one ton current meter array at 3,150', lost for over a year. Deeply moored acoustic tracking arrays and their cables were inspected and photographed from 4,000' to near surface depths. During submerged operations in the West Indies, over eight miles of cable were surveyed for AT&T. It was discovered on one dive that the cables were suspended between two fossil coral heads. To complete the inspection, the submersible dove under the low hanging cable, fortunately without snagging the 5' diameter propeller in the cable. Later, when reminded of the tidal surges and cross currents in the area, the pilots agreed not to do

*Using high resolution sonar, the submersible located the 7' long instrument shortly after arrival on the bottom, and within ten minutes, had it in the manipulator and ready to surface.*

an encore. A contract was made with the Turtle Wax® company in November. A Ford Falcon was carried piggyback on *Aluminaut* to the bottom at 60' and back to the surface near Fredericksburg, American Virgin Islands, while cameras recorded the event and the local population formed yet another opinion of American advertisers.

In May 1967, extensive underwater navigational development tests were conducted for the Deep Submergence Projects Office. Tests and data collection were made on Doppler sonars and a long range multi-transponder system. In September, a series of underwater research projects was conducted for the University of Miami. One of the earliest underwater *in situ* gravity measurements was conducted and the Gulf counter-current was accurately measured near Miami. The Bureau of Commercial Fisheries explored the calico scallops and royal red shrimp near Cape Canaveral utilizing *Aluminaut*. In October, the sunken vessel *Yarmouth Castle* was searched for, but never found in the Bahamas Northwest Passage.

In January 1968, *Aluminaut* made a series of dives for the Atlantic Undersea Test and Evaluation Center (AUTEC) in the Tongue of the Ocean, Bahamas. Inspections and photographs were made of bottom mounted acoustic arrays, deep moors, and cable runs from the equipment to the surface instrumentation. Dives to 6000' for as long as twelve hours were made. Underwater surveys for the Naval Oceanographic Office were made near Puerto Rico. In early June, more deep water navigation equipment tests were made for the Deep Submergence Systems Project Office.

In late June, *Aluminaut* was towed by the R.V. *Privateer* to Cartagena, Columbia, South America. A full month was spent searching for and, indeed, locating the galleon San Jose, which sank during a skirmish with the British in 1708. The galleon was on its way to Spain loaded with over a year's harvest of Spanish gold and treasure. The vessel was reported lost in international waters, so Columbia moved its oceanic boundary to include the wreck. Two and a half decades later, Columbia will still not let any foreigners try to salvage the wreck and its gold.

Beginning in July, a month was spent making biological and geological observations near Cartagena and several other antique wrecks were located. The submersible was then returned to Puerto Rico for more dives for the Naval Oceanographic Office. In August, while at 5500' off of Vieques Island, Puerto Rico, a penetrator in the hull leaked significantly enough to require ballast shot to be dropped and the submersible to be surfaced. The leak was caused by a short in the penetrator. It was redesigned to make a short-out unlikely and certainly not to leak if it did. In October, the stern hemisphere was removed and inspected. The flange of the hemisphere, which was sealed only by a tight fit, was unblemished and showed no sign of water penetration.

In May 1969, *Aluminaut* recovered a 280 lb. scientific package lost by the University of Miami in 1000'. Using high resolution sonar, the submersible located the 7' long instrument shortly after arrival on the bottom, and within ten minutes, had it in the manipulator and ready to surface. Also in May, deep scientific measurements were being made in the Gulf Stream about ten miles from Miami.

In August, under contract to the Navy Superintendent of Salvage, *Aluminaut* dove to 5,000' to locate, photograph and determine if the submersible *Alvin*, lost during a launching accident, could be salvaged. Fortunately, the hatch, left open when it sunk, was still open, permitting a toggle bar to be inserted in the hatch hole. The large sail and fairwater were so much in the way that recovery was questionable at first. Bob Canary, the *Aluminaut* pilot, hacked, chopped, and cut enough of the fiberglass out of the way with the two *Aluminaut* manipulators to get the toggle bar inside the hatch hole. Once connected the surface ship, *Mizar* lifted *Alvin* to the 85 foot depth, where a net was put on by divers. In December the submersible was called on to photograph and recover whole or fragmented pieces of weapons that had exploded or sunk for some reason on the AUTEC Range, Tongue of the Ocean, Bahamas.

In early 1970, the University of Miami lost an expensive instrument array in the middle of the Gulf Stream. In a single dive at midnight, the instrument was located using sonar and recovered from a depth of 1000'. Dive # 251 in 1970 was *Aluminaut*'s last.

The submersible was decommissioned and moth-balled in 1972, when the economic depression in the U.S. caused government cutbacks. The submersible that cost $ 3.5 million originally, would cost $2.0 to $3.0 million just to refurbish a several years later. Part of a required refurbishment would be the replacement of all the rubber and plastic cables that had aged beyond usefulness. The original 15,000' operating depth was never achieved, due to the as-fabricated flaws in the hemispherical ends of the pressure hull which resulted in a certifiable operating depth of 8,000', with a safety factor of 2.8. New hemispherical heads were considered many times, but the expense could never be justified.

In the early 1980s, a Texas oil man named Jack Grimm was seriously considering using *Aluminaut*, if refurbished, for searching and exploring for the *Titanic* at a depth of 13,000'. However, it was about the time the oil boom busted and the expedition never came to fruition. By 1985, Bob Ballard and his associates found *Titanic* and later dived on it, with *Alvin*. *Aluminaut*, in a cosmetically restored condition, rests in the public park in Richmond, Virginia.

The use of light weight aluminum with less strength than steel or titanium required a novel, thick-wall shell design by the Reynolds' engineers. Some future designers who also require the use of low strength materials may find a good example in *Aluminaut's* fabrication techniques and successful operations.

# The Deepstars

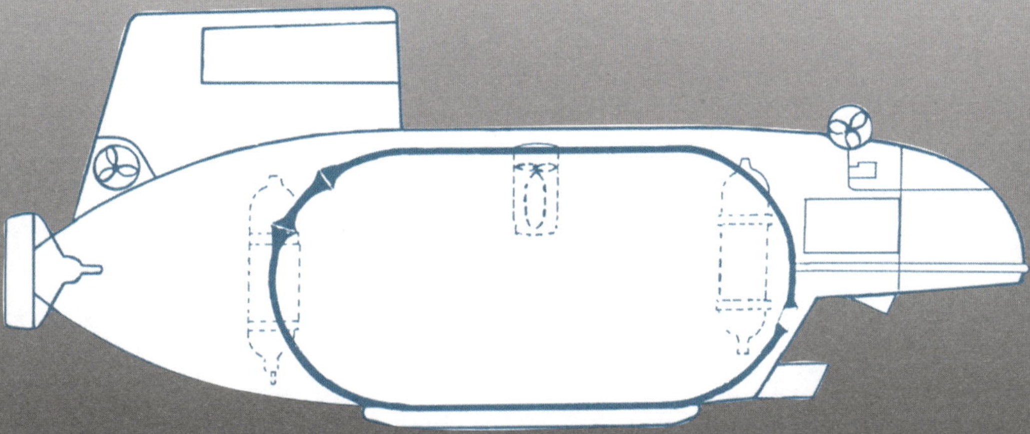

1965–1972

# The Deepstars

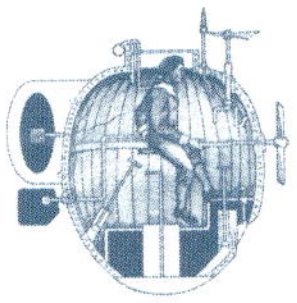

In the early 1960s, the Westinghouse Corporation decided to expand their undersea efforts which, at that time, were limited mainly to torpedo and sonar manufacturing and development. An agreement was entered into, in which the Cousteau group, Office Francais de Recherches Sous-Marine (OFRS) provided technical assistance in the building of a 2000', a 12,000', and a 20,000' operating depth submersibles. The 12,000' submersible was the first to be designed, but it was to be constructed of a brittle steel (Vascojet 90), unacceptable if U.S. Navy was to lease it. Using a lesser strength, more ductile, Navy-acceptable steel (HY-80), the first *Deepstar* submersible was constructed to the same specifications as the 12,000' design, but with an operating depth of 4000'. The French eventually built the originally designed 12,000' submersible using Vascojet steel and called it *Cyana,* which was rated for an operating depth of 3000 meters (9,843'). The 20,000' submersible was canceled due to budget limits and the 2,000' submersible was put on hold until experience with the *Deepstar 4000* was attained.

## Deepstar 4000

Most of the components of *Deepstar 4000* were similar to, or improvements on, those in the 1000' operating depth *Diving Saucer*, built by OFRS and demonstrated in the U.S. earlier in the 60s. The submersible was designed and constructed in Marseilles France under the direction of Cousteau, with guidelines specified by Westinghouse. A series of shallow tests were made near Marseilles before *Deepstar* was shipped to the Westinghouse Undersea Division at Annapolis. After some Americanization of the instruments and a few of the minor hardware components, *Deepstar* was shipped to San Diego. The official launching was held dockside at the Kona Kai Club aboard the Westinghouse R/V *Burch Tide* on a rainy Friday afternoon 17 December 1965.

The steel pressure hull was 79" outside diameter with a nominal thickness of 1.2". Two viewports were located 21 degrees below the horizontal centerline and 16 degrees port and starboard from the vertical centerline, in the forward portion of the hull. The 90 degree conical viewports had an inside diameter of 4.33". Just above the viewports was a 1.69" inside diameter camera port. Similar to the *Diving Saucer*, two fixed position, variable speed, reversible motors and propellers were mounted port and starboard about midships on the external hull. Ejectable, pressure compensated, lead acid batteries provided power to the submersible.

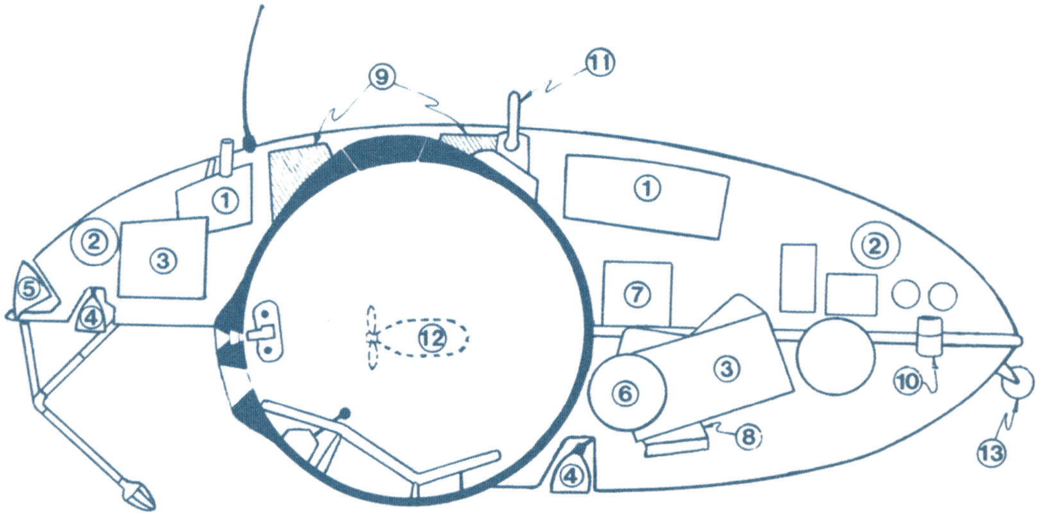

KEY

| | |
|---|---|
| 1 | SYNTACTIC FOAM |
| 2 | MERCURY TRIM TANK |
| 3 | BATTERY |
| 4 | DROPPABLE WEIGHT |
| 5 | LIGHT |
| 6 | VARIABLE BUOYANCY TANK |
| 7 | VARIABLE BUOYANCY BLADDER |
| 8 | BATTERY BLADDER |
| 9 | INFLATABLE ESCAPE TRUNK |
| 10 | TRANSDUCER |
| 11 | LIFT SHACKLE |
| 12 | PROPULSION UNIT |
| 13 | DROPPABLE DESCENT WEIGHT |

## Deepstar 4000

***Name:*** *Deepstar 4000*
***Builder:*** *Westinghouse Corp.*
***Length:*** *18'*
***Maximum Depth Rating:*** *4,000'*
***Capacity:*** *3 Man*
***Year Launched:*** *1965*
***Description:*** *Detailed Cross Section*

Low power emergency batteries were carried internally. Life support provided for 144 man hours of oxygen supplies and carbon dioxide removal. The main ballast tanks provided surface buoyancy and were flooded for diving. On the seafloor, or at a desired depth, a 220 lb. descent weight was dropped to attain neutral buoyancy. For ascent to the surface, a 187 lb. weight was dropped. Variable buoyancy adjustment was made by the transfer of oil between hard tanks and flexible bladders. Incremental trim weights of 3.4 lbs. each could be dropped to increase buoyancy. Pitch trim of 30 degrees above or below the horizontal was made by the transfer of jettisonable mercury, between forward and aft containers. A jettisonable manipulator with 3 degrees of freedom was provided at the bow. Both ascent and descent weights were released automatically if the maximum design depth of 4,200' was exceeded. Overall length of *Deepstar* was 18', weighing 9 tons. The crew consisted of a pilot and two observers.

Sea trials and certification test dives began on 8 April, 1966, including the first deep dive on 11 May, 1966 to a depth of 4231'. The Navy certification of *Deepstar* was completed on 6 June, 1966. This initiated a 13 month, 100 dive contract with the Naval Electronic Laboratory (NEL) San Diego, accommodating 53 scientists.

The scientific dives included:

- The measurement of sediment transport in different areas off the California Coast

- Gravity measurements at shallow and deep depths to determine the variations in the earth's mass along the continental shelf
- Seismic studies of the sea floor using explosive charges set at a distance, while four earphones on long poles on the submersible picked up and measured the signals and echoes
- Geological surveys of underwater canyons and old submerged beaches from of the ice ages, when the huge volumes of ice on earth lowered sea level far below today's sea levels
- The measurement of currents, temperature, salinity, and other chemical and physical oceanographic properties of the sea
- The measuring, observing, and sampling of the Deep Scattering Layer (DSL), which contains a myriad of different small species of sea life and the food chain that feeds on it, as it moves up in the night and down in the day in concert with the light in the sea
- The study and measurement of surface and bottom sound velocity and ambient noise levels
- Bottom core sampling
- Shear vane testing
- Bioacoustic and bioscattering research
- Fish distribution and identification
- Undersea topographic measurements and studies

During a dive on 29 October 1966, pilot Joe Thompson had a "worst case" problem with *Deepstar*. While at 4,000', operating near the bottom, he unknowingly picked up 100 lbs. of silt in the fairing cavities of the submersible. At the same time, the hydraulic powered mercury trim system failed to work, causing the dive to be terminated early. When the 186 lb. ascent weight was released to start the submersible toward the surface, it failed to drop off due to the hydraulic failure that prevented the mercury trim actuation. The mechanical override to the ascent weight release worked, but the weight got hung up in its rack. Due to the hydraulic failure, the 150 lbs. of droppable trim weights could not be released. The 200 lbs. of mercury used for pitch trim was released, but the submersible was still too heavy to rise. If that was not enough, yet another failure had unknowingly occurred: the variable ballast chambers had flooded with 80 lbs. of sea water, due to some faulty pipe joint welding.

In a desperate, but calculated act, Thompson dropped the 450 lb., forward, lead acid battery. As the submersible started up, its ascent weight came loose and dropped off. As the submersible gained speed, most of the silt dropped off. *Deepstar* was now about a 1/2 ton light and, although there was no danger of its going from inner to outer space, it made an unusual high pop up out of the water at the surface – much to Thompson's relief.

In late February 1967, the submersible operations were moved to Morgan City Louisiana aboard the M/V *Search Tide* where dives were conducted off shore until late March when operations were moved to Panama City, Florida for additional dives in the Gulf of Mexico. In late April, *Deepstar* was dropped, accidentally about five feet from the quick release hook on the Birch Tides crane, requiring two weeks of repair in Mobile Alabama. Dives in May 1967, were made near Panama City for organism observations in the course bottom sediments of the 300' contour level. Dives with 11 scientists were made in mid-June for conducting sedimentology, studying ancient ice age terraces and pinnacles, and taking sea floor corings.

*In a desperate, but calculated act,*
*Thompson dropped the 450 lb. foreward, lead acid battery.*

---

In late June, after the NEL contract was completed, a USN Undersea Sound Laboratory (USNUSL) contract for 50 dives was initiated. In July, in the Grand Bahamas, underwater sound and electromagnetic measurements, as well as coral observation and collection were conducted. In August, extensive dives involving underwater sound direction and detection, surface and bottom sound reverberation, and echo location and intensity studies were conducted by USNUSL off Florida. USNUSL conducted similar sound studies to the north at the Georges Bank and at both Block and Corsair Canyons off New London, Connecticut, finishing the contract in October 1967.

Ten dives for the Naval Oceanographic Office were made through November in Hudsons and Norfolk Canyons, as well as in the Bahamas. Physical, chemical oceanographic, geological, and Deep Scattering Layer measurements were accompanied by observations and samplings using customer furnished equipment.

A new 100 dive contract was let to Westinghouse for the use of *Deepstar* by the Navy at the Naval Undersea Center (NUC) San Diego, in late October 1967. Physical Oceanography and Deep Scattering Layer (DSL) Research was conducted off Costa Rica and Nicaragua until early December. Similar (DSL) dives were conducted along the Central American Coast north to the tip of Baja California until late December. Physical Oceanographic and geologic survey dives at the tip of Baja continued until early February 1968. *Deepstar* returned to San Diego in February to complete the remaining half of the dives for NUC in the Southern California area, including dives near Santa Catalina and Santa Cruz Islands, Los Angeles, and La Jolla cities. The last dive of the NUC contract was made on 9 November, 1968, near San Clemente Island to help select a Sea Lab 3 sight in 640' of water. The next dives were made off San Clemente Island during the lowering, monitoring, and raising of Sea Lab 3, including a pictorial record of the gas leaks from the habitat that were being repaired by an aquanaut who died during the event, due to an empty carbon dioxide scrubbing canister.

The last dive made by *Deepstar 4000* was #540, an endurance dive, made over 27 and 28 March 1968. It lasted 24 1/2 hours. *Deepstar* was sold to the Comex company of France in 1974.

## Deepstar 2000

As the 1960s proceeded, the original plans for *Deepstar 2000* still seemed to warrant in-house support for construction, with the prospects of the National Oceanic and Atmospheric Administrations (NOAA) funding it for use as their shallower submersible. *Deepstar 2000* was launched on 20 December 1969 at the B Street pier in downtown San Diego on the M/V *Search Tide*, the former support ship for *Deepstar*

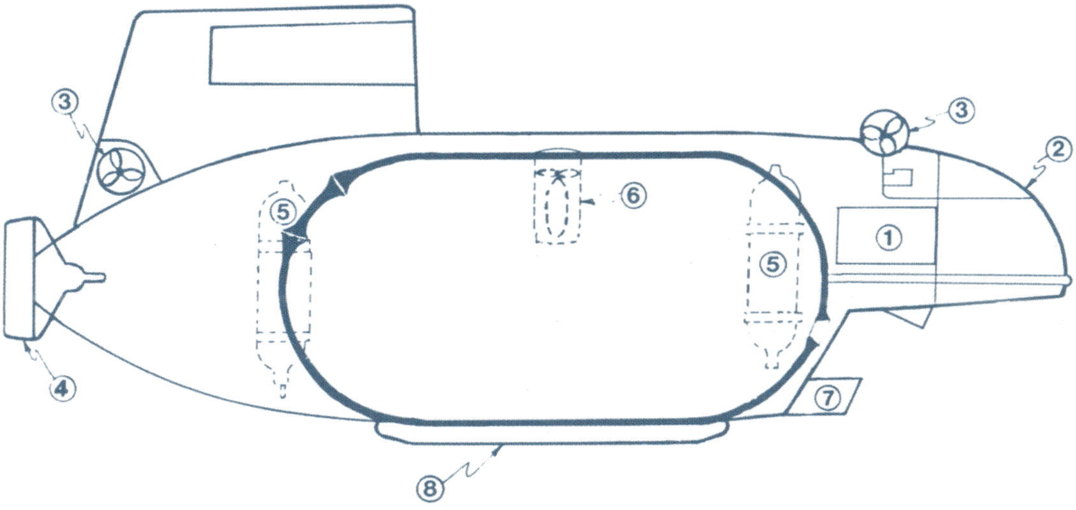


## Deepstar 2000

***Name:*** *Deepstar 2000*
***Builder:*** *Westinghouse Corp.*
***Length:*** *20'*
***Maximum Depth Rating:*** *2,000'*
***Capacity:*** *3 Man*
***Year Launched:*** *1969*
***Description:*** *Detailed Cross Section*

KEY

| | |
|---|---|
| 1 | SYNTACTIC FOAM |
| 2 | INSTRUMENT BROW |
| 3 | LATERAL THRUSTER |
| 4 | MAIN THRUSTER |
| 5 | BALLAST TANK |
| 6 | VERTICAL THRUSTER |
| 7 | SAMPLE BASKET |
| 8 | KEEL BALLAST |

*4000*, just 21 months after *Deepstar 4000's* last dive.

The pressure hull consisted of a 5' outside diameter x 3/4" thick cylinder 5' long with hemispherical end caps of HY 80 steel. The overall length of the hull was 10'. Seven mechanical and three electrical pass throughs were provided, as well as a 15 3/4" diameter hatch in the aft hemisphere of the pressure hull.

Two 4.5" inside diameter x 90 degree acrylic viewports were positioned 19 degrees below the horizontal axis and 15 degrees to port and starboard of the vertical axis, in the forward steel hemisphere. One 2" inside diameter x 90 degree conical acrylic port was located on the center of the horizontal and vertical axis in the forward hemisphere, for still camera photography.

The main ballast system, for surface buoyancy, consists of bellows configured "soft" tanks which provided 500 lbs. of buoyancy when air filled. The variable ballast system consists of four hard tanks containing oil, connected to four bladders to provide buoyancy when the bellows were fully filled with oil. Syntactic foam was installed, external to the pressure hull, to compensate for equipment and battery weight, which exceed the buoyancy of the hull displacement. The batteries could be hydraulically moved fore and aft for pitch adjustment of 25 degrees above or below the horizontal.

Twin reversible and variable speed motors turned 20" diameter shrouded propellers at the stern of the submersible for the main propulsion. Twin shrouded 9"

*Deepstar 4000* Courtesy of Joe Thompson

diameter, fixed position, reversible propeller units on port and starboard provided vertical thrust. Horizontal shrouded 9" diameter, fixed position, reversible propeller propulsion units on the bow and stern provided turning and/or lateral thrust.

One lead acid battery unit at 120 volt d.c. and two at 28 volt d.c., located externally and pressure compensated, provided the total power for the submersible. A small emergency battery was provided internally. The external battery pack was droppable in emergencies.

Life support provided for 144 man hours for the crew. Three one-hour rebreathers were provided for special emergencies such as for fire or smoke inside the hull. The submersible's length was 20', weight in air 8 3/4 tons, normal dive time was eight hours at 1 kt. The crew consisted of a pilot and two observers. The vessel had an operating depth of 2,000' and a safety factor of 2.06, based on crush depth. A manipulator with two degrees of freedom was provided for collections, bottom coring, and the like.

After the launch of *Deepstar 2000* in late 1969, extensive sea trials and new equipment testing were required. Before scientists could employ *Deepstar* in their work, certification was required. The high-pressure test tank located at Annapolis in the Naval Ship Research and Development Center, successfully tested *Deepstar 2000* to a simulated depth of 2,300'.

During an early test in the sea trial period, a rough water launch was in process when the battery pack, normally available for jettisoning in emergencies only, was accidentally dropped. The system was modified sufficiently to prevent reoccurrence and no further problems with the mechanism took place.

On 3 April 1970, *Deepstar 2000* made a dive to 2,242' for certification purposes. During June, 1970, Westinghouse scientists made investigative dives in Scripps Canyon to 1,460'. Extensive bioerosion, primarily by galatheid crabs, was observed on one outcrop. An undercut in the canyon of about 80' with an attending overhang was encountered while the submersible was under the outcrop and near the interior end of the cut. The erosional feature was considered to have been formed when sea level was much lower and surface erosional forces could cause such major geologic deformation.

During 1970, *Deepstar 2000* made approximately 51 dives for testing, sea trials and certification along with geological and biological research programs in the submarine canyons and offshore basins near San Diego.

*Deepstar 2000* was chartered for about 15 dives during June and July, 1972 by the Sandy Hook Laboratory of the National Marine Fisheries to explore the ocean bottom along the New Jersey coast in an attempt to find areas to dump millions of tons of sewage sludge from New York City.

*Deepstar 2000* Courtesy of Hal Palmer, MRJ Inc., Fairfax, Virginia

Instruments and "tracers" were implanted to measure bottom currents and sediment transport. Soluble blocks of dyed sand, about the size of a loaf of bread, called "tracers" were set on the bottom to make visual a synoptic pattern. A two year study of the area concluded that the bottom marine life had become greatly reduced in both species and numbers, due to sewage sludge. Further dumping was not recommended.

During a dive in the Wilmington Canyon off New Jersey, tracking of the submersible was being done from a 16' Boston Whaler with three people aboard. *Deepstar* dived about 1 P.M., followed at the surface by the small boat homing in on the submersible's pinger. The primary surface support ship *State Waves* followed the Whaler as it tracked *Deepstar*. When a storm front suddenly and unexpectedly came into the area, the Whaler and *State Waves* lost sight of each other. The submersible lost all communication with the surface and the Whaler lost track of the submersible. The Whaler was blown out of the area and ended up near a lobster pot buoy to which it tied. The submersible surfaced and gave out periodic radio calls. It was sealed away from the surface splash, but hurled about in the storm.

With a submersible and a small boat lost in a storm miles from shore, the operational director, fighting off insanity on board the surface support ship *State Waves,* called in the Coast Guard. They located the sub quickly and got *State Waves* headed toward *Deepstar* for a hasty join up. The seas were too rough for recovering *Deepstar*, so the sub was tied off from the ship until 3 P.M. the next day, bobbing and rolling for a 26 $^{1}/_{2}$ hour encapsulation. A Spanish fishing boat picked up the Boston Whaler crew and their boat at the lobster pot buoy about 10 A.M. on the day after the dive began. The Coast Guard flew the Whaler crew ashore by helicopter for a check up. They were found to be cold and wet, but OK. The Spanish trawler returned the Whaler to the State Waves with three dozen lobsters from their catch of the day. Hal Palmer, the scientist in the submersible during the ordeal, said that it would be bad enough to be stuck on the bottom in a submersible, but that bobbing helplessly at the surface in the dark of night in a shipping lane was a nightmare of its own. *Deepstar* made about 15 dives on the East Coast of the U.S. in 1972 and then went "on the blocks."

Two of Cousteau's people in France bought *Deepstar 2000*, circa 1975, at a fraction of its original cost, for a proposed trans-Mediterranean pipe line contract. This sale for pipe line work was one of the early indicators of the potential funding available from the offshore oil companies, later to become substantial.

# *The Star Series & Asherah*

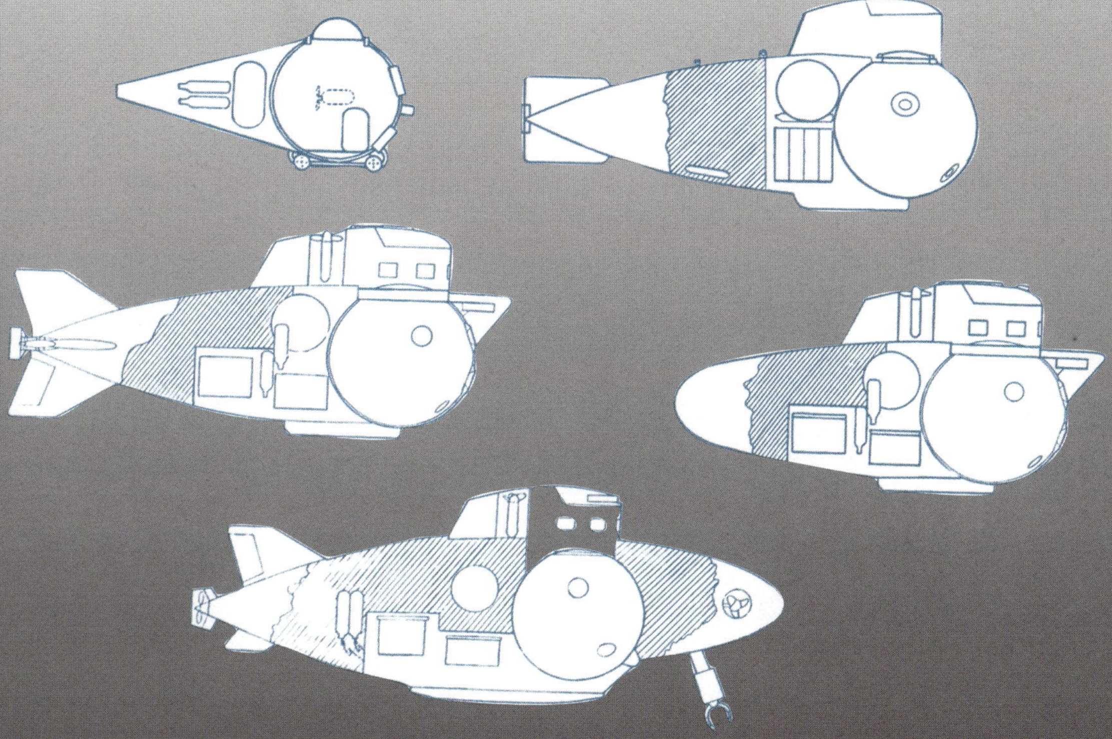

***1963–1987***

# The Star Series & Asherah

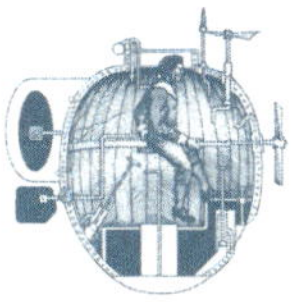

## Star I

Electric Boat (E.B.), a division of General Dynamics, had been a leader in the design and construction of military submarines since building and launching the first U.S. Navy submarine, the *U.S.S. Holland*, at the turn of the century. E.B.'s first entry into the submersible field was a simple, shallow diving submersible that derived its name from the acronym: Submarine Test And Research (STAR) – vehicle #1. *Star I* was a one person submersible with an operating depth of 200 feet and a submerged speed of 1 kt. with four hours endurance and life support for 18 hours. The pressure hull consisted of a 4' diameter by 3/8" thick mild steel sphere, which had a 19" hatch and a Plexiglas hemisphere dome over it for surface use. Two 7" diameter by 1 1/2" thick viewports were provided. One viewport was on the forward vertical centerline just above the equator of the spherical pressure hull. The other viewport was forward on the centerline, about 60 degrees down from the horizontal. Two side mounted, rotatable 1/4 H.P. motors were centered outboard on the horizontal and vertical centerlines of the hull for propulsion and steering. Sea water ballast tanks were

***Name:*** *Star I* ***Builder:*** *Electric Boat* ***Length:*** *10'*
***Maximum Depth Rating:*** *200'* ***Capacity:*** *1 Man*
***Year Launched:*** *1963*

used for buoyancy control. A conical afterbody served as streamlining and for exterior storage of fuel cells, batteries, ballast/buoyancy tank, and miscellaneous apparatus.

The entire assembly weighed 2,750 lbs. and could be mounted on a car-drawn trailer. It was launched in 1963 and operated from the General Dynamics Marine Technology Center (MTC) in San Diego. In July 1964, the submersible demonstrated mock submarine hatch "landings" at Sea Lab I, 30 miles off Bermuda in 192 ft. of water. Aquanaut Manning nearly drowned while photographing the event when his scuba gear malfunctioned; the excitement of the operation delaying his recognition of the problem.

In 1965, an Allis Chalmers hydrazene/oxygen electrochemical reaction fuel cell was installed in the afterbody of the submersible at Groton, Connecticut and operated successfully. This was the first use of a fuel cell in a submersible. In 1966, training dives were conducted at the Marine Technical Center in San Diego. The submersible was deactivated on 29 December 1967 and sent to the Philadelphia Maritime Museum in 1968 for safe-keeping. *Star I* became part of a new display in 1995.

## Asherah

In 1963 George Bass, an archeologist at the University of Pennsylvania, decided that some sunken ships near Turkey could best be explored with a submersible. Named *Asherah* after an ancient Phoenician goddess, it was the second E.B. submersible built and was constructed just after *Star I.* Six viewports provide for extensive visibility: two viewports were located on the forward part of the cabin, slightly above the horizontal centerline; two viewed downward at about a 60 degree angle from the horizontal centerline of the cabin; and two looked sideways a little above the horizontal centerline. The pressure hull was 5' in diameter by 5/8" thick and was comprised of mild boiler steel. A Plexiglas dome or fairwater was located above the 20" diameter hatch at the top of the cabin. Propulsion was provided by two side mounted 2 H.P. pivotable and reversible motors. Fixed syntactic foam buoyancy and external variable ballast tanks provided fixed and controllable buoyancy/ballast. The 4.2 ton submersible had an operating depth of 600'.

On 28 May 1964, some five months after the design program began, the assembled submersible was launched. After two weeks of mostly surfaced sea trials and a dive to 225', *Asherah* was shipped to the Aegean Sea at Guluck, Turkey and then towed to Yassi Ada (flat island), four miles offshore from the Turkish coast. The area Bass chose to explore was west of the island, near a reef which has caused shipwrecks for over 1,000 years. *Asherah* investigated several by passing over them, in effect a photogrammetric "aerial" survey, using a stereo camera. Among the finds was a Fourth Century A.D. Roman wreck which was mapped in a series of 3-D overlapping photos. During one dive, while inspecting the reef up close, the submersible was near enough to the surface that a wave crashed it down on a sharp outcrop of rock. The 5" diameter by 2" thick, bottom viewport was cracked. Only minor leakage occurred and, after recovery, the window was replaced. External guard bars were then installed over the viewport.

Rough weather and political upheaval between Turkey, Greece, and Cyprus, due to the fact that the operations were in the middle of no-man's-land (or water), prevented completion of the dive series. Dr. Bass was later to say that, in the few weeks of operating the submersible, he was able to

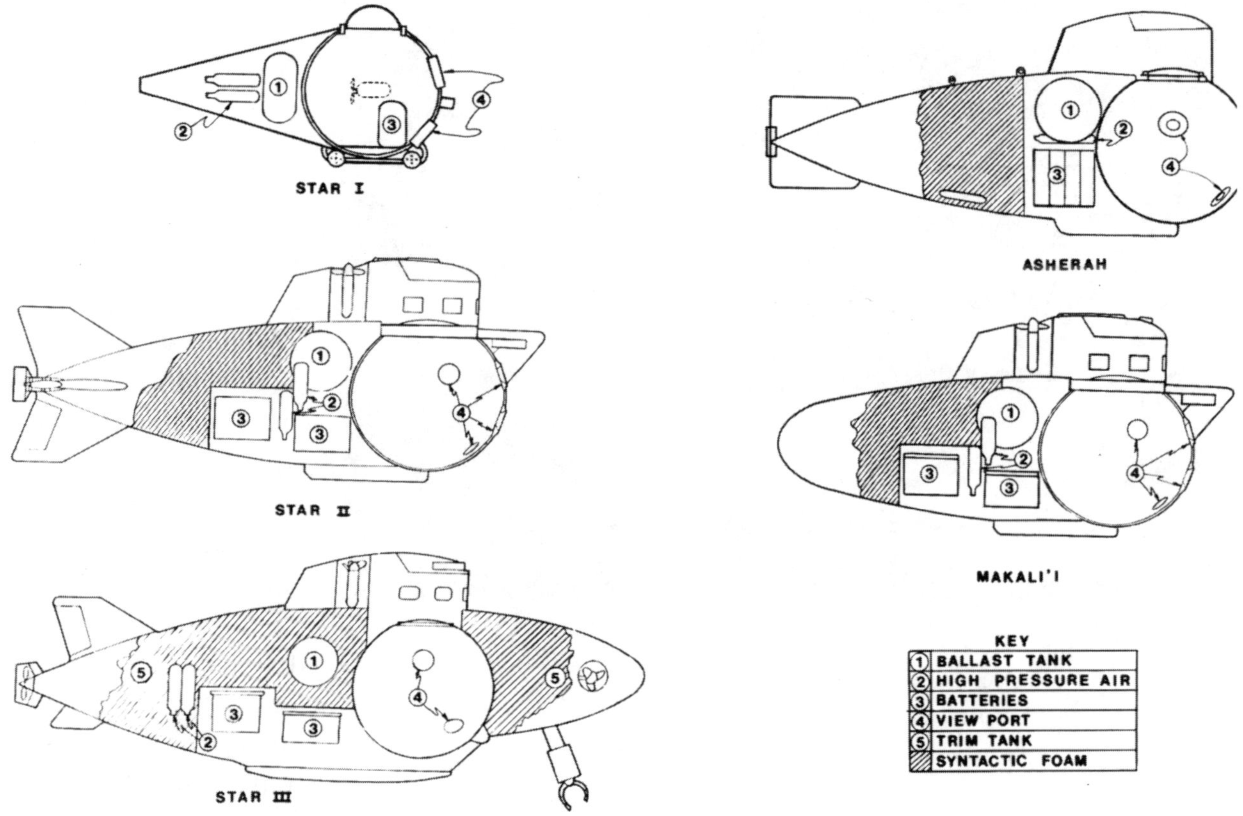


*Electric Boat Submersibles, 1963*

get as much work accomplished as he had previously completed in many months using scuba divers. In June, after refurbishment at E.B., *Asherah* was employed by the University of Rhode Island to explore the floor of Narragansett Bay and Block Island Sound. In July and August 1965, the Smithsonian Institute employed *Asherah* for studying corolline algae growths off Nova Scotia.

During September and October, the Interior Department hired the submersible and crew to dive off the Waimea Coast of Oahu to record fish sounds, photograph fish cleaning stations, survey plankton, and to observe rock and cliff formations. Red coral was seen growing in place for the first time from *Asherah*, deep in the Molakai Straights between Oahu and Molakai Islands. A few weeks later, the coral was being harvested and sold as gem quality jewelry in Honolulu. *Asherah* made 50 dives off Barbers Point, Oahu, Hawaii, for the Bureau of Commercial Fisheries during a month long contract in November and December. At the end of the 1965 operations, the submersible was shipped to the Marine Technology Center in San Diego. Electric Boat conducted various oceanographic projects of their own. These continued until August 1966, when the submersible was packed into the belly of a Navy C-130 at Lindbergh Field and flown to Puerto Rico for "Vieques Salvops," a secret project to recover some lost military hardware. After six dives, the submersible was returned by a Marine C-130 to Lindbergh Field with a "well done" from

***Name:*** *Star II* ***Builder:*** *Electric Boat* ***Length:*** *16'6"*
***Maximum Depth Rating:*** *1200'* ***Capactiy:*** *2 Man*
***Year Launched:*** *1966* ***Description:*** *Star II on LRT*
Courtesy of Makai Undersea Test Range

knowledge from the 1964 dives, Dr. Bass returned with *Asherah* to Turkey and Yassi Ada Island. An elaborate new stereo-photogrammetric mapping system, plus metal detectors and a high pressure water jet system were used. A total of 45 persons made up the crew of diving archeologists and their support. An American (Rosencrantz) pilot and a Turkish observer (Egdemir) made dives during August and September, chasing down likely sonar targets which had been found previously by surface-towed sonar sleds. Starting in August, the side scan sonar search system was upgraded by equipment from Scripps. Several new wrecks were found and stereo-photographed for laboratory analysis. Only twelve useful submersible dives resulted from the 1967 expedition, with the cost burdened this time by many thousands of dollars for insurance coverage. Regrettably, Dr. Bass had to give up his favorite tool.

the Navy Commander of the Caribbean Sea Frontier.

In October 1966, the Bonneville Power Administration of the Department of the Interior hired the *Asherah* and crew for inspecting underwater cable in the Rosario Straights, Washington State, near Anacortes. During nine days of operations, 35,000' of cable were videotaped and photographed. In spite of low visibility water, where unpredictable tides of up to 2 kts. were common, six miles of video tape and 1 1/2 miles of film were taken.

In December 1966, a General Dynamics 64 ton, eight engine XP5Y-1 seaplane was found and photographed six miles west of Point Loma near San Diego. The wreck had been there since 1953, when nine persons aboard had parachuted to safety prior to its crash due to mechanical troubles during a test flight.

In 1967, armed with the experience and

# Star II

*Star II* was designed and built as an Electric Boat second generation research submarine. The general configuration of *Asherah*, its predecessor, was used again. Some improvements over *Asherah* were incorporated. These included: a greater depth capacity of 1,200'; better maneuverability and control; a more reliable and

*A total of 240 dives for pink and gold coral were made before a diving accident on 3 September 1975 occurred, killing two divers and losing the LRT in deep water.*

---

flexible power supply and distribution; and a better internal instrumentation arrangement. Two thrusters of 2 H.P. each were mounted on the stern horizontal stabilizer to provide thrust for fore and aft and turning movement. Another 2 H.P. thruster was located in the sail for vertical thrust. Six 5" diameter viewports provided for visibility. The submersible was capable of 3 1/2" kts. for two hours and weighed 5 tons in air. The 5/8" by 5' diameter hull was made of H.Y. 80 steel (80,000 psi). A pilot and observer made up the crew for the 16 1/2' long submersible.

Both *Star II* and *Star III* were launched at the Underwater Sound Laboratory pier at New London, Connecticut on 3 May 1966. The first operations were sea trials near Groton, Connecticut. The first contract operations were in October 1966 for the inspection of Atlantic telephone cables on the continental shelf off of Cape May, New Jersey, on behalf of the American Telephone and Telegraph Co.

The 1967 dives began with customer relation demonstrations at Nassau, B.W.I., followed by water column temperature measurements off the Bahamas for the New Providence Development Co. The next series of dives were made in Lake Michigan near Beaver Island, using a Coast Guard Cutter as a support ship. During the University of Michigan-sponsored dives, 500 bottom photographs and 7,500' of video tape were taken for the first eye-witness views of the lake bottom. Many unexpected life forms were discovered and the underwater geology was a surprise. During one dive, the bottom was concrete-like with sharp boulders unrounded by glacier action. A 912' dive was the deepest made in the series.

In August, the submersible was taken to the Gulf of Mexico to operate under contract with Continental Oil, Gulf Oil, and Shell Oil, individually, to inspect their offshore oil rigs and pipe lines.

In August of the next year, *Star II* was employed by DIVCON International Divers (for Pauley Petroleum and Humble Oil) in the Santa Barbara Channel to locate downed equipment and assist in its recovery. A 16" riser section had been dropped and recovery was desired. Manipulating and connecting a sling on a lowered drill string, the submersible helped recover a costly, deep-lying riser. Another dive recovered a 200 ton drilling stack after *Star II* cut some cables that were entangling it to the bottom. Also in Santa Barbara Channel, a weather satellite was recovered from the sea floor for the Navy Supervisor of Salvage. Dives were made to 1,200' in the Coronado Canyon, 10 miles south of Point Loma, near San Diego.

Jacques Cousteau had organized a submersible "Jamboree" at Catalina Island during 24-27 April 1969, in which *Star II* took part. A few group maneuvers were made successfully and filmed for Cousteau's audiences. In August 1969, Tap Pryor and Admiral "Chick" Hayward arranged for

*Star II* to be transferred to the Oceanic Institute at Makapuu Point, Oahu, and operated on the Makai Range (later to become Makai Ocean Engineering, Inc). After extensive refitting and refurbishment from storage and shipping, the first instructional dives were made beginning 5 September 1970, for two new Makai pilots.

Pryor, Bill Walker, and I had earlier developed a launch, recovery, and transport (LRT) catamaran for use as a support ship in the ever-rough waters off Makapuu. A submersible is placed on the LRT at a pier by crane and then towed to a dive area. Next, the specially built catamaran is flooded down by several attending divers until it sinks the desired launch depth, normally 75' to 100'. The submersible would then fly off of the LRT and return when finished diving. Many demonstrations were made for the Navy and others, but no paying customers resulted. Instead, the Navy copied the idea and made their own LRT. Some of the 101 dives made at Makai while *Star II* was in Oceanic's possession were scientifically oriented for the University of Hawaii.

When the Oceanic Institute had some financial problems and chose to focus thereafter on aquaculture R & D alone, *Star II* was transferred to Maui Divers Deepex (deep water exploration). A total of 240 dives for pink and gold coral were made before a diving accident on 3 September 1975, occurred, killing two divers and losing the LRT in deep water.

A new LRT, with advanced features, was designed and constructed for *Star II's* first dive on 7 July 1976. Maui divers continued to harvest the deep (1,000'-1,900' ) coral for 217 dives until 26 January, 1979, when General Dynamics transferred ownership to the University of Hawaii. The submersible was sent to Taiwan in early 1979 for 18 dives. Back in Hawaii, dives were made by Maui Divers before they shut down their coral harvesting business. *National Geographic* covered a short dive series with Sylvia Earle being carried in a dive suit to a depth of 1,250', tethered to *Star II*, in October 1979. In another issue, *National Geographic* covered a short dive series, which featured *Star II* harvesting coral. Dive #859 was the last dive made by Maui Divers.

## Makalii

The submersible was then reconfigured and renamed *Makalii* by the new owners, Hawaii Undersea Research Laboratory (HURL). The first dive of *Makalii* was made on 4 June 1981. After sea trials, the LRT and *Makalii* were taken to Eniwetok for scientific dives in 1981.

After returning from Eniwetok, dives were made in Hawaii for HURL with *Makalii* until August 1983. In October 1983, *Makalii* and the LRT were towed to Johnson Atoll to make scientific dives for HURL. In September, the submersible Deep Rover operated with *Makalii* at French Frigate Shoals for a dive series.

HURL's scientific dives continued around the Hawaiian Islands until *Makalii's* last dive, #1271, was made on 29 May 1987, when it was moth-balled at the HURL facility on the Makai Pier at Makapuu.

## Star III

*Star III* was designed to have a high payload/displacement ratio, a high degree of maneuverability, an operating depth of 2,000', and an underwater speed of up to 6 kts. The improvements would allow new systems and components to be tested and deeper and more difficult projects undertaken. For optimum forward thrust, the main thruster was located at the extreme stern of the streamlined outer hull. A horizontally directed thruster was located at the bow to aid turning at zero or low forward speeds. For more accurate depth control, a vertical thruster was installed in the sail. A new manipulator arm allowed more precise handling of external objects. With a weight of only 10 tons in air and 24 1/2' length, *Star III* was serviceable by most marine handling equipment.

The crew, consisting of a pilot and observer, were provided with life support for normal operations of up to 12 hours, with emergency provision for 60 hours. A payload of 2,000 lbs. was easily maneuvered by the 7 1/2 H.P. main thruster and 2 H.P. bow thruster. Located in the 5 1/2' diameter by 1/2" thick steel pressure hull were 5" diameter viewports viewing in four directions: downward; forward; and, to both sides. The 20" hatch at the top of the pressure hull lead to the sail containing three 3" diameter viewports, one looking forward and one for each side view which were used for surface operations. The sea water floodable main ballast tanks when air filled provided for adequate freeboard at the surface. Trim tanks were used for buoyancy and pitch control. The lead acid batteries could supply sufficient power for twelve hours of normal operations.

*Star III* was launched on 3 May 1966, along with *Star II* in Groton Connecticut. During May and June of 1966, the submersible and crew made training and demonstration dives in the Thames River at Groton, and in Black and Hudson Canyons, New York.

July and August were spent making dives in the Atlantic for the Underwater Sound Laboratory, doing a biological survey of the Deep Scattering Layer from New London to Bermuda B.W.I. On one of the Bermuda dives, a power failure occurred when a battery compensation chamber burst. The container was enlarged and strengthened and the dive series continued. During March 1967, dives were made in the Atlantic off Key West, Florida for photogrammetric work, sponsored by the Naval Oceanographic Office. During April, underwater visibility tests were made in the Atlantic near Nassau. In addition, temperature gradient readings

***Name:*** *Star III* ***Builder:*** *Electric Boat* ***Length:*** *24'6"* ***Maximum Depth Rating:*** *2,000'*
***Capacity:*** *2 Man* ***Year Launched:*** *1966* Courtesy of General Dynamics and Scripps Institute of Oceanography

were made for the New Providence Development Co. of Nassau, for environmental protection purposes.

During May and June of 1967, a telephone and telegraph cable survey was made for AT&T in the Atlantic, off Cape May near New Jersey. On several occasions, schools of small lantern fish entered the outer hull filling every crack and cranny until the cabin hatch could not be opened without a clean out of the access trunk. During August 1968, an ocean bottom survey was conducted in conjunction with American and Danish scientists in the waters off Baffin Bay, seven miles southwest of Thule, Greenland. An Air Force B-52 had crashed in January 1968 on seven foot thick ice. Surface cleanup was made immediately, but some of the missing, classified, hardware required recovery after the ice thawed. In ten days of diving, *Star III* recovered "objects" at depths as great as 900'.

In November 1968, *Star III* assisted in the search for a leak in an AMCO (American Oil Co) offshore gas line, in the Haisbrough Sands area of the North Sea. A half-dozen dives in mixed currents and foul weather completed the job.

In April 1969, *Star III's* last dive (#279) was conducted in the North Sea, on a tower 30 miles north of Great Yarmouth, England, which was operated by ARPET (Atlantic Oil Producing Co). The sea-states, tidal currents, and near zero visibility prohibited useful work by the submersible. Shortly afterwards, it was deactivated. Since 1969 it has been in storage at Scripps Institute of Oceanography or on display at Sea World, San Diego.

It is notable that the biggest military submarine builder in the U.S. built one of the smallest submersibles for their initial entry in the field. Electric Boat's *Star II* with dives to as deep as 1,200', operated virtually without modification or redesign during its working life of 1,271 dives, a result that few other deep submersibles attained.

# *The Perry Submersibles*

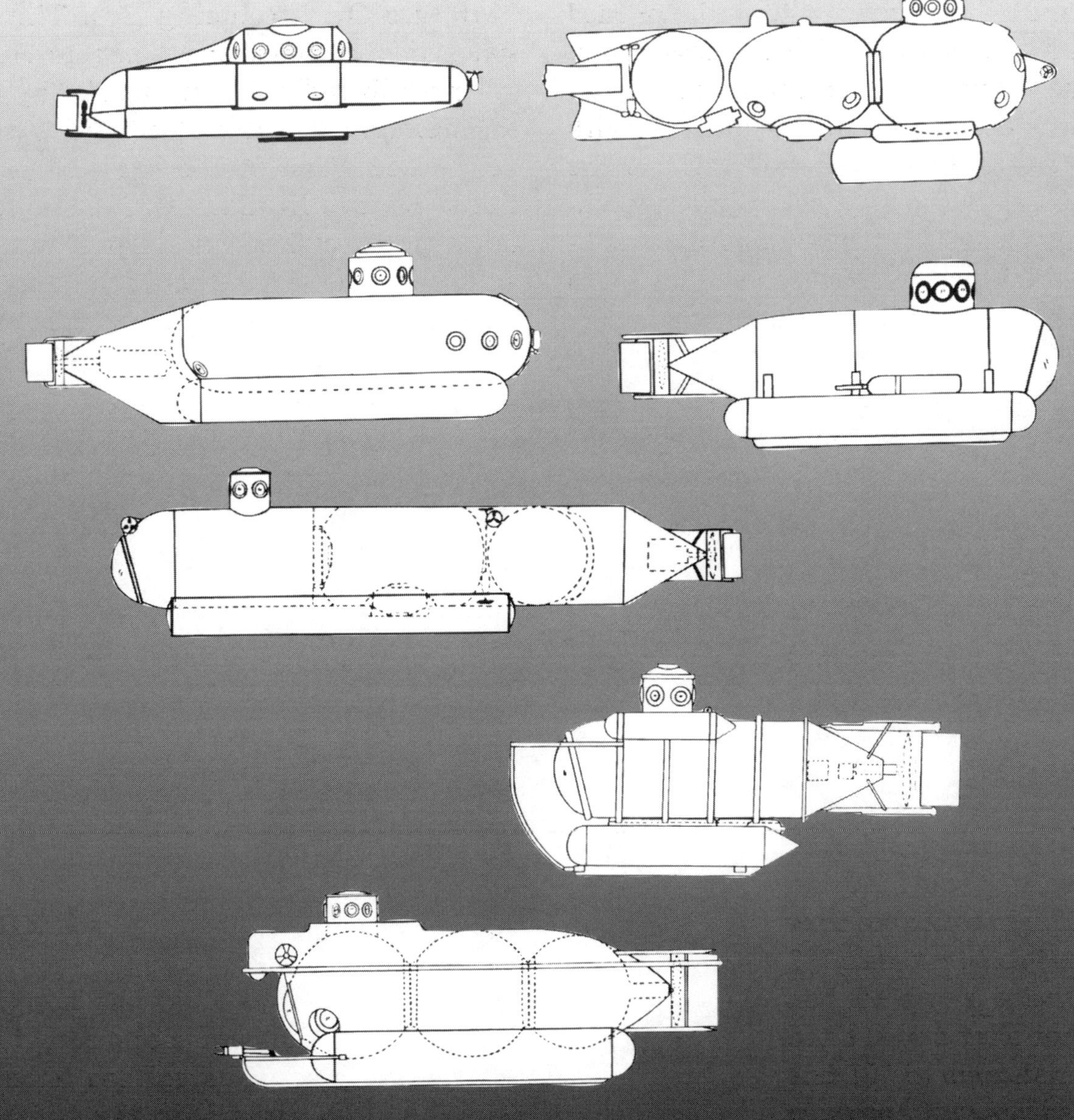

***1962–(       )***

# The Perry Submersibles

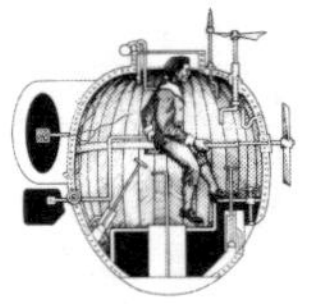

In 1956, John H. Perry Jr. was a successful Florida newspaper publisher who built a submersible in his garage as a hobby. In an interview 37 years later, he commented that it was fortunate his plywood submarine wouldn't sink or he would have drowned. His initial concept was to provide a safe way to go under water without getting cold or being exposed to dangerous marine animals. His thoughts of filling a recreational need were soon overtaken as he learned more of the needs of scientists and for industrial applications.

The hobby became a business when his first production model, with an operating depth of 150', was launched in 1962. It was called the *Perry Cubmarine #3, Experimental (PC3-X)*. Like all of the *PC 3s* to follow, it was comprised of a long cylinder 3 1/2' in diameter, intersected by a long rounded sail or turret containing ten to twelve viewports for the pilot and passenger to look out of from seated positions. *PC3-X* had one hatch at the top of the sail and was powered by a 4 H.P. d.c. motor. With a dry weight of 4,700 lbs. and a 20' length, it was trailerable and could be managed by light marine winches and cranes. It was operated by the Applied Research Laboratory at the University of Texas in Lake Travis — near Austin — until it was retired from active service, and went on display at the Center for Science and Industry in Columbia, Ohio.

The *PC3-B* was similar to the *PC3-X*, but had two hatches in the top of the sail and was built for an operating depth of 600'. It weighed 5,500 lbs. in air, had a length of 22' and a width of 3.5'. The submersible was best known for its 1966 operations in Palomares, Spain, where it was assigned to aircraft salvage and debris collection in shallow areas of the lost bomb site. Later, it was leased by the Bureau of Commercial Fisheries for scallop research near Cape Kennedy, Florida. It was then sold to International Underwater Contractors of New York and named *Techdiver*, for use in fisheries research as well as pipeline inspection. After it was retired, it was put on display at the Planet Ocean Museum in Miami. A twin version was built for Western Oceanographic Systems and designated *WS-2* by that group.

*PC3A-1* and *2* were nearly identical in appearance to the other Cubmarines in the *PC3* class. Operating depth was 300' and a single access hatch was employed. They were powered by 7 H.P. motors and operated for years at Kwajalein and Eniwetok islands and atolls in the South Pacific, recovering missile nose cones. After retirement, one of the submersibles was put on display at the Alabama Space and Rocket Center Museum in Huntsville.

In the summer of 1966, Ed Link started mooring his surface research ship *Sea Diver* at John Perry's berth and submarine shop at Riviera Beach, Florida. Ed and John Perry immediately began collaborating on the design and construction of the first pressurized diver-lock-out, small submersible called *Deep Diver Perry/Link #4* (PL-4). Two elliptical pressure hulls, 54" in diameter, were interconnected by a hatch with each chamber having its own viewports. The two-person pilots' compartment in the bow had an entry trunk at the top with a hatch for surface access. The two-person diver's compartment had its access hatch at the bottom of the chamber. Besides the five bottles of oxygen it carried, the 49" diameter sphere could be pressurized with helium/oxygen mixes for deep or long term use. A total of 21 acrylic viewports were located in the access trunk, divers' and pilots' chambers. The submersible had an operating depth of 1,350' and a lock-out capability down to 1,250'.

In 1967 and 1968, lock-outs were made at 430' and 700' respectively. The first lock-out from any Perry submersible was made on dive #2 shortly after its launch date by Ed Link and Denny Breese in January 1967. The 16,500 lb. submersible, for all its chambers, was only 22' long. A rotating 3 H.P. bow thruster and a 3 H.P. aft vertical thruster gave the vehicle maneuverability even at slow or zero speeds. A 10 H.P. main thruster provided speed in the fore and aft directions.

***Name:*** *Deep Diver* ***Builder:*** *E. Link and J. Perry*
***Length:*** *22'* ***Maximum Depth Rating:*** *1,350'*
***Capacity:*** *4 Man* ***Year Launched:*** *1967*
Courtesy of Perry Trytech

After check out dives and minor modification, Ed Link loaded *Deep Diver* on his ship and headed for the Bahamas. Beginning in March of 1967, a series of dives were made for lock-out and decompression practice. Transits between the submersible's diver chamber and some air filled underwater "igloos" were practiced. The inflatable "igloos" were supplied with air from a compressor on the surface support ship and were used as a source of air and extra tools for some tests simulating the repair of broken underwater power and telephone cables. In June, *Sea Diver*, with *Deep Diver* on board, headed for the Great Stirrup Cay, Bahamas. On board were divers/scientists from the Smithsonian. Link was interested in introducing them to lock-out diving and decompression. A diver/photographer from *National Geographic Magazine* provided film coverage for its January 1968 issue. Shortly after its check out dives, the submersible was sold to Ocean Systems Inc., a company that did underwater engineering and searches all over the world. In conjunction with NOAA's Man in the Sea Program, more lock-out programs were carried out in 1972 in the Gulf of Maine. After retirement, the *Deep Diver* was put on display at the Miami Planet Ocean Museum.

*PC 5* was built in 1967 for Sub-Sea Oil Services (SPA) in Milan, Italy. It was 22' in length, had an operating depth of 1,200' with a total of 26 viewports, nine in the conning tower, 13 in the forward hemisphere of the pressure hull, and four in the stern. Total air weight was 10 tons. The steel pressure hull had a 4.6' diameter cylinder with hemispherical ends. A 400 kg keel weight was mechanically releasable. The crew consisted of one pilot and two observers. The main thruster was a 7 1/2 H.P. motor and propeller mounted astern. Two fractional H.P. thrusters were mounted on the hull, one astern and one forward for maneuvering. Dive planes and a rudder were used as part of the control system. A manipulator with six degrees of freedom was mounted on the bow. The submersible was designated *P51* in Italy.

The *PC 5C* was launched in 1968. When launched, it was nearly identical to the *PC 5*, minus a manipulator. Named *Nai'a* by the owners, Pacific Submarines of Hawaii, it operated near Kaneohe Bay, Makapuu Channel and Hanauma Bay, Oahu Island, Hawaii, on contracts with the Navy, the University of Hawaii, and in-house mineral exploration projects.

*Shelf Diver* (PC4B) was launched in 1968. It was designed for the same functions and depth as *Deep Diver*, including a lock-out chamber. The rotating nose cone with a thruster was replaced with a thruster in the vertical and one in the horizontal direction. The operating depth was reduced from 1,350' to 800', but other statistics and dimensions remained virtually unchanged. The first lease operations were to the Bureau of Energy, Mines, Resources, and Fisheries Research Board of Canada. Using lock-out divers, surveys were made of scallops, herring spawning grounds, crabs, and geological areas south of Nova Scotia and in the Bay of Fundy. The submersible was purchased by COCEAN in 1971, which then sold it to the French Navy.

In 1974, *Shelf Diver* was used to inspect an area inside a 15' diameter water tunnel that carries water from the Alps to Marseilles, France. The inspection required the submersible to travel into the tunnel a distance of 1 km under 1,000' of mountain and return backwards to the opening on its own power. Other work included "Sea-bed exploration of wrecks, recovering materials from the sea floor, intervention, and submarine rescue training." The submarine was sold to the French company INTERSUB in 1979.

The *PC 8* was launched in 1979. It was the first Perry submersible to use an acrylic bow. The main portion of the hull was a steel cylinder 42" in diameter. The overall length was 18'-6" with an air weight of 5 1/2 tons. The classical Perry conning tower contained eight viewports. The main propulsion was provided by a 7 1/2 H.P. motor. The large acrylic bow dome provided the operator with a 140 degree view at a downward angle of approximately 13 degrees from the horizontal. Twin manipulators provided underwater work dexterity for the operator.

In 1973, *PC 8* made a demonstration dive at Loch Ness, Scotland, followed by a

***Name:*** *PC 5C* ***Builder:*** *J. Perry* ***Length:*** *22'*
***Maximum Depth Rating:*** *1,200'*
***Capacity:*** *3 Man* ***Year Launched:*** *1968*
Courtesy of Perry Trytech

lengthy cable laying contract near Long Island, New York. In 1974, an emergency call to Perry brought the *PC 8* to Greenland to rescue a crew in a downed helicopter. The pilot of the submersible was able to place a noose over the aircraft's rotor so that a workboat could hoist it to the surface. The two men in the forward compartment were still alive, but the two in the back did not survive the crash.

In 1974 the *PC 8*, after a refit, was redesignated *PC 8*C. It was purchased by Northern Offshore Ltd. and saw extensive use in the offshore oil fields of the North Sea and the Mediterranean. It was next purchased by Intersub and used for the remainder of its working life in the petroleum fields. It sat on the blocks in Aberdeen, Scotland for several years, until purchased in 1983 by Research Submersibles Ltd. (R.S.L.) for use by the Discovery Bay Marine Laboratory on the north coast of Jamaica in the Caribbean. In mid-1985, the submersible was moved to nearby Grand Cayman Island for 800' deep tourist dives at $175 per hour to an old wreck. During one of these dives, a wedding was performed with witnesses in an adjacent *PC-1203* at 800'. In 1986, the submersible was sold to the Bulgarian Institute of Oceanography in Varna, Bulgaria, after a refit in the Grand Cayman. The Institute planned to do marine research in the Caspian and Black Seas beginning in 1987. Reportedly, the submersible had made 2,000 dives before its transfer to Bulgaria.

***Name:*** *PC 8* ***Builder:*** *J. Perry*
***Length:*** *18'6"* ***Maximum Depth Rating:*** *800'*
***Capacity:*** *2 Man* ***Year Launched:*** *1979*
Courtesy of Perry Trytech

The *PC 15* was launched in 1973 with the same type of large acrylic bow dome on the front of a ring-stiffened cylinder with an inside diameter of 54" as the *PC 8*. The classic Perry conning tower was located on top of the main hull, or pilots' chamber. Attached to the pilots' compartment was the divers' chamber, with a hatch between the two compartments . The divers' chamber had a bottom access hatch for lock-out. Behind the diving chamber was a spherical helium/oxygen tank and, behind that, a short cylinder and cone housing the motors driving the propeller behind the cone. The overall length at launch was 31', weight in air was 15 tons, and the operating depth was 1,200'. The main propulsion propeller was powered by two 10 H.P. motors turning a common shaft. One vertical and two small horizontal thrust units aided in maneuvering, especially at slow speeds. The submersible had a 2,000 lb. payload capacity.

The first owner was Vickers Oceanic Ltd. at Barrow-in-Furness, England, that renamed it *VOL-L1*. The *L-1* operated commercially in the North Sea quite successfully, so it was selected for experiments in submarine and bottom habitat personnel-transfer programs. A skirt was installed, similar to those on U.S. DSRVs, and maneuvering trials were conducted, with satisfactory results. A seabed chamber, like a seafloor habitat, was first used for training, then *HMS Otter* and *HMCS Ojibwa* were practiced on, with men actually transferring from one vessel to another while very deep. Next, a habitat transfer was made, again successfully. Toward the end of the tests, a complete crew could be transferred three times in two hours.

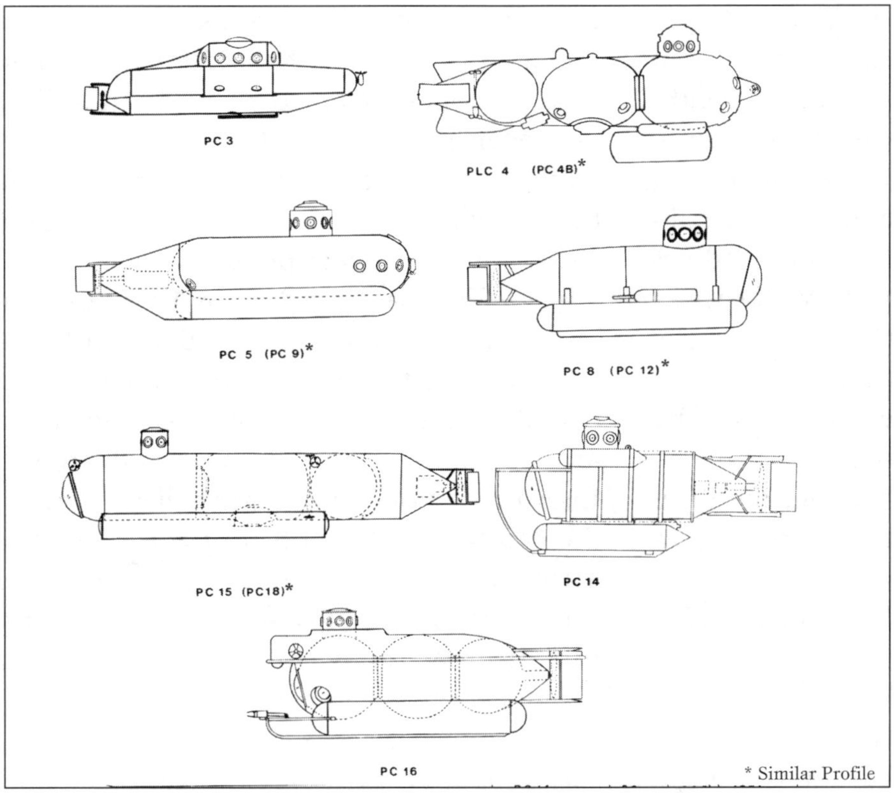


***Name:*** *Perry Submersibles* ***Builder:*** *J. Perry*

| *Name* | *Length* | *Capacity* | *Max. Depth Rating* | *Year Launched* |
|---|---|---|---|---|
| *PC3* | *20'* | *2 Man* | *150-600'* | *1962* |
| *PLC4 Deep Diver* | *22'* | *4 Man* | *1,350'* | *1967* |
| *PC4B Shelf Diver* | *23'* | *4 Man* | *800'* | *1968* |
| *PC5* | *22'* | *3 Man* | *1,200'* | *1968* |
| *PC9* | *26'* | *3 Man* | *1,350'* | *1970* |
| *PC8* | *18'6"* | *2 Man* | *800'* | *1979* |
| *PC12* | *20-26'* | *3-5 Man* | *1,000'* | *1975* |
| *PC15* | *31'* | *4 Man* | *1,200'* | *1973* |
| *PC18* | *25-26'* | *4 Man* | *984'* | *1977* |
| *PC14* | *19'6"* | *2 Man* | *600-1,200'* | *1974* |
| *PC16* | *25'* | *4 Man* | *3,000'* | *1976* |

Five owners later, *PC 15/VOL-L1* was up for sale in Fort Pierce, Florida in 1994.

In 1974, *PC 1401* was launched. The first to buy one of these "Marine Volkswagens" was Texas A&M University's Department of Oceanography. The diminutive size of the *PC 1401* was chosen by many users over the *PC 8* because it was more suited to smaller oceanographic surface ships with light handling equipment. The modular construction permitted optional hull sections to be added after purchase. The 42" diameter ring stiffened cylindrical hull had a conical afterbody with a 36" acrylic dome at the bow. Seven viewports were located in the conning tower. A 3 H.P. Cushman golfcart motor turned the stern mounted propeller. The dive planes and rudder were operated manually. The overall length of 19'-6" and an air weight of 5 tons gave the operator the ability to maneuver in confined quarters and optimize handling around and on oil platforms. The operational depth of 1,200' and a maximum speed of 3 kts. gave the pilot and observer many options in its operations. After acceptance by Texas A&M, the submersible was renamed *Diaphus*. The University's first goals were to track natural oil seepages in the Gulf of Mexico and elsewhere. After its use for oceanography, the submersible was returned to Perry for fuel cell development. In 1994 it was for sale for $200,000 and was ABS certified.

The *PC 1402* was launched in 1975. It is nearly identical to the *PC 1401*, but with an operating depth of 600'. It replaced the two previously operated Perry Submersibles at Kwajalein, Marshal Islands, Southwest Pacific. A construction contract with the Army's Ballistic Missile Defense Systems Command, Huntsville, Alabama, included the requirement of an operational depth of only 600'. The lagoon is a missile target at the downrange end of the Pacific Missile Range. Rockets are usually fired from Vandenberg AFB, California. The submersible is used to locate the missile components as well as to maintain underwater instrumentation in the lagoon. The submersible and crew average about 30 dives per year. A civilian company, Kentron Hawaii Ltd., operates the submersible under contract to the government.

*PC 1201* was launched in 1975. The modular construction made it possible to obtain this model with or without the diver lock-out chamber. Even after leasing or purchasing, the lock-out chamber could be added to the standard or observation model. Operating depth for the submersible in either form was 1,000'. The lock-out version had an air weight of 9 tons, while the observation version had a weight of 7 tons. Length of the observation type was 20' with another six feet added for the lock-out. The cylindrical hull had a diameter of 44" with a 36" diameter bow dome and a conical after-body containing the 10 H.P. d.c. motor for the main propulsion. Eight viewports were located in the turret. A three person crew, usually consisting of one pilot and two observers, was located in the forward chamber. The capacity of this submersible fell somewhere between the *PC 8* and the *PC 15*. Twin battery packs under the hull could be varied in size for different power needs.

The first client for the *PC 1201* was Intersub of Marseille, France and London, England. It was used in the North Seas for pipeline inspection, which included clearing nets, setting explosive charges, checking platforms, making cathodic protection measurements, and the like. Frequently, clients and pipeline experts were taken on the dives as observers with video and photo coverage as well. In 1986, the *PC 1201* was sent to the Island of Rota, south of Guam in the Mariana Islands, for use as a tourist submersible catering to Japanese tourists and honeymooners. Approximately 1,100 dives had been made by 1989. Typhoons, overhauls, and off-seasons make the business in the Mariana Islands sporadic. On 4 March 1991, the *PC 1201* was put up for sale.

*PC 1202* was launched in 1975. Its construction details are nearly identical to *PC 1201*, except that it has the diver lock-out chamber as part of the hull. Also included are two manipulators. It was delivered to Intersub of London and Marseille for use in the North Sea gas and oil fields. There, it allowed divers to lock out, repairing pipelines, doing close-up rig inspections, installing & recovering pingers, making video as well as camera recordings, and performing deep water search and recovery. It was supported by a specially designed ship with a decompression chamber that connected to the diver's chamber on the submersible.

In March 1978, one of the *PC 9s* of a competitor of Intersubs, called P&O Subsea became entangled in some steel cables near a well head. Since the propeller was badly fouled, no available surface systems could free the submersible with the two men inside. With a three man crew and two divers, the Intersub *PC 1202* dove on the entangled submersible. Once on the bottom, the rescue submersible's divers took only 15 minutes to free the *PC 9*. The crew of the *PC 9* were back on the surface after a 24 hour dive and back with their families in nearby Aberdeen, Scotland that night.

*PC 1203* was launched in 1976. It was nearly identical to the observation type *PC 1201* except that it had two manipulators. The first organization to use the submersible was COMEX of France. COMEX is a worldwide French offshore company. *PC 1203* was used for support of oil industry operations, in pipeline surveys; and observations in the North Sea. In 1985, Research Submersibles Ltd. (RSL) obtained *PC 1203*. The submersible, along with a *PC 8B* now makes three to four dives a day to a wreck at 800' with a pilot and two paying passengers, totaling an average of 60 dives per month. In 1977, a *National Geographic* crew photographed the geologically famous 1,000' underwater wall in the area. Over 600 dives were made in the first nine months of 1988. Perry's records show it to be still active in 1995.

*PC 1204* was launched in 1976. It was initially used by Intersub Ltd. of London, England for the oil fields in the North Sea. In 1990 it was advertised for sale by the Subsea Works of Seattle, Washington. It is currently (1995) very actively engaged at the Harbor Branch Oceanographic Institute of Fort Pierce, Florida under the name Clelia.

*PC 1205* was launched in 1978 and sold, or leased, to Intersub (London). In 1979, *PC 1205* and *PC 16* were involved in the deepest pipe laying program ever made at that time. In June of 1987, the *PC 1205* had a second acrylic dome installed on the aft end of the pressure hull with side mounted thrusters. The work was done by RSL in the Grand Cayman.

The *PC 1601* was launched in 1976. It was the deepest operating submersible Perry built. Many features of the construction were unique. Three 6' diameter modular spheres were connected together by bolts, providing a 22" diameter hatch between the pilots' sphere and the divers' sphere. The forward sphere had a large acrylic dome, which was 4" thick and 30" in diameter, plus two 6" diameter viewports, for viewing to the side and downward. The conning tower, which is part of the forward pressure hull, had six viewing ports around its cylindrical surface, plus one in the hatch. The conning tower could be occupied and used submerged or surfaced. The divers' sphere had a hatch for dry transfer to habitats and other submarines, with a 6" diameter viewport in the hatch. The aftermost sphere houses the 10 H.P. motor for main propulsion and other machinery, equipment, and electronics. The operating depth of 3000' with dry transfer capability was exceptional for the time, and has not been duplicated in the private sector in the 18 years since its launch. The 25' length and its 15 ton weight made a compact vehicle for all of its capacity. Two manipulators provided additional capabilities to the submersible. The rudder and bow planes operate by hydraulic power. Two small thrusters attached to the dive plane and another, directed laterally at the bow provide low speed maneuvering.

The *PC 1601* was purchased by Intersub of London. They began using the submersible immediately after delivery, following testing in the Bahamas, for route surveying the Trans-Mediterranean pipeline project. For this project the seafloor topography was being mapped to depths of 2,200'.

The *PC 1602*, identical to the *PC 1601*, was launched in 1980. Saipem of Milano, Italy purchased it to lay the deepest offshore pipeline at the time. Sea trials and training dives were made in the Mediterranean, in addition to test dives by the manufacturer, prior to shipping from the U.S. The pipe-line worked on runs from Sicily to Tunisia, Africa to supply Algerian gas to the Italian mainland.

In 1980, *PC 1602* became entangled in cable and was disabled at a depth of 1,700' while surveying a pipeline. By prearrangement, a diving bell with a manipulator was on standby status. It was lowered and divers cut some soft line connected to a massive bottom clump of concrete that had entangled in the propeller. The diving bell, called the OMB-1, had been built by Perry for the Italians for just such an event. As of 1995, *PC 1602* is still in use by Saipem.

*PC 1801, 1802*, and *1803* were all launched in 1977. The series was specifically designed for the petroleum industry for pipeline, rig, and tower inspection, as well as rig repairs in place. With only a 984' operating depth, its design is smaller than the *PC 15* which it emulates. Economy and maneuverability were the paramount design criteria. The diver lock-out capability was designed for 656' maximum, which was as deep as future needs indicated at the time. The pilots' cylindrical pressure hull was 54" in diameter with a 36" acrylic dome forward, a conning tower and hatch on top, and a hatchway, centered aft, to the divers chamber. The divers chamber was also 54" in diameter, with a lock-out hatch at the bottom, which could be used for mating to a deck decompression chamber on a support ship.

The submersibles' over-all length of 25 1/2' gave the pilot good maneuverability, using a 180 degree pivotal 10 H.P. motor and propeller at the stern. Eight viewports girdle the conning tower, another viewport is in the internal hatch between the divers' chamber and the pilot's chamber. The third viewport is in the divers' lock-out hatch. Forward mounted thrusters, for lateral and vertical thrust, augment the rudder and dive planes. The submersible's weight in air is 12 tons; it can carry four people, usually one pilot, one observer and two divers. Additional thrusters were added externally on some units. The Perry work-system on this submersible provided unusual capabilities by way of the heavy duty manipulators with grinders, cutters and diver plug-in tools to the power take-off on the submersible.

Both *PC 1801* and *PC 1802* were tested in the Bahamas before delivery to Intersub in Aberdeen, Scotland. *PC 1801* made a video/profile survey, marine growth, and structural survey on a Shell Oil Co. 36" pipeline during the summer of 1981.

*PC 1802* was purchased by Research Submersible Ltd. (RSL). In 1985, the submersible was active, making tourist dives near Provo in the British West Indies. The operation was moved to the Grand Cayman Island next and, in the first half of 1985, over 100 tourist dives had been made. In 1986, RSL used *PC 1802* for a scientific survey in the Gulf of Mexico. In 1987, the submersible was back at the Grand Cayman Island, diving tourists to the 800' deep wreck of the *Kirk Pride*. In the first nine months of 1987, *PC 1802* made 540 tourist dives.

*PC 1803* was launched in late 1977 and delivered to Super Pesa Transportes Maritimo LTD of Rio de Janeiro, Brazil. It was purchased for the development and maintenance of undersea oil installations off the coast of Brazil. In June 1981, a team of British submersible pilots were operating the submersible off Brazil. In 1994 the submersible was for sale in Brazil.

The *PC 1804* was launched in 1982 and sold to Shell/ESS for underwater work in the North Sea. This was the fifth and largest in the 1800 series, with an air weight of 12 tons and a length of 25'. It was demobilized in 1986 and put up for sale in England. In 1994, it was up for sale at Perry's facility in West Palm Beach, Florida for $750,000.

The end of the oil boom also ended Perry's submersible construction. It was an extraordinary run of excellence beginning with John Perry's garage model that "wouldn't sink." As the hobby grew to a business, John's son, "Hap" Perry took over operational responsibilities and continued the family commitment for a safe product, with continuing improvements to meet both safety and customer demands.

# *Deep Quest*

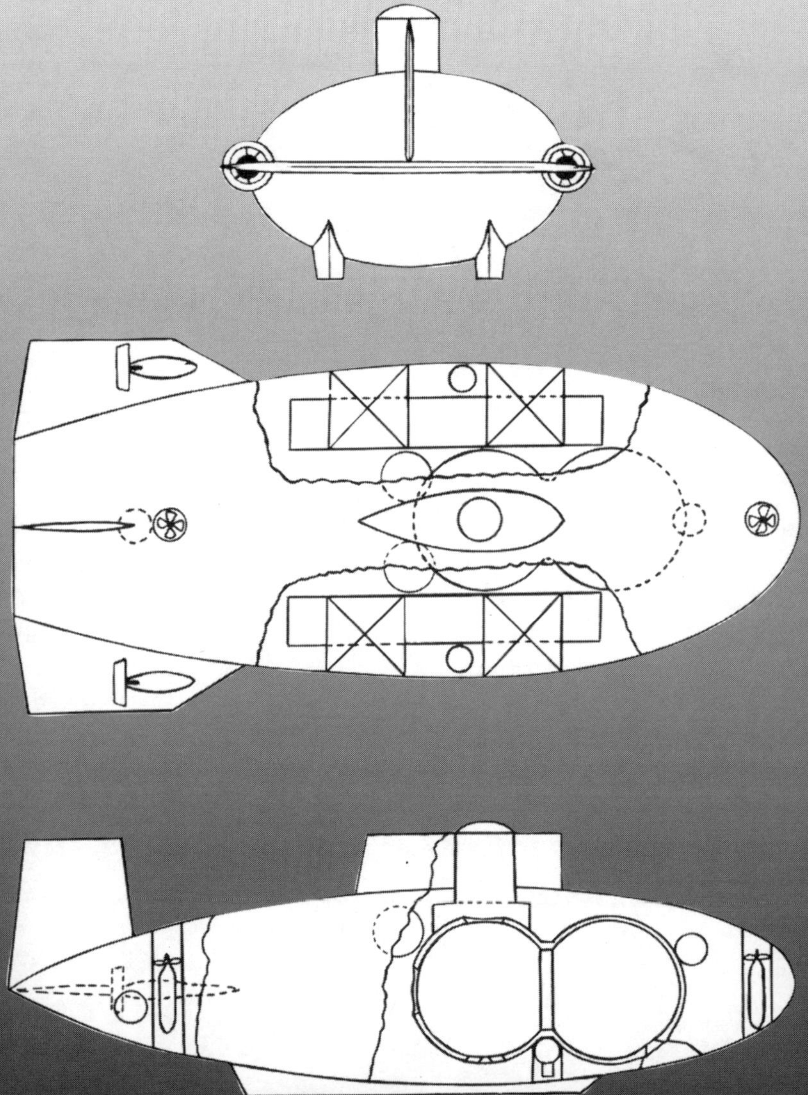

## *1967–1980*

# Deep Quest

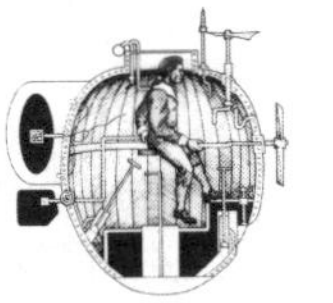

The first priority in the Navy's deep submergence program, established after the loss of the submarine *Thresher*, was the Deep Submersible Rescue Vehicles *(DSRV-1* and *DSRV-2)*. To position itself to obtain design and construction contracts, Lockheed Missile and Space Co. (LMSC) needed to prove its ability. It was decided to build not only a DSRV-like submersible, but a support ship to demonstrate LMSC's ability to conduct operations at sea and, perhaps, obtain the operational handling contracts for the future DSRV's as well.

The company elected to use their newly developed maraging steel used in rocket boosters for the pressure hull. Their brilliant metallurgist, Dr. Shapiro, would reduce the brittleness, work out the difficult welding procedures, and still have a steel 2 1/2 times stronger than the steel used for nuclear submarine hulls.

The DSRV prototype was called *Deep Quest*. The pressure hull was comprised of two 7' diameter steel spheres welded together at the equatorial centerline. The spherical connection formed a passageway between the spheres. A hatch in the top center of the forward sphere provided access to the pressure hull when surfaced. A hatch at the bottom centerline of the aft sphere was provided for access to a skirt, which could be attached below the pressure hull as required, for making submarine rescues. The skirt was never constructed, but the design concept was successfully used by the DSRV's in later years. A third hatch at the rear and on the equatorial centerline of the aft sphere was provided for a divers' lock-out chamber, to be added later. This chamber, which was never built, was to permit divers to lock in and out under water from a special and separate pressure chamber, with access to the main pressure hull, when appropriate. A viewport was provided on the centerline of the forward sphere at approximately 51 degrees downward from the horizontal. A viewport in the lower hatch of the aft sphere was also included.

Since demonstrations of maneuverability were primary requirements of the operational capabilities of *Deep Quest*, the propulsion and buoyancy control were critical systems. Two 7 1/2 H.P. main thrusters were located aft in the horizontal stabilizers, with two vertical thrusters

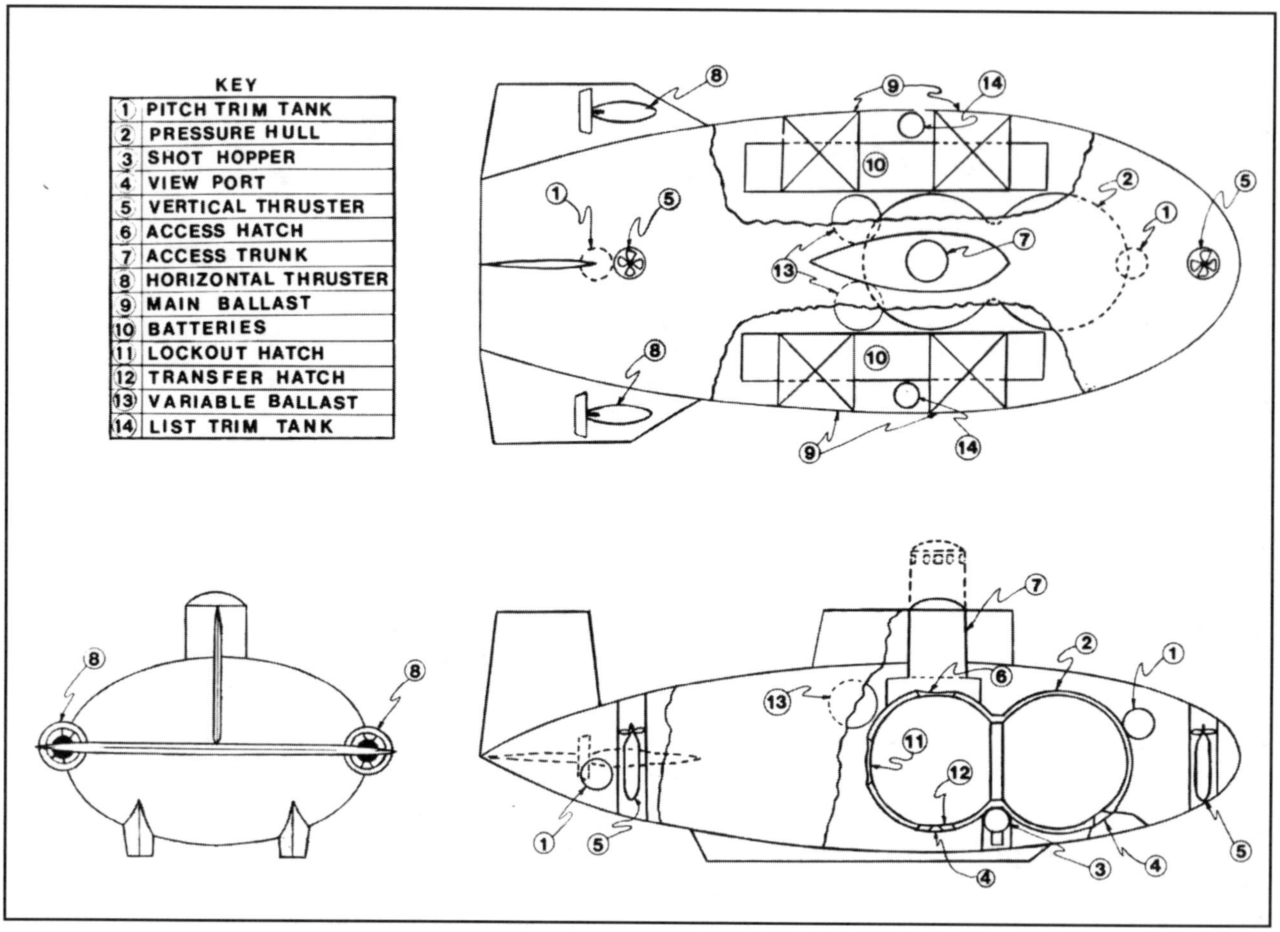


***Name:*** *Deep Quest* ***Builder:*** *Lockheed Missile and Space Co.* ***Length:*** *40'*
***Maximum Depth Rating:*** *8,000'* ***Capacity:*** *4 Man* ***Year Launched:*** *1967*
***Description:*** *Top, Side, and Rear Views with Cutaways*

in tubes, one located at the stern and one at the bow. Mercury-filled pitch trim tanks were located forward and aft. List trim tanks, containing mercury, were located outboard on either side of the aft sphere. All of the mercury used for trim was droppable for a total of 220 lbs. A shot hopper, containing 1,656 lbs. of droppable shot, provided the primary lifting capability when released. The main ballast tanks were used for buoyancy and freeboard at the surface and were flooded for diving. The forward batteries were droppable in an emergency, for an additional 3,000 lbs. of lift. Two manipulators and a bow rack were also releasable. In total, 7,500 lbs. of ballast and equipment could be dropped, if required. Lateral jet thrusters, fore and aft, completed the maneuvering systems. State-of-the-art instrumentation, control system, and communications set-up gave the cockpit the appearance of that in a 747 superjet.

*Deep Quest*'s overall length of 40' and displacement of 50 tons made launch and recovery a significant job in rough weather for the 108' long support ship, *Transquest*. For launch or recovery, the submersible was positioned on a 23' wide by 31' long elevator platform at the stern of *Transquest*. When lowered, the remaining portion of the ship has the configuration of a squared off "U" shape when viewed from overhead. The platform was positioned 12' below the surface of the water for launch or recovery.

*Deep Quest* was maneuvered on or off the platform by guide lines and/or its own power, with *Transquest* headed into the wind and the existing seaway to shelter the operation. The 19' width of *Deep Quest* required a close and careful fit inside the 23' wide elevator. In spite of the large volume of the pressure hull, which housed all the instrumentation, controls, and up to five people, the submersible had a design depth in excess of 8,000'.

The gamble paid off. Contracts were awarded to the Lockheed Missile and Space Company and the Draper Laboratory for the design and construction of the *DSRV-1* in June 1966, based in great part on Lockheed's work on *Deep Quest*. *Deep Quest* made its inaugural dive on 30 September 1967 in San Diego Bay. The aerospace company was now ready to test a submersible the size of a humpback whale, comprised of sophisticated systems, as yet untried under water. Maneuvering and control of the large submersible was comparable, in accuracy, to that of a helicopter in an underwater environment, using sensors, television, and instrumentation as substitutes for direct viewing. The new and complex systems would require test after test, sometimes to failure, with subsequent repairs during a period of test piloting in its early operations. Pilot training and testing, as well as failure analysis, were paramount goals and accomplishments during dives in 1967.

All dives made in 1968 were made in the San Diego area. They included mostly pilot training and demonstration dives in an effort to find customers. *Deep Quest's* deepest dive was made in February. During the eight and a half hour dive, a depth of 8,310' was reached off the Coronado Roads, near San Diego.

The 1969 operations began in early January in the San Diego area, initially for pilot training. However, a DC-8 Scandinavian Air Service (SAS) jet plane crashed six miles west of the Los Angeles Airport (LAX) on 13 January, with 30 survivors, but leaving 15 others in the plane. The *Deep Quest* sub and crew were called on to recover the flight recorder in about 350' of water, in order to find the cause of the crash. On 18 January, the same day that *Transquest* and *Deep Quest* arrived at the dive site, a United Airlines 727 crashed 11 miles west of LAX only five miles from the SAS crash site. None of the 38 passengers and crew survived in that crash, and the plane lay in 1,000 feet of water.

*Artist's Rendition of Deep Quest*
Courtesy of Lockheed Ocean Systems

*In the process, some of the line got sucked into the port horizontal thruster. Efforts to clear it failed. The thruster became inoperable and severely fouled.*

On 20 January, an hour after the first dive began, the tail of the SAS, DC-8 was spotted in 355' of water. The worst storm of the season delayed further operations for two days. Other parts of the plane were located and recovered, but no recorder could be found. Weather delayed operations another few days. On the next dive, the flight recorder housing was located, but it held no recorder. Two more dives failed to locate the recorder, but other hardware was recovered. On 4 March, in heavy seas, during the fifth dive in the SAS series, the crew spotted the recorder and captured it in one of the two manipulators on board. The seas were unusually rough for recovery at the dive site, so *Transquest* towed the submersible to quieter water before bringing it back on board. On 22 March 1969, *Transquest* launched *Deep Quest* 9.7 miles west of LAX in 855' of water to search for the United aircraft. Using sonar, the crew was able to locate the main fuselage, but no recorder. On the third dive the voice recorder was recovered. On dive #5 in the series, on 4 March, the flight recorder was found and recovered to complete the operations.

Local dives for pilot training, a search for a WWI military submarine hull, and demonstration rides occupied the next dive series. Four dives were made at Catalina Island for Cousteau's group-submersible gathering. Unfortunately, the squid were in "bloom" and the newly born fingerlings filled the water so much, that photography was limited.

During a dive in the afternoon of 7 October, at a depth of 430', about five miles west of Point Loma (San Diego), *Deep Quest* was maneuvering near a sonar target clump which was connected to a surface buoy by 3/8 diameter polypropylene line. The connection was comprised of excess lengths and loops, which draped about in the water outside of the view of the pilot who was directing the co-pilot on various courses around the target. In the process, some of the line got sucked into the port horizontal thruster. Efforts to clear it failed. The thruster became inoperable and severely fouled. No real emergency existed, since batteries or mercury could be dropped, thereby snapping the line as *Deep Quest* rapidly ascended. However, these alternatives were expensive choices that required replacement and/or recovery, or which would cause severe pollution.

The *Nekton* submersible group in Los Angeles was contacted and found to be available in two or three hours. A Navy torpedo retriever boat met the group at a nearby pier and towed *Nekton* five miles to the dive site. After a briefing aboard *Transquest*, the two man crew of *Nekton* dove on *Deep Quest*. They found no other lines except the one on the port motor fouling the sub. Using underwater phones, *Nekton* had *Deep Quest* deballast enough to tighten the line for easy cutting. *Nekton*, with a knife in its manipulator, cut the line. Surfacing by both submersibles ended the operation at 0400 in the morning on 8 October. Newspapers at the time played up the aspect of David rescuing Goliath. The remaining dives of 1969 were for pilot training, bottom coring and surveying, as well as module and pinger recovery, for a year's total of 37 dives.

The dives in 1970 began on 27 January. Pilot and crew training dives were made in preparation for coming operations with the new National Oceanographic and Atmospheric Administration (NOAA). During a DSRV pilot training dive in March, a WW II Navy F6F was spotted and the location was recorded. After the NOAA dives, 19 additional dives were made for the Naval Oceanographic Office, averaging five or six hours a piece. Many basic salinity, temperature, sound velocity, chemical analysis, and similar tests were conducted in the Santa Barbara, San Nicholas, and San Clemente Island areas. In July, sheer vane tests were practiced in preparation for sediment testing operations that were scheduled to begin in October with Lehigh University. On 10 September, the F6F spotted in March was dived on with a 5,000' reel of nylon line located outboard on *Deep Quest*. The aircraft was located at a depth of 3,360'. After several hours of work with the manipulator, wire straps were attached around the aircraft. After attaching the nylon line to the steel straps, the nylon was payed out as *Deep Quest* ascended to the surface where the line was transferred to the *U.S.S. White Sands*. The plane was slowly lifted on board over an eight hour effort. The plane was in unexpectedly good condition after being on the bottom for 26 years. After refurbishment, the plane was displayed at the San Diego Aerospace Museum. On 21 June, *Deep Quest* received Navy Certification, which was the result of inspection and conformity to a large number of strict safety rules and conditions.

On 21 October, James Parks, the head of Lehigh University, made a dive in *Deep Quest* and wrote a book about it. The dive was the first in a series to be made by *Deep Quest* in a University-Industry partnership. Parks was one of several scientific investigators funded by the National Sea Grant Office, U.S. Dept. of Commerce under NOAA, in partnership with Lockheed. The dive was part of a project to study the engineering properties of the deep ocean floor. When the Sea Grant program was completed in 1974, after running for four years, over 20 engineering and scientific papers were published as a result. Some of Park's comments deserve mentioning.

The icy cold steel that he lay on to peer out of the 6" diameter window seemed to curve the wrong way, no matter what position he was in. The view through the viewport was of a world he'd never seen before, with strange and weird objects moving across his field of vision. He thought the submersible instrumentation panels and controls looked like those of a "moon rocket," with the submersible having the additional requirement of repeated use and wear, that required continual maintenance when out of the water. Parks concluded (correctly) that the most dangerous parts of the whole operation were the launch and recovery phases. He found the deep scattering layer to be far less densely populated with biology than he would have suspected, from the way it reflects and scatters sonar signals from the surface. Parks estimated that *Deep Quest* must cost $1,000 per underwater hour, which compares favorably with another estimate at the time of $6,000 to $10,000 per dive *(Los Angles Times)*.

He was surprised, that at a distance of twenty miles from shore, at a depth of 4,000', to find so much man-made debris. One of his frustrations was to see unknown or unusual organisms and not be able capture them, partly due the limited sampling tools and partly due to the organisms not being able to stand the change in pressure at the surface. In spite of the discomforts and inconveniences of the dive, his conclusion was that he "definitely wanted to dive again if he had a chance, as there was much to be learned by direct observation on the sea floor." The Lehigh and Lockheed investigators made up the last dives of the year 1970 in December, for the year's total of thirty-nine dives. The *Los Angles Times* reported that all but 28 dives of the sub's four years of diving were funded by Lockheed as a research expense.

The first six dives of 1971 were spent in pilot training and searching for a pre-WWI submarine hull thought to have been abandoned in the area off San Diego. The next dives were made in conjunction with Lehigh University, for bottom sediment, infrared transmission investigation and survey, as well as sheer vane tests (to determine the strength of the bottom sediments, at various depths, of the sea floor). During a dive in November, a torpedo was discovered and avoided. Through the years, several torpedoes have been seen and avoided by various submersibles operating in the San Diego area. Two demonstration dives were made for NOAA in November, in combination with pilot training dives. The last two dives of the year were made to recover a shroud from the *DSRV* and a pinger. The last dive of the year was made on 22 December.

The first dives in 1972 were for pilot training and bottom coring practice tests near San Diego. The remaining dives of the year were all made in conjunction with Lehigh University, sponsored by NOAA. In June, while diving at 6,285' in the Santa Cruz channel to inspect atomic waste barrels, *Deep Quest* landed in the middle of an unmarked ammunition dump area amidst tens of thousands of 50 caliber machine gun bullets. Dumping had been barred at the site since 1953, but sea water has various unpredictable affects on old, wet ammunition, sometimes causing it to explode when disturbed, so the pilot (Saner) "slowly" hurried away from the live ammunition. Dives were made to 2,900' in the San Pedro Channel to inspect the dumping of industrial and chemical waste. Dives to 2,400' were made southeast of Catalina Island to observe the garbage dumping grounds used by ships and the City of Los Angeles. In all areas, bottom cores and water samples were taken for later lab analysis. A set of three dives were made at the Mission Beach Tower in 60' of water for a special project, sponsored by the Defense Advanced Research Projects Agency (DARPA). A few more pilot training and equipment practice dives were made until October, when *Deep Quest* went into overhaul.

On 17 July 1973, after coming out of overhaul, *Deep Quest* began a series of dives for Lehigh University. The dives were once again a joint University-Industry effort and took place in the San Diego Trough, averaging around 4,000' in depth. Ocean bottom sedimentation and general sea floor research were conducted. In November, a three dive series was conducted to retrace the track and to make comparative evaluation of core samples and bottom sediment measurements, made by a bottom crawling ROV, called RUM, owned by Scripps. The last dive of the year was made on 30 November.

Three years of inactivity for *Deep Quest* followed, while *Transquest* was in use by the Navy as a support ship for the *DSRV-1* and *2*. In November 1976, *Deep Quest* made three dives after overhaul in the San Diego bay at 40'. Two more dives were made near San Diego in December for core sampling purposes.

In January 1977, a dive was made to a depth of 5,112' for PMS 395 (a Washington division of the Navy Department) to demonstrate the capability for doing fuel cell tests, as a potential power source for the *DSRVs*. Originally, the *DSRVs* were to conduct the tests, but it was decided to keep them on line for submarine rescue practice or for emergency availability.

From May to November 1978, *Deep Quest* conducted dives to as deep as 5,040' and for as long as nine-plus hours, operating a new Navy fuel cell as its source of electrical energy. The submersible was the first underwater test platform for the new cell, which was to be the primary source of electrical power for the DSRVs. The 30 kw. fuel cell power plant, with tankage, required the removal of one of two sets of batteries on *Deep Quest* to make room for it. The system was completely self-contained and required no outside power source for start up or operations. The fuel cell makes electricity by consuming gaseous hydrogen and oxygen electrochemically.

A new life support system was installed in 1978. A potassium super-oxide system replaced the former high pressure oxygen source and lithium hydroxide "scrubber." The new system more than doubled the previous life support time of 48 hours, without increasing the size or weight. Both a closed system, with face masks for emergency use, and a standard open system circulating air throughout the submersible were installed. The potassium super-oxide absorbs both carbon dioxide and moisture and gives off, in return, oxygen, thereby nicely fulfilling the basic requirements of a submersible environmental system.

In January 1979, dive #199 ended the first series of fuel cell tests for the Navy. The remaining three dives of the year were made in November and December for the filming of the movie "Raise the Titanic." The movie cost eight times as much to make as the *Titanic* had cost to build in 1912. The *Deep Quest* scenes were all made near the surface, with many retakes. The three days of diving are listed as three dives, but the pilot (Saner) stated that at one point they made three shallow dives in ten minutes. A full scale mock up was made for movie purposes that "looked better than the real submersible." One of the stars of the movie had both a fear of water and a weak stomach, which made him an "interesting" passenger during the submersible operations.

All of the 14 dives in 1980 were made for the Navy fuel cell sea trials and operations. The new series of tests consisted of dives of increasing duration, with the purpose of testing the fuel cell power system to its maximum capability and to evaluate the physiological and psychological effects of long endurance dives on crew performance and operational safety. The dives began in April and were completed on 9 September 1980. A depth of 3,760' was the deepest attained in the series during a 24 $1/2$ hour round-the-clock dive in August. The final dive for 54 miles from San Clemente Island to La Jolla, California took 42 hours at a maximum depth of 3,920' to successfully complete the series.

*Deep Quest* was "put on the blocks" after 1980. Its ultimate resting place was the Naval Undersea Museum in Keyport, Washington. Besides Lockheed's initial goal of doing the exploratory work for the *DSRVs*, they accomplished all of the oceanographic research that was assigned as well as a remarkable string of practical operations.

# *DOWB*
# *(Deep Ocean Work Boat)*

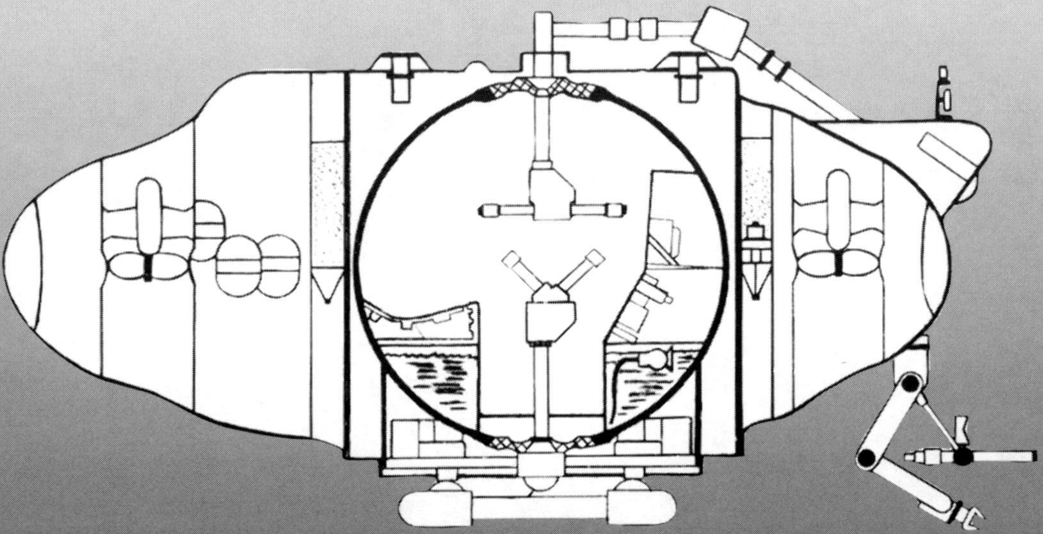

***1967–1970***

# DOWB
# (Deep Ocean Work Boat)

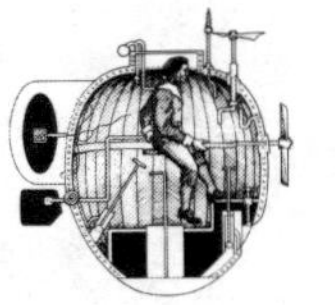

I was at China Lake, California on a very hot day in the summer of 1963, sitting on top of the Deep Jeep submersible running the motors and controller for the first time, when a voice called out from below me. The loading dock where we were was a non air-conditioned area we referred to as the skunk works, and visitors were rare. Dr. Dave Potter introduced himself and told me about the Deep Ocean Work Boat *(DOWB)* that his group at the A.C. Electronics Defense Research Laboratory (ACE-DRL) planned to build. He asked me if I would like to use their research ship *Swan*, as a surface support craft during my manned tests for $2,000 per 24 hour day so that they could get submersible handling and operation experience before building their own sub. It was so perfect a solution at the exact right time that, if he had been female, I could have kissed him on the mouth.

In the 1960s, the AC Electronics-Defense Research Laboratory (ACE-DRL) at Goleta, near Santa Barbara, was no stranger to operations and equipment testing at sea. Department of Defense contracts for laboratory research and/or at-sea development and tests were its forté.

The Deep Ocean Work Boat *(DOWB)* submersible was an effort to expand the existing ocean engineering capabilities of the ACE-DRL and increase its oceanographic vessel capability. The operational goals were to provide a three person, highly maneuverable submersible for a large variety of assignments. *DOWB* was designed by Scott Dobbins, a retired Navy Captain with a PhD in physics, who had specialized in submarine design and construction prior to his Navy retirement and employment by ACE-DRL shortly after the laboratory was established in 1960.

*DOWB* was 17' long with a dry weight of 9 1/2 tons. The pressure hull was made from HY 100 steel, and had an 82" outside diameter with a nominal wall thickness of .935". A 20" diameter hatch was located at the top and bottom of the sphere with all pass-throughs located in the thickened portion of the upper hatch. An operating depth of 6,500' was based on a collapse depth of 10,000'.

Propulsion and maneuvering were provided by four ducted, 18" diameter

***Name:*** *Deep Ocean Work Boat (DOWB)*
***Builder:*** *A.C. Electronics Defense Research Laboratory* ***Length:*** *17'*
***Maximum Depth Rating:*** *6,500'* ***Capacity:*** *3 Man* ***Year Launched:*** *1967*
Courtesy of Dick Hegeman

propellers driven by 2 H.P. motors. The main propulsion thrusters were located near the stern and positioned parallel to the fore and aft axis of the vehicle, but angled 15 degrees inboard, toward the after part of the submersible. The angling of the motors had the effect of increasing the turning moment significantly at very little loss to the fore and aft thrust. The cruising speed of 1 kt. could be maintained for 26 hours.

Instead of conventional viewports, *DOWB* was equipped with periscopic optical systems, one through the top hatch and the other through the bottom hatch. Both observers, as well as the pilot, viewed their outside surroundings through separate eye pieces from their seated positions. The systems provided a full 180 degrees of visibility in the horizontal plane in a forward looking direction from two locations. To accomplish this, the periscopes were each terminated with an optical dome. One was located on the outboard side of the lower hatch at the termination of the lower optical system, the other, which passed through the upper hatch, was on the bow of the submersible.

The submersible had several ballast/ buoyancy systems. To attain surfaced freeboard and stability, two ballast tanks were blown free of water with compressed air before surfacing. This provided 3,330 lbs. of buoyancy. When diving, the main vents on the tanks were opened to let air escape and water to flood them. Two tanks were filled with 900 lbs. of iron shot used as ejectable ballast. Descent was achieved by the flooding the surface buoyancy tanks and then dumping enough shot to attain approximate neutral buoyancy. Twin bladders, located at the bow and stern, could be filled with oil from a reservoir inside the pressure hull to provide minor, more accurate, buoyancy control. By transferring the oil between the fore and aft bladders, the pitch angle could be altered 2 1/2 degrees up or down.

Electrical energy for the submersible was provided by lead acid batteries located in the skirt of the vehicle, beneath the pressure hull. The power available from the batteries was converted to alternating current through two inverters. Silver zinc batteries inside the sphere provided power in an emergency.

The environmental system could provide a total of 65 hours of life support for three people. Sensors for oxygen and carbon dioxide, humidity, and temperature provided the environmental readout and status. A closed circuit breathing system was provided for emergency use.

*DOWB* was equipped with a number of instruments and tools for scientific use: a manipulator with six degrees of freedom; a constant transmitting frequency modulated (CTFM) sonar for searching the terrain and obstacle avoidance; a fathometer which could be pivoted upward or downward for determining the distance to the bottom or the sea surface; and an underwater telephone for communication with the surface ship. For special circumstances, a 7,000' reel of nylon could be attached to the submersible at the rear for recovering objects. The manipulator, shot tanks, and batteries were all jetisonable in emergencies.

*DOWB* and all of the supporting gear for the submersible could be transported on land by low boy trailer or by air in a C-133. The support ship for *DOWB* was the R/V *Swan*, a 250 ton, 135' long ship with a monorail over the stern, fantail and afterdeck for launch and recovery.

*DOWB* was first launched, pierside, at Santa Barbara, in 12' of water on 13 October 1967. A test made at the completion of the vehicle's check out included a practice personnel escape through the bottom hatch into the cold harbor water. The first open ocean dive was made off Santa Cruz Island, California from the R/V *Swan* on 26 October, 1967.

The first deep dive was made on 24 July, 1968. Earlier in the year, a $50 million Nimbus-B nuclear-powered experimental weather satellite had plunged into the ocean near San Miguel Island, California after its launch rocket was destroyed, intentionally, during a faulty launch. The satellite and its booster were located on 27 September 1968, while *DOWB* was under contract to NASA and the AEC. A conflict between the Navy, NASA, and the AEC then took place over the administrative responsibility for the recovery of the $700,000 worth of plutonium 238 dioxide in the satellite wreckage.

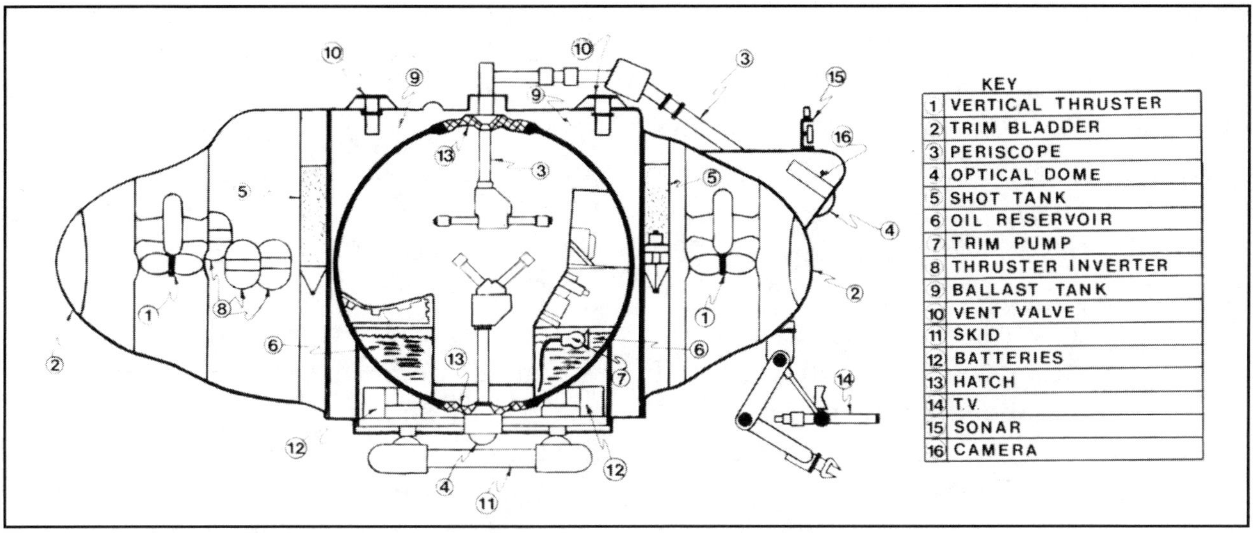


*DOWB Detailed Cross Section*

The Navy won the recovery rights, but could not find the wreckage. *DOWB* was then rehired to locate the wreckage once again. Using deep water divers, the Navy then recovered the undamaged atomic batteries in 310' of water.

In December 1968, *DOWB* was brought into the search for the *Alvin* submersible lost in October off Nantucket Island, east of Massachusetts. Dives to over a mile deep were unsuccessful, and foul winter weather prevented any further chance of success.

Most of the dives in 1969 were made near Santa Cruz Island on the Santa Cruz Island Acoustic Range Facility (SCARF), operated by General Motors doing DoD contract work. During the month of May, dives between 1,000' and 4,150' were made, including one dive with *DOWB* acting as the mystery submarine in a Cousteau movie. During the last two weeks in June, *DOWB's* mission was to locate, photograph, and place upright four large sonar array structures. These weighed 400 lbs. each, lay in water depths as deep as 4,385' and had numerous cables draped about them. During one relocation move, it was necessary to lift a 700 lb. hydrophone array over a 30' reef and reposition it.

My opportunity to dive *DOWB* in November 1969, provided a chance to evaluate the optical system as well as the operations in general. The observers and pilot are all strapped into comfortable seats and use optics for external viewing. The upper optic (periscope) provides a view ahead and upward, while the lower optic provides a view primarily downward and ahead. The parabolic effect of the optical system makes the flat bottom appear slightly like the surface of a very large sphere. The eye and brain quickly adapt to the distortion and soon it is unnoticed. Viewing ahead on the Straza CTFM-500 sonar at a long distance and then with the periscope up close, in the lighted zone, provided almost identical images. The bottom optic was useful primarily for very close up inspection and when "flying" over the sea bottom at some distance above it. The viewing by optics and sonar was quite satisfactory and the general maneuvering and buoyancy control was good.

The 31 dives made in 1970 ended with the last dive the submersible made on 21 July 1970 (#137). Almost all the dives in 1970 were made on the "SCARF" range for G.M. projects.

*DOWB* remained "on-the-blocks" until its donation in the late 1970s to the Marine Technology Department of Santa Barbara City College. During a brief ceremony at the college, Dr. Dave Potter said that, "*DOWB* had performed all of the visualized missions by her designers including search and recovery of objects lost on the bottom, inspection and maintenance of Delco Electronics underwater range off Santa Cruz Island, repair and precision implantment of bottom mounted equipment and some geological research missions."

In the late 1970s the submersible was moved to Southwest Research Institute in San Antonio, Texas for remodeling into an undersea tractor. The new owners, a corporation in New York, failed to provide enough money for the modifications so the outer shell was acquired by Sea World Florida for display, and the pressure hull was still at Southwest Research Institute in 1995.

So as to not penetrate the main pressure hull, the challenge of having all observations made through the hatch viewports was met by an innovative and unusual solution. The *DOWB* optical method, aided by the sonar system, was a good substitute for the eye ball. Scott Dobbins' innovation provided the nearest thing to direct viewing.

# *Trieste 2*

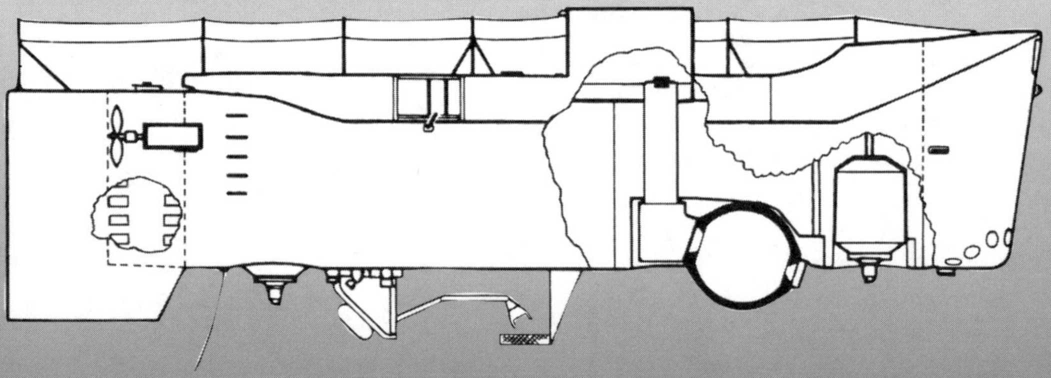

## *1964–1984*

# Trieste 2

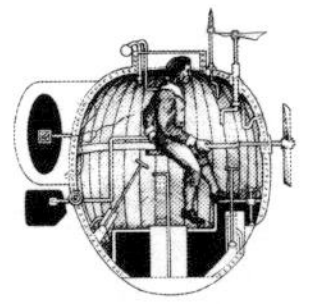

After the last dives for the Thresher search in 1963, the original Trieste float, enlarged for the Mariana's dives, was in such bad shape that a new one was indicated. With a pointed bow and flat sides, the new float – sometimes called the "boat float" – had the original but refurbished Terni sphere recessed into it for a better towing combination. Larger, externally located lead acid batteries were substituted for the silver-zinc batteries providing 8 times the power previously carried inside of the sphere. The two propulsion motors were upgraded to 5 H.P. with 48" diameter propellers. Sonar, other search systems and instrumentation were all vastly improved in preparation for undersea search, salvage, and photography. By launch date, 17 January 1964, the modifications had been so major that the vessel deserved a new name: *Trieste 2*.

***Name:*** *Second Trieste 2* ***Builder:*** *Mare Island* ***Length:*** *78.6'*
***Maximum Depth Rating:*** *20,000'* ***Capacity:*** *3 Man*
***Year Launched:*** *1968*

The first half dozen dives in 1964 were spent testing the new equipment and training new pilots. Oceanographic-oriented dives were then made for the scientists at the Naval Electronic Laboratory. In June, the submersible was moved to Portsmouth, Rhode Island in preparation for the second series of "*Thresher* search" dives. During dive #1 at a depth of 8,400', a severe short occurred in the bathyscaphe's main propulsion motors, causing the external batteries to drain to zero voltage. Arcing from the shorted circuits occurred only 4' from the gasoline in the float, causing

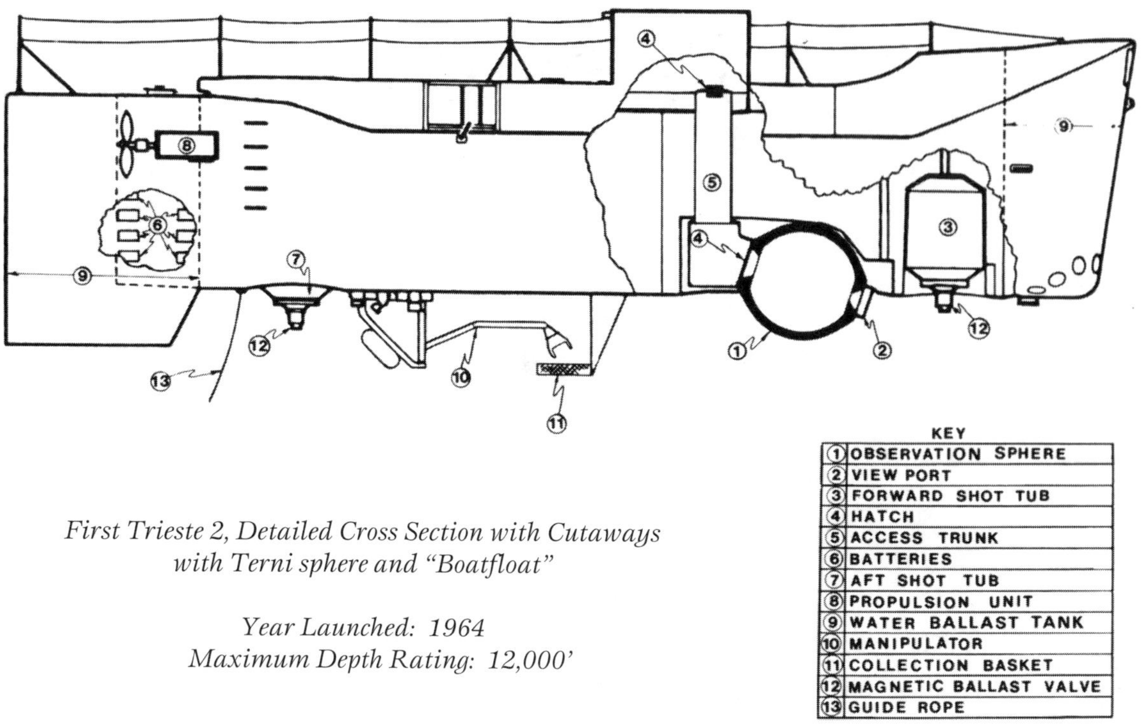


*First Trieste 2, Detailed Cross Section with Cutaways with Terni sphere and "Boatfloat"*

*Year Launched: 1964*
*Maximum Depth Rating: 12,000'*

great concern that, at the surface, gasoline fumes might be ignited from the sparks. Fortunately, this didn't happen. Photographs taken on five dives and from a camera array towed on a sled from the *USNS Mizar* showed the tail section of *Thresher* with stern draft numbers, stern planes, and the sail with the side numbers "593". After the last dive in the series, *Trieste 2* was returned to San Diego for maintenance and overhaul. In March, 1965, *Trieste 2* was transferred from NEL to the Deep Submergence System Project (DSSP).

The first dive of 1965 took place in May, in the San Diego area. Testing of the overhauled submersible equipment took place in the first few dives, but most of the rest of the year was spent training pilots and crews in developing search, operation, rescue vehicle, and object recovery techniques. All dives in 1965 were made in the San Diego area, mostly at 3,400'-3,500' depths. The last dive of the year took place in August for a year's total of nineteen dives.

Most of the 1966 dives were made in La Jolla or Scripps Canyon. All dive depths were 2,000 ft. or less. The dives were virtually all training dives.

The first dive of 1967 took place in La Jolla Canyon. All nine dives in 1967 were made in the San Diego-La Jolla area with four pilots qualifying. The last dive of the year and for the submersible was #54 in March, after which it was retired from service.

In early 1968, there emerged from the Mare Island Shipyard, a completely redesigned submersible confusingly also called *Trieste 2*. The second *Trieste 2* was the first to be designed and fabricated entirely in the U.S. It had a blunted torpedo shaped float containing 65,830 gallons of aviation gasoline and extended legs forward for setting on land, seabeds, or shipboard. The "X" shaped stabilizer fins aft provided two points of support at the stern, facilitated sea keeping ability, and provided contact points during docking and undocking.

The pressure sphere was made of two hemispheres of HY-120 steel designed by Southwest Research Institute of San Antonio, Texas, with an operating depth of 20,000' and a safety factor of two. The first dive of the new submersible was made in April, 1968. More than a half dozen new pilots were qualified during the dives made in 1968. In July, a six hour dive was made to 12,720', all other dives were in the San Diego area and did not exceed 6,300'.

When the nuclear submarine *U.S.S. Scorpion* was declared "officially missing" in early June 1968, an exhaustive search of the sea bottom was immediately initiated. In the fall of 1968, five months after the search began, the wreck of *Scorpion* was located by submerged cameras, remotely operated from surface ships. The location, 400 miles SW of the Azores at 10,800', indicated that the sub had sunk on 22 May, a day after the last radio report.

At the time, Dr. John Craven was head of the Special Project Office in the Navy that dealt with Navy deep submersible operations. Once the *Scorpion* wreckage was located, Craven called Dr. William B. McLean, the Technical Director of the newly formed Naval Undersea Center (NUC) in San Diego to where McLean and I and a few others had recently transferred from China Lake. McLean then dropped by my office and said "Craven wants us to make an ROV to look inside the torpedo tubes of *Scorpion*!"

Basically our assignment was to do whatever was necessary to write up the specifications with all of the design parameters detailed for a contractor to build the ROV in a very short time. (As it turned out, most of the contract work was completed before the contracts were signed.) Our device was to deploy and operate on an umbilical at depth from *Trieste 2*. If it found any empty torpedo tubes, that would indicate that there had been a hostile engagement.

I hooked up some electrical solenoids, valves, and a control box full of on/off switches to a garden hose to simulate propulsion jets and the controls for them. About 6 P.M., McLean came back over to see how it was going and we played with the hardware and made notes, while we brainstormed away at the problems.

The next day, Thanksgiving, was the start of a four day weekend. We spent much of Thanksgiving day with our families, but lined up hardware later in the day for experiments in the pool at McLean's house during the rest of the weekend. By Friday, when we met at the pool, Dr. McLean had located some submersible fountain pumps that would take the place of the garden hose as a pressurized water source. George Hatchet, who was the main innovator and head of new product development at Hydroproducts (a San Diego subsea company), brought over a company-owned video camera in a pressure-resistant housing. To compensate for all the weight of the equipment and hardware in our makeshift ROV, we attached package-type styrofoam until it barely floated. Later syntactic foam, especially formulated to withstand very high pressures, would be substituted.

We took turns running the control box and improving our hardware, until we got our "gizmo" working as well as we could. Most of the time was spent fixing and adjusting things or running to my shop at the base, a neighborhood hardware, or to some other store in town for parts. By late Sunday night, we had a vehicle with a TV camera that was working with acceptable reliability.

Monday morning, McLean went to a meeting in Pasadena and while there, had a mild heart attack. The medical prognosis was good, but I felt partly to blame for letting him overwork. He was a diabetic and older than George and me. We got caught up in one of those rare times of demanding and useful creativity that was highly productive and I had neglected to take care of my director and friend.

The specifications were written, Hydroproducts got a contract from Cravens and Co. and everyone at Hydroproducts worked long hours for six weeks to produce the first ROV for the Scorpion search. The unit, named Tortuga (Spanish for turtle) by Hydroproducts, looked like a small, 3 gallon barrel. Besides the control jets, there were lights, a TV camera and an umbilical line with flotation balls on it to compensate for the weight of the cable. Later, these balls were to cause problems due to snagging.

On 3 February 1969, *U.S.S. Apache*, towing *U.S.S. White Sands* with *Trieste 2* on board, left the West Coast of the U.S. for a 14,000 mile trip to the Scorpion dive site by way of the Panama Canal and the Azores. Enroute to the Azores, launching tests of *Trieste 2* were conducted. During one of the tests, the tow line on Trieste got loose and it was adrift for nine hours. Later, two days of storms caused damage to the submersible so no more tests were conducted until repairs could be made. Another storm encountered en route caused additional damage to the equipment on *Trieste 2*, including the hanger that housed "Tortuga." Repairs to all the equipment were completed by arrival time in Terceira, Azores, on 21 May. Departure for the dive site was made a few days later with the on-site commander Captain Bob Gautier (pronounced "go-shay") who was the COM SUB DEV GRP ONE (Commander Submarine Development Group One) on board. The dive site was reached on 2 June. Three days were spent implanting DOTs (deep ocean transponders), small devices resting on or implanted in the sea floor that give directional or homing signals. Another five days were spent waiting out the weather.

The first *Scorpion* site dive was made on Friday, the thirteenth of June, for 8 1/2 hours to survey the navigation network and to test equipment. The second dive was a night dive starting in the evening of 18 June and remaining over twelve hours to survey the entire navigation network of DOTS. All DOTS were in position and working. Dive #3 lasted for over fourteen hours and the debris field was located. After the dive, there was a delay while a dead battery was replaced on *Trieste 2*. The delay was caused by the need to drain the aviation fuel (AVGAS) to get at the battery compartment inside the gas-filled float.

On 2 July, during the fourth dive, Scorpion and its entombed crew of 99 were located during a 14 hour-17 minute dive. On all five additional dives, *Scorpion* was located accurately and timely. During those five dives on *Scorpion*, (the average length of dive was just under 15 hours), many photographs were taken. At last, Tortuga was launched. While the ROV worked as intended, the flotation balls on the umbilical constantly got snagged in the debris, preventing a survey of the torpedo tubes. The last of nine dives was made in July. The submersible was returned to San Diego in October and an overhaul began in December.

Back in San Diego, after the pictures had been developed, Captain Gautier invited me to lunch at Submarine Development (SUB DEV) one day. Afterwards he showed me a large table full of 8x10 photos taken at the crash site. He then asked, "what do you see in the pictures?" I recognized quite clearly the modes of structural failure that were evident, because they looked identical to pictures of submarine models that had been purposely failed by over pressurizing for research purposes. I said so, and referred Bob to the works of Krenzke and Kiernan at David Taylor Model Basin, where the research work had been done. In the photos, ring stiffeners had buckled; between stiffening rings the hull plates had caved in; and, in addition, an accordion collapse was evident lengthwise of the hull. It seemed apparent that the section of hull being shown had failed in all three classic failure modes at the same time, indicating that the area photographed had passed its crush depth very rapidly. After hearing my summary, Capt. Goutier smiled and thanked me. About twenty-seven years later, at a luncheon, I asked him if anything new had been discovered and he shook his head to indicate no. Just about every anniversary of the tragedy, there are one or more new theories proposed on the cause of the accident, but none so far have seemed at all conclusive based on stills and videos taken in the area in 1969.

Most of 1970 found *Trieste* in overhaul. In the fall, after the overhaul, the *Trieste 2* was designated *DSV-1* by the Navy. Only two dives were made during the year, one in November and one in December at 13,200'. Both were in the Southern California area.

The first dives in 1971 were all made in the San Diego area at depths of 4700' or less, for post overhaul sea trials, vehicle certification, navigational network training, and one five hour dive to recover a Scripps instrumentation package that was lost. The next two dives were made 300 miles NNW of Kauai Island, Hawaii to recover military instrumentation packages at 16,400' depths during two 10 hour dives.

In April 1972, *Trieste 2* returned to the dive site near Hawaii for an additional instrumentation package recovery dive at 16,400', lasting nine hours, 19 minutes. The next dives were made near San Diego at 4,575' in search of an F-4 aircraft crash site in mid-July, averaging 7 hours per dive. Three dives were made in late September, searching for the hulk of a sunken Navy Ship at 4,700'.

The first dives in 1973 were made near San Diego at the F-4 plane crash site in January, February, and May at depths of 4,600', averaging five hours per dive. In July *Trieste 2* (alternately now known in Navy logs as *DSV-1*) was deployed to an area NW of San Francisco. At a depth of 10,699', she recovered a lost and submerged, unmanned sled carrying oceanographic equipment with a replacement value estimated at $250,000. Two dives were made in the San Diego area for pilot certification. The last dive of the year was made in October for a manipulator test near San Diego. The dive duration was reduced to four hours due to a battery melt-down. In late 1973, *Trieste 2* went in for major overhaul that lasted all through 1974.

After leaving the shipyard on 4 May 1975, three dives were made pierside, while tethered, for post-overhaul tests. The fourth and last dive of the year was made in August near San Diego.

Major new electronic systems installed during the 1974 overhaul continued to be tested in early 1976. A "10 K certification dive" was made in February to 10,004' for seven hours near San Diego. Four dives were made to locate a lost helicopter and for pilot training near San Diego. Two dives were made to recover F-14 debris near Ensenada, Mexico. A dive was made to recover a DOT at 3,946' near San Diego in November. A lost Sea Stallion Helicopter from Brown Field was located during an eleven and a half hour dive. The rotor assembly was recovered in November, and on the last dive of the year in December the tail rotor was snared and pulled loose after a 13 hour dive.

The most dives per year, the deepest series of dives, and the most important scientific dives made by *Trieste 2* were all made in 1977. The first of these were in January to 4,000' near Ensenada, Mexico to recover DOTS. They averaged 13 hours per dive. The next dive series was made near Catalina Island for sea trials and pilot training in conjunction with the support ship *Point Loma*. A dive was made west of

*The most dives per year, the deepest series of dives, and the most important scientific dives made by Trieste 2 were all made in 1977.*

---

San Francisco in an attempt to certify *Trieste 2* for 20,000 feet, however *Trieste* reached only 17,246' at the bottom of that site in the Mendocino Fracture Zone.

*Trieste 2*, on board *Point Loma*, departed San Diego in May for a 16 dive series of geological investigations in the Cayman Trough, Puerto Rico Trench, Blake-Bahama Plateau, and Mid-America Trench, in conjunction with several universities and research institutions sponsored by the Office of Naval Research (ONR). The ships reached the Cayman Trough operations area in mid-June. The Cayman dives were made in conjunction with Dr. Bob Ballard of Woods Hole Oceanographic Institute, to investigate tectonic plate spreading and volcanism.

The first dive in the series was made on 23 June to a depth of 20,236' for ten hours to certify the submersible. While on the bottom, *Trieste 2* recovered a piece of lava the size of a softball, still crackling from contact with sea water. It was the deepest object recovered from the sea floor at the time. The Cayman Trough is unusual in having all three tectonic geological zones: subduction, spreading, and strike or slip zones. The second dive was in the same area at 19,310' for 16 1/2 hours under water. The third dive was in the same area, but was cut short to three and a half hours when a vertical cliff was bumped while blindly sinking to the bottom during descent. The bow planes and forward pan and tilt were damaged, but repairable. Since bubbles in the water resulting from the impact might have been AVGAS used for buoyancy in the float, the dive was prudently aborted. Upon surfacing it was determined that no damage to the float had occurred and that the dive series could continue.

Dives 4, 5, and 6 were made in August at 9,200' or less, with Naval Ocean Science Center (NOSC) scientists as observers. The dives lasted from 14 1/2 to 20 1/2 hours. Dives 7 thru 11 ran from 20 September to 1 October in the Blake-Bahama Outer Ridge. These made it possible to study the formations and structures of furrows and mud waves, as well as the currents that caused them. The smaller mud waves were a yard wide and a yard deep. Others were over 30' wide and deeper than the height of *Trieste 2*. Arctic currents and tsunamis may have caused the "ripples." Dives were 16,446' or less in depth, and were of five to seventeen hours in duration. One dive had to be aborted after five hours when bottom sand got jammed into one of the shot tub outlets, preventing its ballast dropping valve from working.

Dives 10 and 11 were made to prove Puerto Rico had rotated 90 degrees during its geological history. Dives 12 thru 15 were made in support of scientists interested in the structure of calcium carbonate platforms, their modification by plate tectonic forces, and the nature of plate-tectonic subduction in the area of Mona Canyon. Fossil coral was discovered during the dives which were made between 16,500' and 20,200' depths for 12 1/2 to 21 1/2 hours per dive. Dive 16 was completed on 28 October for the last dive of the series. The submersible was returned to San Diego in November and the crew was awarded the Meritorious Unit Commendation for the deployment.

LCDR Les Parsons relieved Kirk Newell as O-in-C for the 1978 dives. The first three of four dives made during the year were made in Southern California waters for training and DOT retrieval. A major deployment was made to an area near Midway Island in late August. An Air Force navigation system with recorder had splashed down in open ocean at a depth of 16,000'. Upon sighting the sunken hardware, at depth, it was apparent that the one ton capsule had buried itself deep into the seabed. After a manipulator was broken trying to free the heavy object, the Chief of Naval Operations (CNO) called the project off and *Trieste 2* returned to San Diego.

The first dives in 1979 were made between San Diego and San Clemente Island at depths of 12,700' or less for pilot training and certification. In June, *Trieste 2* and the *U.S.S. Point Loma* departed San Diego for the Panama Canal and the Mid-Atlantic Ocean. Two dives were made on the Deep Ocean Station #1 and #2 to pick up specimens for Woods Hole Oceanographic Institute, due to *Alvin* being out of commission and in overhaul.

Three dives were made on the crash site of the Scorpion for CNO at 11,300' to 11,700' depths. On one of the several long dives to see if radio activity or its migration existed (it did not), 25 hours were spent submerged. During another of the dives, the surface ship lost all power. Communications from the ship to *Trieste* were made by battery power. After surfacing, the *Trieste* was left surfaced, unmanned, for three days while the ship's power was restored. On a dive in the AUTEC Range in the Bahamas in October at 5,000', a mobile submarine simulator was recovered in record time due to the fact that, after the launch and travel to the sea bed, Trieste had accidentally landed on the bottom only a few feet from the simulator. After this last dive of the year, *Trieste 2* was returned to San Diego for post-deployment analysis and upkeep.

The first two dives in 1980 were made in Southern California waters in February and March. The next dives were expeditionary in nature, requiring transit to an area off of the coast of Guatemala known as the Mid-America Trench. The deployment was to assist Scripps Institute of Oceanography in investigating plate tectonic dynamics within the trench. The dives took place during April at a 16,140' depth. Earlier in the year, a Navy-owned RUWS (remotely operated underwater work system) was lost at sea near the big island of Hawaii. *Trieste 2* was asked by CNO to find and recover it. In July a dive was made in the area, but the bottom was so full of massive blocks of rock bigger than *Trieste* – or, more to the point, bigger than RUWS) – that it made SONAR nearly useless, as all targets looked alike. During the next two dives in August to 16,330', the vehicle was located and recovered.

After the eighth and last dive in 1980, all future dives by *Trieste 2* were for training. Chief Warrant Officer Burt Tharp became Officer in Charge until the submersible was decommissioned. *Trieste 2* was kept on line due to a letter from CNO ordering that a 20 K capability be maintained until *Sea Cliff* could be outfitted with its new 20 K titanium hull.

Four dives were made during 1981 at 5,000' or less, followed by a brief overhaul of the sphere in July. The six dives in 1982 were made in the "reduced operating status."

They were routine dives to 4,000' operating from a barge stationed at San Clemente Island. In 1983, two routine dives were made, followed by the 122nd and last dive of her career. During the last dive on 7 October, a malfunctioning San Clemente Island Range hydrophone was recovered. *Trieste 2* was deactivated on 18 May, 1984 and arrived at the Naval Undersea Museum in Keyport, Washington on 3 August 1988, where it was placed on display.

From Dr. Piccard's first vision of a balloon-in-reverse through to the historic Picarrd/Walsh voyage to the bottom of the Marianas Trench, and the numerous scientific and military missions that were to follow until the wind down of the *Trieste* series in 1984, all of those involved ran a truly varsity operation.

# *The Nektons and Delta*

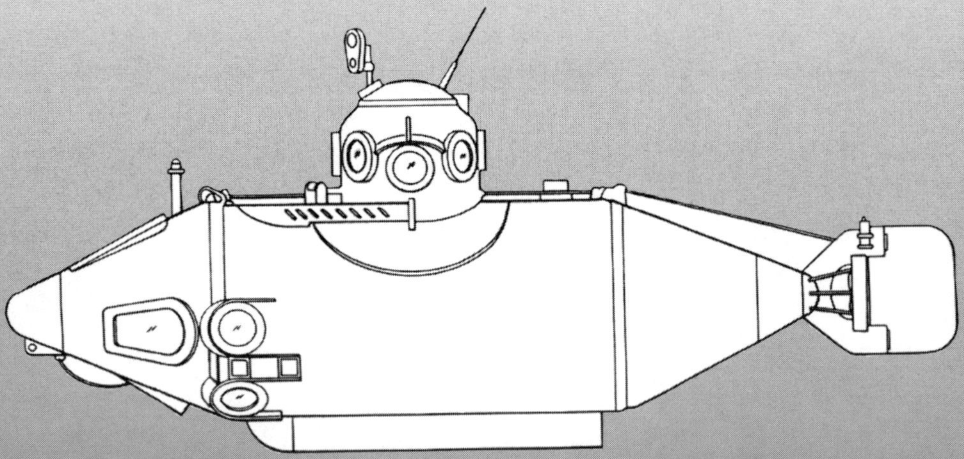

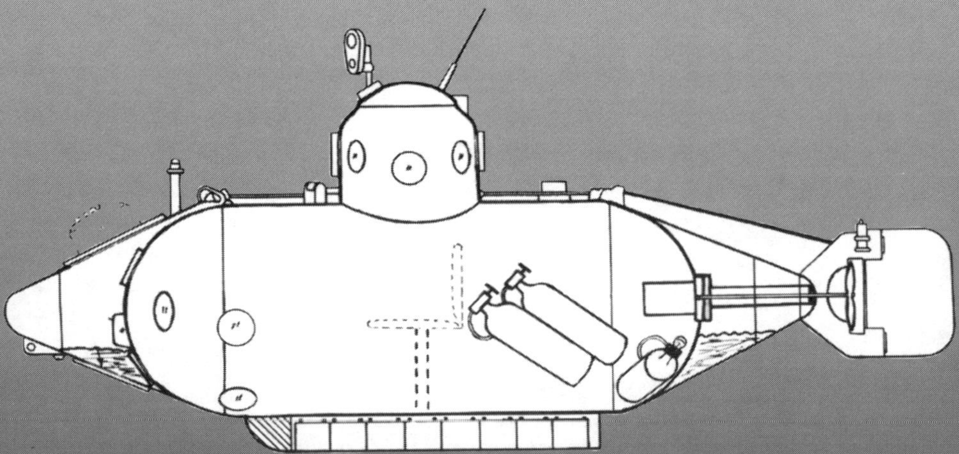

## *1968–(       )*

# The Nektons and Delta

As John Holland was to submarines in the late 1890s, so Doug Privitt is to submersibles in the 1990s. Privitt built small sports

*John Holland and Doug Privitt*

submarines for clients starting in the mid-1950s, first with the *Aquasub*, later four *Sportsman* subs and, finally, the *Submaray* in the early 60s. The experience he gained led to the building of the first of his own submersibles, *Nekton Alpha*. In it, he included some unorthodox design criteria which later proved more than adequate by testing many models to failure.

His first break with design tradition was to have the long box containing the batteries running the entire length of the main cylindrical hull. In this configuration the battery box became part of the pressure hull and the batteries ended up inside of the submersible. Hemispherical ends with a small cylindrical turret and hatch completed the mild steel pressure hull. Models designed by engineering hand book formulas were fabricated and pressurized to failure in a test tank he built. The failed areas were reinforced and retested over and over again, until they were found strong enough.

*Nekton Alpha* was the first deep submersible tested in the new pressure tank at the Naval Civil Engineering

***Name:*** *Nekton Alpha* ***Builder:*** *D. Privitt*
***Length:*** *15'* ***Maximum Depth Rating:*** *1,000'*
***Capacity:*** *2 Man* ***Year Launched:*** *1968*
***Description:*** *Nekton Alpha Aboard Dawn Star*
Courtesy of Marfab, Torrance, California

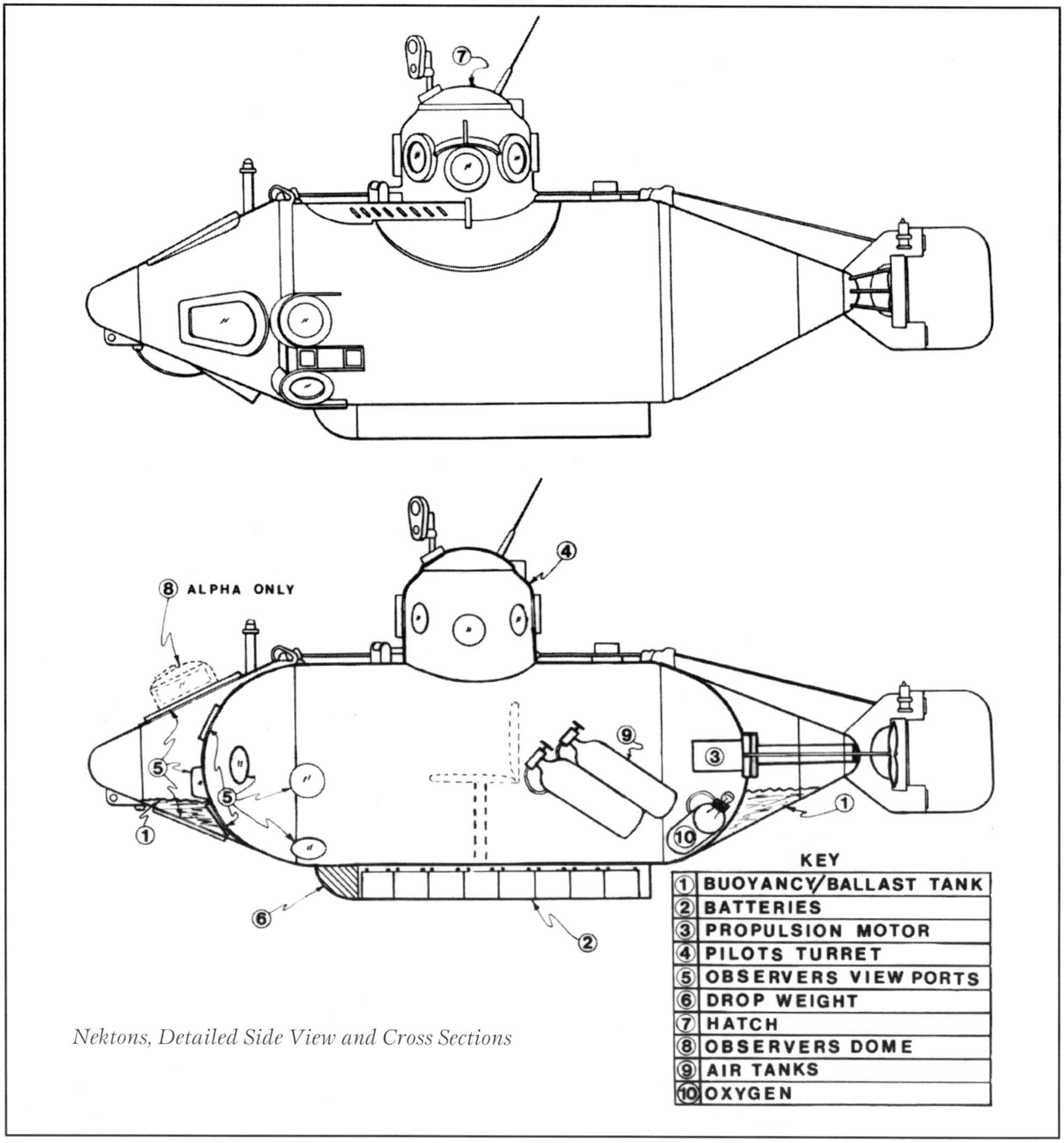


*Nektons, Detailed Side View and Cross Sections*

Laboratory at Port Hueneme, California in February 1968. The 15'-0" vessel weighing a little over two tons was launched in May. The main pressure hull was a cylinder 8' in length x 42" inside diameter made of mild steel 9/16" thick. The conning tower, or pilot's turret, was a 24" diameter cylinder. Collapse depth of the *Alpha* was 2,500' with an operating depth of 1,000'.

Viewing by the observer in the main pressure hull could be done through the forward overhead turret, through four viewports in the forward pressure hull hemisphere or two ports in the adjacent pressure hull cylinder. The pilot's turret contained eight viewports at 6 $^1/_2$" in diameter. A forward and aft tank provided an open-to-the-sea or "soft" ballast/buoyancy tank system. The chambers were completely blown full of air for surfacing after a dive for 1,500 lbs of buoyancy and flooded for diving, except for minor addition of air at depth for buoyancy control. A 3 1/2 H.P. reversible, direct current motor powered a 10" diameter propeller. The motor could be operated at 24 volts for cruising at 1 1/2 kts. or 48 volts for fast operations at 2 1/2 kts. The rudder, located aft of the propeller, provided lateral control and steering. A dive plane was located on the starboard side of the main hull, forward of the pilot's turret. Batteries were housed in the pressure resistant skid box at the bottom of the main hull. The observer and pilot were provided 144 man-hours of life support. A pinger on board allowed for the surface ship to track and locate the submersible under water.

The most simple and effective manipulator ever seen on any submersible extended out the starboard side of the pressure hull, up forward. A billiard cue ball in a spherical socket with "O" rings and concentric shafts through its center, provided a telescoping arm that could be extended and/or pivoted through about a 120 degree cone of operation. The claw on the end could pick up 100 lbs. at a time.

The droppable 75 lb. lead keel provided for an emergency jettisonable weight, it also served as a good bumper. A large mat of old tires tied together provided a fender between the submersible and the support ship for launch and recovery. Some of the first dives in 1968 were for testing and adjusting the various new systems. There was time during the year for several pipeline surveys and sewer outfall inspection contracts off Southern California, as well as pipeline dives in Lake Michigan. The submersible was found to tow easily on its trailer during the first of many round trips across the country.

In 1969 *Alpha* made demonstration dives for several oil companies early in the year, followed by gas seepage inspection dives in the Santa Barbara area for the Union Oil Co. and the U.S. Geological Survey (USGS). In April, *Alpha* made 24 dives in six days of filming for Cousteau's TV program. Sixty days of geological mapping and sampling in the Gulf of Alaska began in June, for a total of 122 dives sponsored by Conoco, Phillips, Union, and the Tenneco Oil in the Gulf.

Off San Diego on 7 October, Lockheed's submersible, *Deep Quest* tangled a polypropylene line in its port screw while checking out a large object recovery module under water. A call from the Navy at 4:00 P.M. started *Nekton Alpha* from Goleta, California. *Alpha* arrived at San Diego on a trailer towed by a truck full of equipment and was able to dive shortly after midnight. The pilot/builder Doug Privitt and the manipulator operator, Larry Headlee, had *Deep Quest* tighten the fouled line and holding a knife in the manipulator, cutting it. With *Nekton Alpha* less than an hour under water, *Deep Quest* was free.

A total of 96 dives were made during the rest of the year for geological mapping, sampling gas seepage areas, well head inspections, and drill site survey off Southern California.

During the first part of 1970, *Alpha* conducted sonar tests near San Pedro, California, a week of geological survey diving off Cortez Bank, California, a week of diving on the Makai Range off Oahu Island, Hawaii, and 94 geological mapping dives for five different oil companies in Alaska and Southern California.

On 17 August 1970 *Nekton Beta*, with the observer's dome omitted and 6" more length (15'-6") than *Nekton Alpha*, was launched in Longbeach Harbor. Training dives and searches for sunken sports boats off Catalina Island took place until 21 September when *Alpha* and *Beta* converged at Catalina to search for and recover a sunken speed boat. The 27' long speed boat had flipped over doing about 55 mph. and sank undamaged. *Alpha* accompanied *Beta* to get photographic coverage of the recovery. After finding the speed boat, both *Alpha* and *Beta* dived on the sunken hull in 225' of water, 500 yards off

Jew Fish Point on Catalina Island.

The *Beta* was piloted by Larry Headlee, who was giving pilot training to Rich Slater. Slater, in the observer's compartment, using the submersibles manipulator, began tying a line lowered from the overhead support ship *Oil City*, to an existing bow line of unknown strength on the sunken speed boat. When Headlee traded places with Slater to finish attaching the lines to the manipulator, Slater became the pilot with his head in the conning tower. Communications between the two subs and the surface were good and they could see each other and the wreck clearly. Instructions were given to the winch operator to take up the slack and put a strain on the line in order to test the knots and line strength. After the line was tight and didn't break, Privitt, in the *Alpha*, ordered *Beta* away from the wreck and he did the same. Slater moved about 100 yards off and rose to mid-water to watch the lift.

The wreck was raised almost to the surface when the bow line parted. The speedboat hull glided rapidly at a shallow incline towards *Beta*. One of the steel skegs on the speedboat struck the pilot's viewport, crushing it and slamming the turret and viewport hard against Slater's head, spattering him with plexiglas shards and leaving him semiconscious. After *Beta* sank to the bottom, tilted to a 45 degree angle and flooded enough to equal the pressure at depth, Slater opened the hatch and floated like a drunk without panic to the surface. He had just enough sense to blow the relatively high pressure air from his lungs as he ascended. The semi-consciousness probably saved his life, since it allowed him to overcome the instinct to hold his breath, which usually occurs during a panic, bursting lungs.

Slater came up next to a small boat containing a friend and his son from the *Oil City*, who had gone fishing in the ship's dinghy, also about 100 yards from the ship. The son saw the bubbles escaping from the wreck and rowed to where they were just as Slater surfaced. Had he not come up next to the dinghy, he probably wouldn't have been noticed and would have sunk back down. The father and son, with the help of an *Oil City* engineer, pulled him into a nearby water-ski boat and sped away for the city of Avalon five miles distant as fast as they could go. The hard smacking of the boat bottom against the sea apparently acted something like CPR, because Slater gained consciousness, vomited, and began to breathe. Meantime, when visibility returned, Privitt in *Alpha* saw Larry Headlee's body on the bottom near *Beta*. Privitt inserted the manipulator in Headlee's belt and raced for the surface where Headlee was declared dead.

Slater ended up with ruptured ear drums, numerous cuts and abrasions on his neck and head, and a concussion. Shortly after the accident, he read his obituary in one of the undersea monthly publications. Six weeks after the accident, he dived again, this time in *Alpha* for his 36th dive in a submersible. Since the accident, he has made 2,500 dives (up to 1995). The incident of the escape is in the *Guinness Book of Records*. Slater says of the record, "I sure as heck wouldn't want to try to break it."

*Alpha* made 31 more dives for geological mapping off the Southern California coast during the rest of the year.

*Beta* made 63 dives during 1971, most of them for the University of Miami, off British Honduras for geological sampling and training dives. *Alpha* made only 14 training and demonstration dives during the year. On 21 December *Nekton Gamma*, similar to *Beta*, was launched and made a few fun dives.

*Alpha* was inoperative and in the shop during 1972 and 1973. *Beta* made demonstration dives during the first half of the year for the University of Miami, followed

*The incident of the escape is in the Guinness Book of Records. Slater says of the record, "I sure as heck wouldn't want to try to break it."*

by two months of diving off the coast of Spain and Morocco for geological survey work. The University of Miami hired the services of *Beta* for 86 dives over a period of 2 1/2 months, for coral reef photography and surveying off British Honduras. In March, *Gamma* was successful in locating a lost tug off Vancouver, British Columbia, and photographing it for insurance purposes. During April, while searching for a sunken racing boat in *Gamma*, Privitt spotted a huge abandoned propeller and shaft on the bottom off San Pedro, California. He returned weeks later with salvage pontoons which were sunk, flooded, then attached to the propeller and shaft by line. The pontoons were then air filled and 14,300 lbs. of bronze propeller and 12,500 lbs. of steel shaft were recovered and sold for scrap. The project was called "get the shaft and screw up."

In June of 1972, *Gamma* spent eight days and 30 dives in the Gulf of Mexico, 120 miles from Freeport, Texas, carrying geologists and biologists to the Flower Garden coral reefs, to study and photograph the coral there. During a week of dives in Lake Michigan, at the Straits of Mackinac, pipeline engineers were taken to the lake floor to observe and videotape 2' diameter pipelines over a twelve mile total length. The previous inspection by divers cost $100,000 which compared to *Gamma's* cost of $5,500. The next 126 dives were made in August in Discovery Bay, Jamaica for the National Science Foundation, University of Texas and the Jamaican Government, for biological and geological surveys and photography.

*Beta* operated only in the last half of 1973, making dives for the State of California, for offshore oil seepage observations, geological mapping in the Channel Islands for Mobile Oil, as well as biological observations at Catalina Island. *Gamma* made 54 dives in the first half of the year doing pipeline and oil platform inspections for four different oil companies in the Santa Barbara area. Another 54 dives were made from June to October for NOAA, doing biological and fisheries surveys. It took three dives to find a Navy instrument package lost off Boddeck, Nova Scotia, Canada.

*Alpha* operated briefly in September 1974 for training, testing, and a few dives off the Cortez Bank near Southern California doing geological observations for Union Oil. *Beta* made demonstration and training dives at Long Beach, California, Catalina Island, and in the San Pedro, harbor. *Gamma* made 105 dives doing pipeline inspection and repair in the Gulf of Mexico over a two month period. *Gamma* spent a week and made 56 dives off Santa Barbara, making pipeline inspections, followed by a torpedo recovery for the Navy off Anacapa Island. During the next three weeks, 39 dives were made for the National Marine Services off the N.W. Atlantic Coast for biological surveys, followed by 24 dives for bottom and dumpsite surveys offshore from the mouth of the Delaware Bay, to evaluate the effect of sewage dumping in the area. Three dives were made for the Army Corp of Engineers inspecting Libby Dam in Montana. Arco Oil hired *Gamma* for a week of dives inspecting platforms and wellheads in the Santa Barbara Channel. Dives for Mobile Oil were made on the Tanner Banks off Southern California for bottom sampling.

*Alpha* made two dives off the Tanner Banks in September 1975, doing geological observations for Union Oil. *Beta* surveyed a damaged Mobile wellhead and a Texaco underwater construction project. *Beta* made 100 dives over a three week period diving for NOAA, USGS, and the EPA (Environmental Protection Agency), evaluating the effect of toxic and sewage waste dumping at several offshore locations from Maine to New Jersey. Eleven dives over a three day period in September were made, recovering $23,000 worth of scientific instruments for NOAA and USGS south of the Georges Bank, off the east coast of the U.S. *Gamma* was employed by the University of North Colorado for 33 oceanographic orientation dives in the Santa Cruz-Santa Barbara Islands area in March. Training dives for new pilots took up 42 dives in April and May. Twenty-three dives off Santa Barbara were made on pipeline and platform inspections for Union, Sun, and Phillips Oil. The California Department of Water Resources employed *Gamma* for 10 dives to inspect The Castaic and Pyramid Dams.

*Alpha* was operational during the last half of 1976, making 63 dives for pipeline and platform inspections in the Santa Cruz-Santa Barbara area for Exxon, Phillips, and Union Oil. *Beta* was employed during the first half of the year in the Bahamas for 34 coral reef inspections dives followed by 15 dives for pipeline inspection and repair for Mobil and Sun Oil.

*Gamma* made 15 dives in January to search and survey a shipwreck at Catalina Island, and to recover oceanographic equipment for the University of Southern California (USC), off Santa Barbara Island. Twenty-three dives over a three day period were made for the California Department of Water Resources, inspecting the San Louis, Orville, and Del Valle Dams in the Sierra Mountains. Forty-seven biological survey dives were made off New York and in the Gulf of Maine for NOAA and USGS. Twenty-three Geological survey dives were made in the Hudson Canyon off New York City and 23 dump site inspection dives were made in the Rhode Island Sound in July, followed by several dives for pipeline inspections in the Gulf of Mexico in October.

*Alpha* made 66 dives in the Gulf of Mexico doing pipeline inspections during a week in June 1977 and from mid-October through mid-December. *Beta* was hired by Global Marine and Gulf Oil to retrieve well head fittings during 31 dives on the Tanner Banks off Southern California in March. The California Department of Water Resources required an inspection of the inlet pipe at Orville Dam in May. A wellhead base was removed at Tanner Bank in July during seven dives. A pipeline survey was conducted for one week in the Bahamas for the William Brothers Engineering Co.

*Alpha* was involved in only one project during 1978. Sixty dives were made for pipeline inspections during nearly six months of operations in the Gulf of Mexico. *Beta* operated only three months of the year in Southern California waters, primarily for training, recreation, certification and coral surveying during 103 dives. Another Navy carrier plane of unknown identification was spotted and photographed three miles off Catalina Island during one of the training dives. *Gamma* was operational for only a month during mid-year. The American Bureau of Shipping certification was completed during three dives. While on an expedition to Alaska, a crab and coral survey was completed for NOAA and the National Marine Fishery Services (NMFS) as well as an environmental survey for U.S. Borax.

In 1979, *Beta* was trailored to Michigan for eight dives over a four day period in the Straits of Mackinac to inspect oil and gas lines. *Gamma* made a three day power cable inspection off Des Moines, Washington (south of Seattle on Puget Sound) followed by a two dive series, testing underwater gear and demonstrating the submersible in the Gulf of Mexico. NOAA and the NMFS employed the submersible for 44 dives for biological surveys and benthic ecology studies in Hudson and Oceanographers Canyons, off the east coast of the U.S. Union Oil employed *Gamma* for 26 dives in October, for pipeline inspection.

In 1980, *Beta* made 36 dives during a one month summer period, making pipe line inspections in the Gulf of Mexico. A month later, a six dive series was completed for inspecting a platform and riser system at High Island, in the Gulf of Mexico. On 1 July, *Gamma* recovered some U.S. Geodetic Survey equipment in the Russian River in California. A month of operations followed in which I participated as one of the pilots, near Sitka, Alaska. The first part of the Alaska operations was to analyze the local long line halibut fishing techniques sponsored by the National Marine Fisheries Services (NMFS). The second part, sponsored by Alaska Department of Fish and Game, was to locate and identify Pacific Ocean Perch fingerlings thought to be schooling in the area.

All efforts failed to accomplish the goal until a young female biologist suggested catching the fish that ate the fingerlings and opening up their stomachs for examination. Thereupon, the submersible *Gamma* became a fishing boat with lures strung between the pilot's turret and the manipulator at the bow. Just as we were closing the hatch to dive someone asked permission to put more lures near the buoy that was towed at the surface above and behind the submersible. I gave a quick OK as we were going over the side of the ship to dive. We would try to catch fish on the bottom next to the submersible and at the surface near the floating buoy. Within a few minutes after arriving on the bottom

***Name:*** *Delta* ***Builder:*** *D. Privitt* ***Length:*** *15'*
***Maximum Depth Rating:*** *1,200'* ***Year Launched:*** *1982*
Courtesy of Marfab, Torrance California

hundreds of 18" long cod circled us like Indians around wagon train. Once a brave one took a lure and was hooked, the others followed in a frenzy to get at the other lures and hooks. Soon all of the lures in front of my viewport on the submersible, as well as all the ones a few hundred feet above and near the surface, had caught large fish. We immediately surfaced. There were so many hooked fish with their mouths wide open that they slowed the forward speed of the submersible to a crawl. Upon recovery and examining the stomach of our catch, we found many small Pacific Ocean Perch which were photographed for later use in preventing commercial fishing in the area.

Early in 1981, *Gamma* was operated off Oceanside, California to locate and help recover a light plane that crashed there in January. Gas seep studies took place in the Santa Barbara Channel, supported by the State Land Commission, in an effort to locate and determine their overall environmental impact. Arco Oil employed *Gamma* for platform, pipeline, wellhead, and seep area inspection. An expensive ship's anchor was searched for, located, and recovered near Santa Barbara. Faunal studies were made in Scripps Canyon, the Coronado Escarpment, and off Catalina Island.

In 1982, *Gamma* made six dives in July to inspect pipelines in the Straits of Mackinac, Michigan, on a contract with the Lockheed Pipeline Co. *Gamma* made 22 dives in March and April in a continuing oil and/or gas seep study in the Santa Barbara Channel. A Kodiak Island, Alaska investigation took place during three weeks in July, when *Gamma* made 53 dives to investigate halibut, king crab, and rock fish populations and conditions for NMFS.

In 1983, a long term biological survey was continued in the Santa Barbara Channel, where *Gamma* made 72 dives over a ten month period under contract to Texaco. There were 43 additional dives for training, certification, demonstrations, and equipment testing.

*Gamma* made eleven dives off San Diego for demonstration purposes for the Navy and the San Diego *Evening Tribune* to observe the sewer outfall off Point Loma in 1984. Fifteen dives were made in September in the Santa Barbara Channel for the State Land Commission as part of a thirteen month continuing drill site environmental impact study.

*Gamma* made only nineteen training dives during 1985 prior to the submersible going to inactive status. No dives were made in 1986.

After a few training dives in the Los Angeles harbor in 1986, *Gamma* was shipped to Wishon Dam high up in the Sierra Mountains about 70 miles east of Fresno. Tim Caffarella and I made 28 dives over a two week period in July. We repaired the leaking dam using the submersible's manipulator to stuff rolled up burlap bags filled with fast setting Portland Cement into cracks and holes in the concrete, water side facing of the rock-filled dam. The last dive of the series before *Gamma* was sold to out-of-state investors was #1442.

In 1990, all of the Nektons were sold; *Gamma* was shipped to Lee Stocking in the Bahamas; *Alpha* to the East Coast of the U.S.; and *Beta* to the Great Lakes. Prior to leaving San Diego, *Alpha* had made 828 dives, *Beta* 842 and *Gamma* 1442, for a total of 3,112 dives. In August 1995, the husband and wife team of Don and Sally Tondro piloted *Nekton Gamma* to its seven hundredth dive for the Caribbean Marine Research Center at Lee Stocking Island in the Bahamas.

# Delta

In 1974, Doug Privitt had started a new commercial enterprise called Delta Oceanographics, so as to completely separate the new operations from those of the Nekton submersibles. Privitt wanted to get the maximum operating depth for his new sub, so we reviewed all of the academic formulas and systems for calculating pressure hull designs used by the Navy and industry. The results were that his empirical model testing method could not be significantly improved on. He was able to add another 200' to the operating depth mostly due to using a stronger steel.

A major problem was getting *Delta* certified using a grounded hull system similar to virtually every system used on American autos, airplanes, boats, golf carts, and fork lifts, as well as many deep submersibles certified by the U.S. Navy at the time. The maximum voltage available aboard *Delta* was 48 volts, direct current. The shock factor for 48 volts d.c. is nearly imperceptible, even in the unlikely event of a handling accident while diving. The safe records of the Nektons over many years and many dives all over the world, using the same system, were not accepted for several years as proof. Letters from an ad-hoc committee in the Deep Submersible Pilots Association (DSPA) to ABS (American Bureau of Shipping) justifying — with volumes of references — the grounded hull system were not even acknowledged. Finally, in 1982, after a change in personnel at ABS and a show and tell presentation by Delta Oceanographics partner Dr. Rich Slater to new, open-minded officials, the *Delta* electrical system was accepted.

On 21 December 1982, *Delta* made its first dive off Redondo Beach, California. The life support system was increased in capacity to 216 man-hours. The total battery power available was more than doubled. The overall length is the same as *Alpha's* at 15', with a large slanted viewport at the nose of the ballast/buoyancy tank and an added viewport in the bow of the pressure hull. The cruise speed was increased to 2 kts. and the maximum speed to 3 1/2 kts. The operating depth was increased to 1,200' with an increase of the crush depth to 2700'. The human powered manipulator that has never failed was used again on *Delta*, as was the droppable tail assembly and external ballast weight. Models of the pressure hull were pressure tested to failure at 3,000' of sea water pressure. Drawings of the full scale submersible were made on computer, then printed out and/or manufactured from computer discs directly, by numerically controlled machines in the shop. In summary, *Delta's* design provided improved instrumentation, operational capabilities, visibility, and more comfort for both the pilot and observer.

Most of the 45 dives made in 1983 were made in the first six months at Point Conception, California for geological surveys for the oil companies.

A series of dives for environmental studies was made in January 1984, at Gaviota, California. In July, a lengthy series of dives at Eniwetok Atoll, Marshall Islands in the South Pacific was made to explore a few of the 40 nuclear bomb craters, including the one made by the world's first H-bomb, exploded by the U.S. in the barrier reef surrounding the lagoon in the early 1950s. The diameter of the largest crater, Koa, was about three miles but its maximum depth was only 180 feet. (One observer, viewing the crater from a helicopter said, "it looked as if a giant tooth had been pulled.") Two selected craters were completely mapped out in bathymetric grids. The few dives made outside the lagoons indicated that extensive cracks existed in the reefs from the detonations. October dives were made at Bodega Bay, California for the oil companies for geological surveys.

The Eniwetok dives were continued in June 1985, for monitoring and mapping the craters for the USGS. Dive visibility was 200' at a depth of 1,000' outside the lagoons.

ABS certification and biological dives were made in April, 1986. A pipeline inspection series off San Diego was completed in May. On 5 June, a summer dive program was initiated with the first dives made off Long Island Sound. The program was funded by NOAA, the U.S. Geological Survey, and several universities and colleges for biological and geological programs using the Perry Foundation's *Undersea Hunter* as a support ship. In June, a scientific instrument package was installed on the bottom in the Gulf of Maine for pick up later, with its accumulated data, by the *Johnson-Sea-Link 2*. Dives were next made off Six Mile Reef, Connecticut, Long Island, New York, and Cape Cod, Massachusetts. On 22 July, dive #500 was made in the Florida Keys during one of the 47 dives in the South Florida portion of the project. The project research team's primary goal was to learn about the general health of the local coral reefs, in shallow water as well as in deep. In August, underwater studies were made for the Caribbean Marine Research Center, involving sedimentation studies for the University of Miami in the Bahamas near the Tongue of the Ocean.

The 1987 dive series was a repeat of the series done in 1986 for the National Undersea Research Program and the Caribbean Marine Research Center. Three hundred eighty-five dives were made during the year, which included dives off Cape Cod, Atlantic City, as well as dives off St. George's Bank, Hudson's Canyon, New York, Palmer's Point in the Bahamas, and the Tongue of the Ocean. These dives that began in January 1987 in the Bahamas, were completed in 1988 for the Caribbean Marine Research Center. A lobster study in June involved eighteen dives in one day in the Gulf of Maine. A sediment survey, as well as a fish and crustation count, was made in the same area over a several week period. In September, dives off the Oregon Coast were made for Oregon State University for a fish study and an underwater gas vent study. The fish studies were continued in September off Newport, Oregon with hagfish studies off Monterey, California in October.

The 377 dives of 1989 began in Auke Bay, Alaska for a king crab survey, then proceeded to Bristol Bay for the University of Alaska. Additional dives were made off Baranger Island, Alaska and in the North Bearing Sea above the Arctic Circle for rock fish and whale pit surveys. Oil spill investigations were made off Alaska in August for the National Marine Fisheries. In September, dives in the Gulf of Maine and off Cape Cod were conducted for the Universities of Maine and Boston for investigating sediment transport, oceanography teacher training, and rockfish surveys. In October, the University of Connecticut made a series of nighttime biological survey dives. During November and December many dives were made for the Papua New Guinea Government to analyze the local lobster migration patterns.

The Papua New Guinea dives were continued through February 1990 as part of the year's 451 dives. The submersible then dove on the Great Barrier Reef off Australia with a Papua New Guinea biologist as observer during a lobster survey.

Dive #1800 was made on 12 June off Bimini in the Bahamas as part of a National Underwater Research Program. In July, a series of dives were made off Charleston, South Carolina on the Blake Plateau for a geological investigation program. A wolf fish survey combined with a high school student indoctrination program was conducted off Pigeon Hill, Massachusetts in late July. In September, a fish survey and invertebrate collection was made off Daisy Banks Oregon.

The 553 dives made during 1991 began in April, with inspection of a commercial fisheries dumpsite at Kodiak, Alaska followed by the collecting of 50 mating Tanner crab pairs for NOAA and NMFS off Kodiak, Alaska. In May, a hagfish survey was made for the Monterey Aquarium and Clemson University in Monterey Bay, California. In June, a king crab survey was conducted near Sitka, Alaska for the University of Alaska and the Alaska Department of Fish and Game. In July, an invertebrate study was made in the North Bearing Sea for the University of Alaska. In August, the National Undersea Research Program sponsored a high school aquanaut program off Pigeon Hill, Massachusetts. In late August, anesthetizing wolf fish for recovery purposes was done during a series of dives. Gas vent seepage inspections were made off Coos Bay, Oregon in September, followed by fish migration recordings for Oregon State University, off the Oregon and Washington Coasts during October. November dives in the Gulf of Mexico were made for the U.S. Geodetic Survey at oil well test sites.

The 1992 dives began in May off Kodiak, Alaska in surveys of Tanner crabs followed by Rockfish surveys off Sitka, Alaska for the National Marine Fisheries. The June dives began in Puget Sound investigating the shipwreck of a 1914 luxury liner, the 214' *Admiral Sampson*, and collecting artifacts from it. A seismic fault survey was conducted in August for the University of Washington and Oregon State University off the Oregon Coast, followed by thermal vent inspections and methane gas sample collecting off Eureka California. *Delta* dive #3000 was made on 29 November, 1992 in Puget Sound searching and collecting additional items from the 1914 shipwreck. A seven hour-plus dive was made in November during one of the dives on it. *Delta* resumed the *Admiral Sampson* dives in March, 1993.

During August, 21 dives were made on the *Lusitania* shipwreck. In 1915, the *Lusitania* was torpedoed by a German U-boat, just off the southern coast of Ireland, while only a few miles from its destination of Liverpool, England. Her coal bunkers were close to empty after leaving New York a week earlier. After many dives by *Delta*, Bob Ballard, conducting the 1993 expedition, concluded that a second explosion was due to coal dust exploding in the near-empty bunker areas and that massive damage to the starboard side of the ship caused the luxury liner to sink in 15 minutes, killing 2/3 of the 2,000 passengers and crew on board.

In April 1994, filming of the *Admiral Sampson* shipwreck continued with recovery of over 100 lbs. of equipment such as portholes and engine telegraphs. *Delta* was requested to return in 1995 for more work. In September, geological surveys were conducted for Oregon State University off Astoria Oregon which completed twelve years of operation, using anything from a 50' fishing boat to 300' drilling vessels as support ships.

*Delta* performed crab and rockfish surveys for National Marine Fisheries and the Alaska Department of Fish and Game near Kodiak and Sitka during April, May, and June, 1995. A series of dives was made in August in Lake Michigan on the wrecks of *Cederville* and the *Carl D. Bradley*. Biological sampling in Scripps Canyon near San Diego in late August was followed by fault zone rock sampling near Coos Bay, Oregon in September, for the 3687th dive in less than 13 years.

Like John and Hap Perry, Doug Privitt is a private businessman. He spans the entire spectrum of skills from design through fabrication, following these with successful commercial submersible operations. While the Perrys, for the most part focused on manufacturing and sale, Privitt concentrates on fulfilling operating contracts of all sorts — literally thousands of them in every corner of the globe.

# *Dolphin*

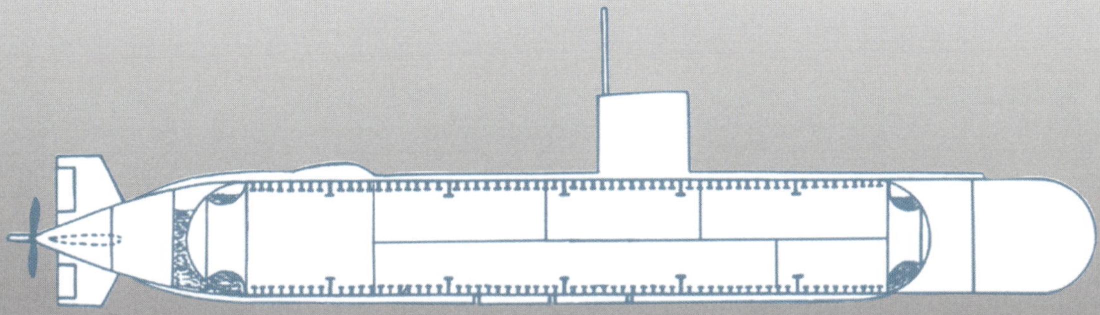

**1968–(     )**

# Dolphin

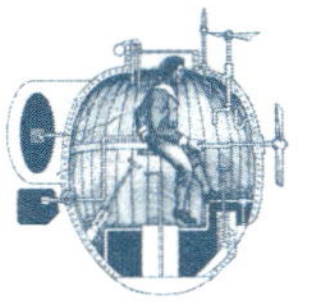

*Dolphin* is a U. S. Navy diesel-electric research submarine. Although experimental torpedoes have been fired from the submarine for test purposes, its functions are non-combatant. With a length of 165', a beam of 18', a capacity to operate with 35-40 people on board and an underwater displacement of 950 tons, *Dolphin's* size and complement makes it the largest of its kind in the U.S. submersible inventory. Officially, the operating depth of *Dolphin* is given as 3000'. Unauthorized, published literature alleges a 4000' capability. Either figure is impressive for a submarine of its type. Having the capacity to operate at sea unattended with a full crew for 10-12 days, it can be just as accurately classed as a research "submarine" since the majority of submersibles are capable of only brief periods of immersion.

To a structural engineer, *Dolphin*'s pressure hull represents the epitome in a ring-stiffened cylinder design. The 1960's designers used the latest Navy approved steel (HY 80) and produced the deepest operating, large-scale, cylindrical, steel vessel of the day. Similar size U.S. combatant submarines of the same period, using the same steel, had operating depths of less than half that of the *Dolphin*. The submarine is also distinctive as a manned research tool in that it has no viewports and no manipulator. Electronic viewing and/or sensing by television, sonar, and laser substitute for the usual direct eyeball-to-object viewing. In the Navy, *Dolphin* straddles the line somewhere between the usual Navy commissioned ship and a research vessel. A commanding officer, executive officer, and two lieutenants head up the 26 member crew with space allotted for five scientists.

The uniform diameter, cylindrical pressure hull, with hemispherical end closures, is approximately 115' long. Rings about 18 inches apart stiffen the steel shell. In addition, five deep frames are proportionally distributed throughout the hull to act similar to bulkheads, making the hull structurally the equivalent of six short, very strong cylinders out of an otherwise long, weak cylinder. The effect is that the pressure hull is less compressible than sea water, which is unusual for ring stiffened, thin shelled, cylindrical pressure hulls.

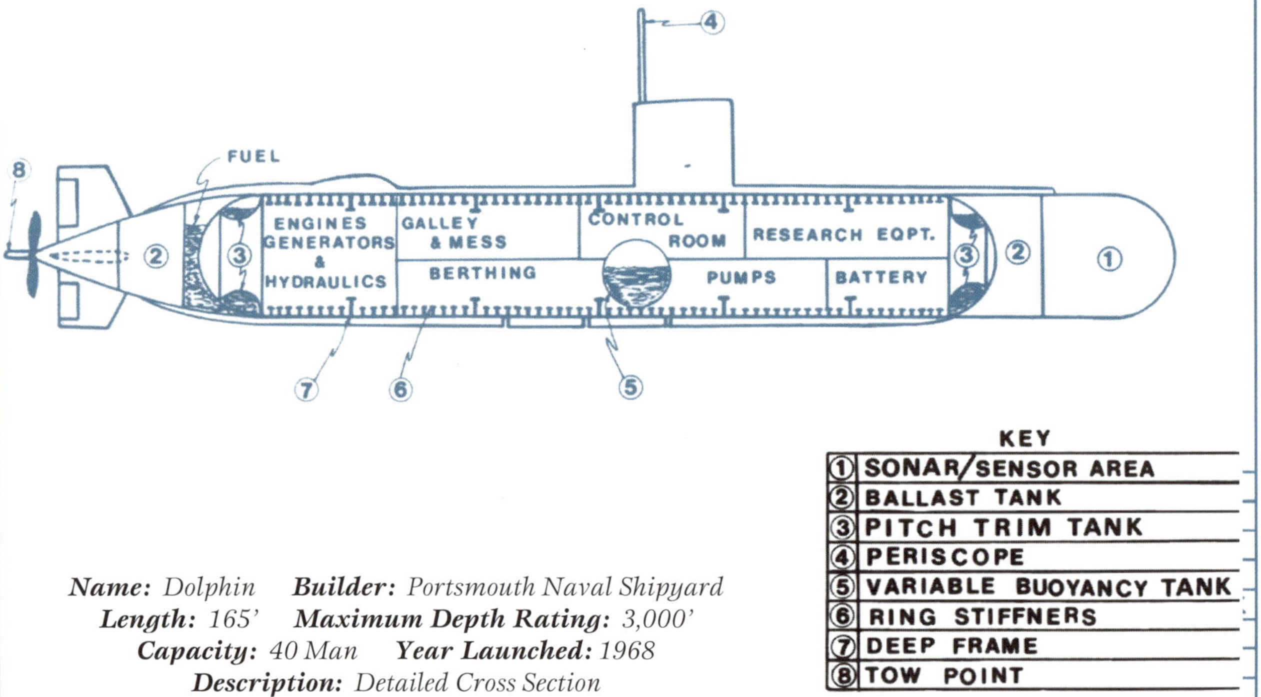


***Name:*** *Dolphin* ***Builder:*** *Portsmouth Naval Shipyard*
***Length:*** *165'* ***Maximum Depth Rating:*** *3,000'*
***Capacity:*** *40 Man* ***Year Launched:*** *1968*
***Description:*** *Detailed Cross Section*

A periscope, 25" hatch, propeller shaft, and the diesel exhaust line are the few mechanical penetrations in the hull. Various electrical pass-throughs are provided for the external equipment. Propulsion power is provided by two General Motors 425 H.P. V-12 diesel engines. They power the 5' diameter, 2-bladed propeller, which thrusts the submarine at up to 7 kts. at the surface. The diesels also power an electric generator for charging two lead acid, 126 cell Gould batteries for underwater cruising speeds of up to 5 kts. with burst speeds of 10 kts. Combustion air for the diesels, which are only operated at the surface, is obtained through the single access hatch, while exhaust fumes are discharged overboard via piping.

Control and variation of ballast/buoyancy can be done in several ways. The primary buoyancy adjustment is done by taking on or pumping out sea water from two large pressure resistant spherical tanks with 11,000 lb. capacity, located amidships. A 5,000 psi air supply is used to eject water from the tanks when desired at any operating depth. Torroidal shaped trim tanks are located inside the forward and aft hemispherical end closure of the pressure hull. Freshwater can be pumped between the two torroid tanks for pitch control. A cast-in-place fixed ballast of lead is located internally at the bottom of the pressure hull between the ring stiffeners producing about a 5' wide chord across most of the bottom of the cylindrical hull. Two 11 ton lead weights under the sub can be dropped in an emergency, by actuating explosive bolts that secure them to the hull.

The helmsman controls *Dolphin* by a wheel on a stick control which actuates hydraulic pistons to the rudder and dive planes located at the stern. There are no dive planes on the sail or bow. During hovering modes, all personnel must maintain their position on the sub for the sake of the sub's stability in pitch. When surfacing or diving, all personnel available are directed to move forward or aft, to help get a dive or climb angle on the sub. If more personnel than the normal compliment are on board, "hot bunking" and/or passageway area sleeping is required on extended tours. The airline type galley has a microwave and conventional oven. One table that seats six is the extent of the messing space. TV dinners are not served, but fresh food type buffets are prepared by crew members who take turns as cooks away from their regular duties for a week at a time. Audio-visual entertainment is available to off-duty personnel. Cellular phones for both official and private use keep the sub crews in touch with the command and with their homes ashore. The twelve day limit on duration at sea is primarily due to the space available for storing food. Submerged times typically average from six to ten hours due to battery power limitations. Navigation is conducted by both inertial dead reckoning and satellite geopositioning. Collision avoidance sonar can be directed from 40 degrees upward through 40 degrees downward.

*Dolphin* has racks and cargo carrying capacity for 12 tons of scientific equipment. All material carried inside the submarine must pass through the 25" diameter access hatch. A cooling system for all heat emitting equipment is provided by low pressure, water flowing through a heat exchange system located between the ring stiffeners, against the shell of the pressure hull. One million BTU of heat dissipation is available. Five low-light-level TV cameras and one electronic still camera are mounted externally. Electrical power is available in 120 volts or 250 volts. at 60 hertz or 400 hertz 3 phase. Direct current power is available in most voltages. A trailing bar aft of the propeller is provided for towing sonar arrays. Most physical oceanographic parameters are measured routinely and can be recorded with on-board equipment. *Dolphin* is normally available thirty weeks of the year for scientific research.

*Dolphin* was launched at the Portsmouth Naval Shipyard in Kittery, Maine on 8 June 1968 and made its first dive in September, 1968. The first dive to test depth was made in November, 1968 near Nantucket Island, Massachusetts. A record for the deepest launching of a torpedo was set in August, 1969 on the torpedo test range near Fort Lauderdale. From October 1969 to August 1970, the submarine underwent major modifications at the Portsmouth Naval Shipyard. In October 1970, *Dolphin* transited the Panama Canal and arrived at its new home port of San Diego, California on 7 November 1970.

*Dolphin Courtesy of Navy Deep Submergence Group, San Diego, California*

*A collision with a University of Hawaii surface ship on 30 October near San Diego, caused the submarine to be dry docked for damage assessment.*

---

From January through March 1971 *Dolphin* conducted sea tests of its underwater sonar and electronics at the Naval Research and Development Center (NRDC) [NB: NRDC, NUC, and NOSC respectively became different designations for the same place] in San Diego, CA. In April, the submarine was deployed to the Gulf of Alaska on a sonar and oceanographic development expedition. On the way north, shaft seals were repaired at Mare Island and Puget Sound Naval Shipyards. Basic oceanic acoustical path research conducted at sea confirmed predictions made by Navy scientists. From September to mid-December, *Dolphin* underwent battery replacement in dry dock at Mare Island Naval Shipyard, then returned to San Diego in December.

From January through March 1972, local operations near San Diego were conducted to test a new sonar system and for crew training. During April, two dry dockings at National Steel & Shipbuilding Company were required for the installation of hydrodynamic test equipment. Hydrodynamic tests were then conducted on the hull in local waters. An acoustic range system for submarines was installed, calibrated and tested at sea in September. Dry docking and installation of a deep, experimental, towed array system was initiated in late September, with sea trials conducted in November and December.

From January to April 1973, *Dolphin* conducted extensive operations in the Southern California area evaluating the Deep Towed Sonar Array System. During the regular shipboard overhaul which began on 1 May, major modifications and routine maintenance procedures were conducted, including another replacement of the silver zinc main battery. Several research projects with Scripps Institute of Oceanography, involving underwater acoustic studies, were accomplished along with a week of project service to the Naval Undersea Center (former NRDC).

From January to August 1974, *Dolphin* operated in the San Diego area in support of various research, development, test, and evaluation (RDT&E) projects oriented toward underwater acoustics and ocean thermal microstructure for NUC and the Naval Research Laboratory. From August through December, only surface operations were conducted due to the demise of the silver zinc main storage battery.

From January through September 1975, *Dolphin* was limited to surface operations, while waiting for a replacement for the silver zinc main battery. Pierside replacement of the battery was accomplished during the first week in September. The new battery was also found to be flawed and operations were limited to shallow depths. A collision with a University of Hawaii surface ship on 30 October near San Diego, caused the submarine to be dry docked for damage assessment. The remainder of the year was spent pierside awaiting repair materials for the port stern plane and its shaft.

After repairs were completed in early 1976, *Dolphin* resumed its services to the Naval Undersea Center, San Diego. The services provided from February through April, included optical communications to aircraft, mobile submarine simulation, towed array operations, and passive ranging and tracking tests. In April, the silver zinc battery was replaced by a new lead acid main battery. In June, operations resumed with low frequency communication tests. The main shaft seals were refurbished in dry dock in November, followed by sea trials to test them.

During January through May 1977, *Dolphin* provided services to the Naval Ocean Science Center (formerly NUC) in support of projects involving submarine slow scan television, passive sonar ranging, and experimental sonar. During a brief dry docking in May, an improved sonar array tow point was installed. In July, a downed F-14 was located and marked for others to recover. Port calls to Alameda and Eureka, California were made in October. During the remainder of the year, a towed array project was conducted.

From January through April 1978, *Dolphin* provided services to the Naval Ocean Systems Center, San Diego, California for classified sonar and acoustics projects. In early May, the number 2 diesel malfunctioned, so *Dolphin* was towed to Mare Island Naval Shipyard, remaining there in overhaul the rest of the year.

*Dolphin*s overhaul was completed in June 1979, after which acoustic tests were conducted near Fox Island, Washington. *Dolphin* operated the remainder of the year in support of submarine research projects for the NOSC.

During January 1980, the number 2 diesel failed again and a new engine block had to be installed at Mare Island Naval Shipyard. During repairs, a new cryptographic communication system was installed. The engine blocks were passed through a hole cut in the pressure hull, which was then repaired. A silver zinc main battery was installed to replace the lead acid unit installed four years earlier.

In May, operations were conducted in the Southern California area to evaluate the effects of deep submergence on sonar hydrophones, built for conventional submergence depths. *Dolphin's* conventional dead reckoning system was replaced by a computer controlled plotter. In July, *Dolphin* transited to Astoria, Oregon to conduct a sonar research project, returning to San Diego in August. In September, another sonar research project was conducted in the Southern California area. Sonar and wake turbulence studies were made at San Clemente Island in November. In December, *Dolphin* located a lost Air Launched Cruise Missile near San Miguel Island, California.

In January 1981, *Dolphin* went to sea for crew training, but returned early due to problems with the silver zinc battery installed only a year earlier. In February a new silver zinc battery was installed pierside in San Diego by Mare Island personnel. In March, sea trials for testing the batteries were "completely satisfactory." In April, crew training dives and a deep test dive were conducted near San Clemente Island. In May and early June, laser communication tests were conducted at San Clemente Island. Messages from submarine to aircraft were exchanged in spite of cloudy weather. In June, *Dolphin* went to Eureka, California for communication tests in the nutrient rich water in that area in order to study the effect of the bioluminescence on the blue/green laser signals. July and August were spent testing an expendable vertical array, as well as the temperature of thermal gradients in the open ocean near San Diego, California. At pierside in San Diego from September through December, new lead acid batteries replaced the silver zinc units tested in March.

The lead acid battery installation lasted from January into February of 1982. Sea trials in late February indicated that the main ballast tank blow valves needed repairs, which took four weeks in port. During April and May, transient ocean acoustic detection measurement tests were conducted for the Defense Advance Research Projects Agency. In June, data was gathered near San Clemente Island on ocean front currents and eddies for analysis. In August, *Dolphin* conducted anti-submarine training services in conjunction with aircraft. In September, a submarine satellite information exchange system was installed, as an augmentation to the radio room. A seven ton winch was installed on the bow of *Dolphin* (externally) to handle 6,000' of deep tow sonar array cable. October through December were spent testing deep tow arrays.

During January 1983, local operations for crew training and on-board equipment testing took place. In February, a deep towed array test was made for the Naval Ocean Science Center. Anti-submarine exercises were conducted in April, and *Dolphin* demonstrated a new towing system. In June, *Dolphin* was used as a target for range tracking exercises off San Clemente Island. In July, *Dolphin* was towed to Port Angeles, Washington, where technical evaluation was made of a reverse osmosis desalinization plant. Minor repairs were made to the submarine at the Trident Repair Facility, Bangor, Washington, followed by remote sensors operations nearby. *Dolphin* was towed back to San Diego in October, where acoustic navigation system evaluation was made by a deep submergence support ship. Preparations were made for dry docking at Mare Island Naval Shipyard.

From January through March 1984, crew training and dry dock repairs took place. During April, oceanographic services to the Naval Postgraduate School in Monterey, California, were conducted. Significant data was collected concerning "upwelling" in the Monterey underwater canyon. Submarine laser communications to aircraft were conducted in May and June. Acoustic experiments for the Marine Physics Laboratory, San Diego, was conducted in the shallow water of San Diego Bay. Services for the NOSC for conducting ocean color changes, and their measurements due to biological agitation, were carried out in August. Additional current upwelling measurements in Monterey Canyon took place in September.

During November and December of 1984, replacement of the main battery was initiated and completed in February 1985. Submarine laser communication to aircraft operations took place in March and April 1985. A demonstration of a deep ocean transponder field was made in June. Regular overhaul at Mare Island was begun in mid-June and continued through the rest of the year. Overhaul was completed in April 1986, although the installation of several new equipment systems took place at dockside during the rest of the year.

Sea trials following overhaul were completed in January 1987. While conducting upper ocean turbulence experiments in February, the number 2 diesel generator arced and required new brushes. Repairs were made and the upper ocean turbulence experiments were continued. Following a port call to Long Beach, California in April, a fast cruise drill damaged the #1 diesel engine requiring another engine block replacement. The large size of the V-12 engine block required dry docking and a hole to be cut in the hull again. In July and August a non-acoustic anti-submarine program was executed. A high frequency Obstacle Avoidance Sonar System and a Wide Angle Imaging System were installed dockside in September and October, followed by a successful fast cruise test. A navigation evaluation exercise was completed satisfactorily in November. During January through June 1988, *Dolphin* conducted open ocean thermocline observations using the Upper Ocean Turbulence System off San Clemente Island for NOSC, San Diego. During July and August, *Dolphin* conducted vorticity meter testing for the University of Washington, Applied Physics Laboratory, in the Southern California Operating area. During October and November, High Frequency Obstacle Avoidance Sonar testing was accomplished near San Diego.

Sea trials in January 1989 were followed by mine detection evaluation of the Obstacle Avoidance Sonar tests. In June, the new main battery was removed and replaced with a successful fast cruise test afterwards. From July through September, at-sea testing of the Obstacle Avoidance Sonar and the Wide Area Imaging (sonar) System were tested. Extensive gravity anomaly experiments at test depth were conducted for Scripps Institute of Oceanography during September and October. A Supply Management inspection was passed in December.

In January 1990, *Dolphin* served as a test platform for a very low impedance antenna system that processed minute magnetic signals for use in communication and detection. A sunken recreational boat was searched for and found near San Diego, as part of a police investigation. Components for integrating target data with navigation and velocity log data were installed, to improve their management. Testing of the components and systems lasted until mid-June. In late June, an advance optical system for generating visual images by scanning underwater objects with blue-green laser was tested. Two weeks of testing very low frequency background noise levels and frequencies were conducted to establish a data base for the attenuation of their signals through the air-sea interface. From August through December, testing of the target management of obstacle avoidance and high resolution sonar was carried out.

During January 1991, the Laser Line Scanning System was under test. In May, two way laser communications between submarine and aircraft were accomplished, in spite of turbid water and a cloudy overcast. During October, *Dolphin* served as a platform for testing and evaluating a new acoustic Doppler current profiler. During November and December, the submarine was a target for a new type sonar detection system which uses frequency dispersion for better performance. In July, *Dolphin* completed dive #1200 while in the Southern California area.

During January 1992, *Dolphin* was again a target for ASW tests. Sound channel research was conducted off Mexico in February. In March, *Dolphin* acted as a target for active sonar detection from shipboard underwater transducer arrays at San Clemente Island and La Jolla. During April, periscope detection tests were run off Point Loma, California with *Dolphin* at periscope depth as the target. In June *Dolphin* left San Diego for lengthy overhaul at Mare Island Naval Shipyard.

*Dolphin* returned from the Mare Island overhaul in mid-August 1994. A test program was conducted in mid-

September off the coast of Washington state. Calibration of onboard equipment was conducted in the San Diego area in December.

Support for fleet amphibious exercises off Camp Pendleton, California, sonar tests and calibrated high alloy corrosion testing took place off the Southern California Coast in the first half of 1995.

*Dolphin's* need to undergo frequent updating of its equipment caused many dockside and dry dock repairs. Equipment exchanges also caused many interruptions of its chartered research and development mission. In part, these resulted from the long at-sea tours when the sub personnel work a 24 hour day, thereby lengthening maintenance time ashore. Equipment exchange was in itself an R & D function. Furthermore, *Dolphin's* many classified projects placed extra demands on its overall performance. *Dolphin's* record shows that it has performed and been managed well for both civilian and military purposes.

# *Beaver MK 4*

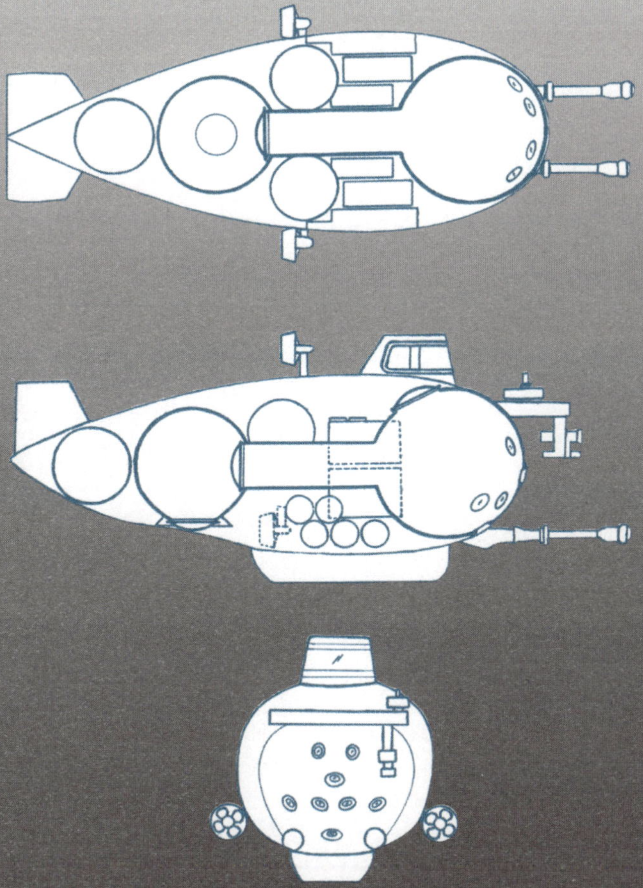

*1968–1976*

# Beaver MK 4

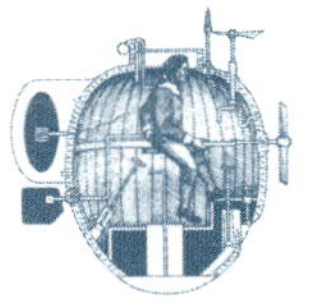

Following his assignment as Chief Scientist on *Trieste*, Dr. Andy Rechnitzer accepted a position at North American Rockwell in their Research and Development branch of their Autonetics Division in 1961. He immediately initiated the development of a submersible for use as a tool for offshore oil field work and oceanographic research. At first called "Roughneck," it became *Beaver MK 4*. *Beaver's* overall length was 26'-4" with a weight of seventeen tons. It held a maximum crew of three in the pilot's sphere and two in the diver's sphere. It had a payload of 2,000 lbs. The pressure hull had the configuration of a lopsided dumbbell with a pilot's 7' diameter sphere forward, and a 5'-6" diameter diver's sphere placed aft of a 25" diameter tunnel between the two spheres.

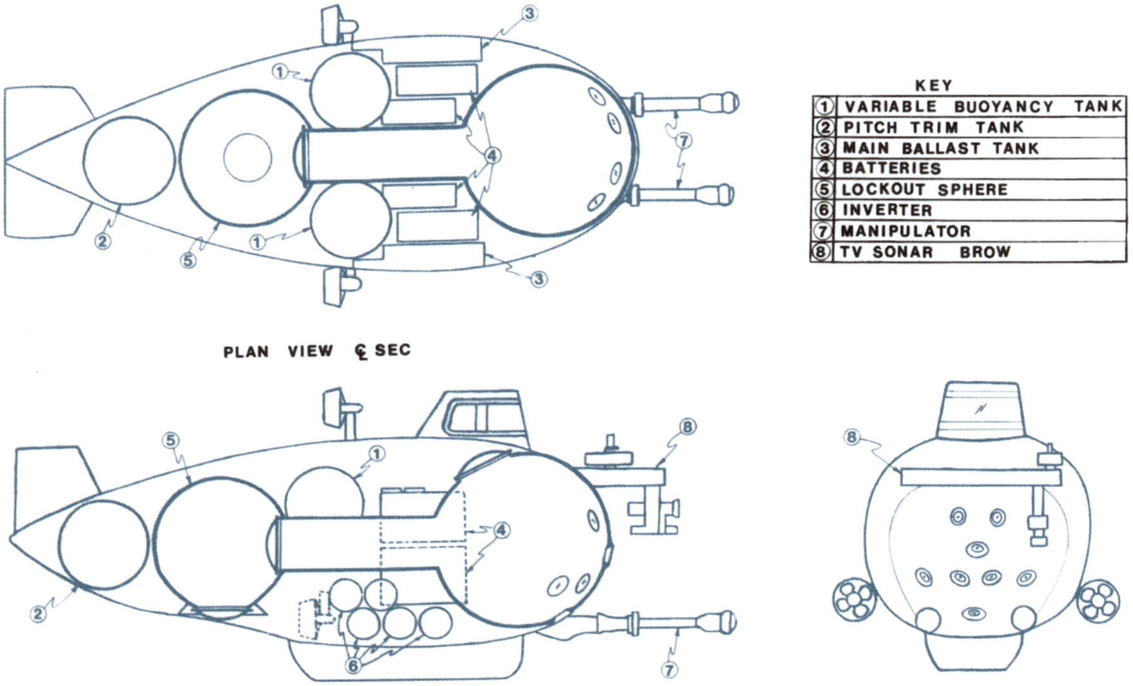


***Name:*** *Beaver IV* ***Builder:*** *Autonetics* ***Length:*** *26'4"* ***Maximum Depth Rating:*** *2,000'*
***Capacity:*** *5 Man* ***Year Launched:*** *1968* ***Description:*** *Detailed Top, Side, and Rear View*

All material in the pressure hull was HY100 steel. There were three hatches in the pressure hull. A 25" diameter hatch was in the upper part of the pilot's forward sphere. Two 22 1/4" diameter hatches were in the diver's aft sphere, one for lock-out and one going to the tunnel between the two spheres. The entry hatch to the pilot's sphere had a 2" diameter viewport; the lock-out hatch in the diver's sphere had a 1" diameter viewport. In addition to the two viewports in the hatches, there were nine 5" diameter viewports in the pilot's sphere. The submersible had an operating depth of 2,000' with a crush depth of 4,000'. The diver lock-out capability was 1,000'. The pressure hull was fabricated at Hahn & Clay in Houston and hydrostatically tested at Southwest Research Institute in San Antonio, Texas.

There were several ballast/buoyancy control and/or trim methods. The main ballast tanks, for surface buoyancy, were located approximately amidships and outboard of the pressure hull. Their volume provided 1,540 lbs. of buoyancy at the surface when blown free of water using compressed air. Two spherical tanks outboard of the pilot's sphere provided a ballast/buoyancy trim of 942 lbs. A third spherical tank at the stern of the vehicle could provide up to 1,240 lbs. ballast/buoyancy. All three ballast/buoyancy spheres were operated in conjunction with one another using venting to flood and compressed air to blow out water.

By various combinations, the trim tank system could provide 30 degrees up or down angles in pitch or 12 degrees left or right angles in roll. The tanks were also used for varying the net buoyancy in minute amounts to maintain a stable vertical status. Inside the diver's sphere there was a winch with an anchor weighing 100 lbs. secured to it by 500' of cable. The anchor system was used to stabilize the submersible during diver lock-out or lock-in as well as to stabilize the submersible at given distances off the sea floor.

Three propulsion units provided thrust for the submersible. They were located near midships. One unit was at the top of the submersible. The other two were on either side, 120 degrees apart around the perimeter of the outer hull. Each thruster consisted of a 5 H.P. d.c. electric, reversible, oil flooded motor with a 4-bladed 18" diameter propeller. All three thrusters could rotate 360 degrees. Pulse width modulation electronic controls provided variable motor speeds up to 550 R.P.M. and 6 degrees of freedom in maneuvering. Limited only by the energy capacity of the batteries, the cruise speed of 2 1/2 kts. could be maintained for twelve hours. The electrical power was supplied by oil flooded, lead acid batteries. Emergency batteries in the forward sphere provided for such emergency functions as jettisoning appendages and a float marker.

Life support in the forward sphere and diver's sphere was provided for 72 man-hours. Two manipulators were at the bow; each had a reach of 9 feet with 8 degrees of freedom. The emergency systems included the ability to jettison the main batteries, manipulator, propulsion pods, anchor, and cameras by explosive bolts for a total of 3,400 lbs. Emergency breathing air was provided by compressed air tanks and full face masks.

Nancy Reagan, wife of then-Governor Ronald Reagan, christened *Beaver MK 4* on Thursday, 26 September 1968 at Long Beach California. Twenty-five pool tests followed for weight and balance adjustments. The first dive series was made in March 1969, from the marine railway at Big Fisherman's cove, in the isthmus area of Santa Catalina Island. In April 1969, *Beaver MK 4* took part in Cousteau's

*Beaver 4* Courtesy of Ed Cargile

submersible jamboree at Catalina Island. In May 1969 the first operational depth dive to 2,094' was made.

A contract with the Naval Undersea Research and Development Center (NURDC) began a series of scientific dives. Fifteen of the these were diver lock-out. The rest consisted of the following:

- Midwater plankton observations and Deep Scattering Layer studies
- Geological and biological surveys of Scripps and La Jolla Canyon, areas off Santa Catalina Island, and in the Santa Barbara basin
- Physical Oceanographic analysis

A few in-house dives were made for North American, some of which were lock-out experiments. On one occasion, the lock-out diver was leaving the diver's sphere at 120' when the submersible suddenly became heavy and settled on his air line connected to his air supply aboard the submersible. Just as he prepared for free ascent by removing a lot of gear and a full-face mask, a diver from the surface dove down and gave him air by sharing his regulator and taking alternate turns at breathing while ascending to the surface. In another operation, *Beaver* was mating with an underwater habitat operated by North American at a depth of 200' near the isthmus at Santa Catalina Island. After pressurizing and blowing out the water in the diver lock-out trunk, a transfer of personnel was made by opening the habitat hatch.

*Beaver's* batteries and electrical breakers were all oil flooded and external to the pressure hull. A special oil-filled distribution box contained the breakers next to the main batteries. Toward the end of a dive on 17 June 1970 at a depth of 2,000', an overload which failed to disconnect the breakers occurred in the electrical system. A short in the starboard propulsion motor caused a major short circuit in the distribution box. The arcing and resulting fire burned a large hole in the box which let seawater in, shorting out all of the other breakers. A major modification in the electrical system was indicated, but the vehicle's redesign was not undertaken because of the existing poor market for submersible services.

*Beaver MK 4* was deactivated by North American on 10 July 1970 and was idle until purchased by Andre Galerne's International Underwater contractors Inc. (IUC) of NY, NY. About 20 dives were made by IUC in the Block Island, Rhode Island, area after refurbishing the submersible. The dives which began in July 1973 were mostly for AT&T for burying and/or inspecting transatlantic communication cables. A few of the dives in the series were made for the National Oceanographic and Atmospheric Administration (NOAA) for underwater pollution studies in the area.

The submersible was then leased to Oceanwide Submersible Services (OSS) of England. Dive #102 was the first dive made in the United Kingdom. It was made in Loch Linnhe in December 1974. Following about eight checkout dives in the United Kingdom, the submersible was loaded onto a cargo plane and flown to Cairo, Egypt where it was trucked to Ras Shukeir and launched in the Gulf of Suez. There, 132 miles of submarine pipeline from the El Morgan oil field platforms to an onshore terminal were inspected and videotape recorded under water by *Beaver*, every day for 50 days, beginning in January 1975. After the Suez operations, the submersible was returned to IUC in NY.

In July 1976, the pressure hull was sent to Southwest Research Institute in San Antonio, where the middle three viewports were replaced with a single 36" diameter viewport. A spare pressure hull for *Beaver* was fitted with a 36" diameter viewport at the same time and later sold to the Canadians for their submersible *SDL-1*. Pressure tests on the pressure hulls after the modifications showed that the original operating depth of 2,000' was still valid. After the pressure hull modification, IUC had plans to replace the aerospace instrumentation and equipment with approved and tested underwater hardware and to simplify the instrumentation and control system. As part of this program in 1981, brushless d.c. motors were installed. About 80% of the planned refurbishment was complete when the Canadian submersible *Pisces* was acquired by IUC. Since the *Pisces* best met IUC's operational requirements, *Beaver MK 4* was sent to the Man in the Sea Museum in Panama City, Florida after approximately 160 lifetime dives.

Planned originally for offshore oil field work, the *Beaver's* initial nickname was "Roughneck." When contracts of that sort did not materialize, it was put to other uses. By 1970, the Nixon Administration made it clear that they would not support a wet-NASA. North American Rockwell then sold *Beaver* to a commercial underwater contractor. Later, much of the off shore oil demand for submersibles developed. *Beaver 4* was just a little ahead of its time.

# *Ben Franklin*

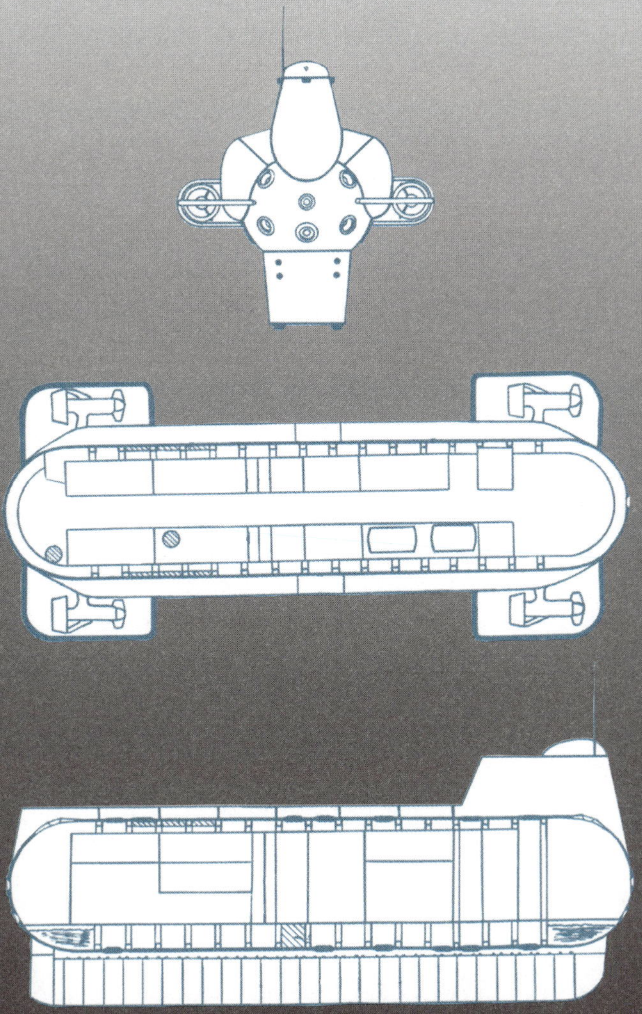

***1968–1971***

# Ben Franklin

Although Spanish ship masters had learned to take advantage of the Gulf Stream in the mid-1500s, the first charts of this river within an ocean were prepared at the direction of Ben Franklin who was Deputy Postmaster General at the time. A Nantucket sea captain had told him that whalers and merchant ships all sailed the Gulf Stream going east and avoided it coming west. Whereupon, Franklin had it drawn in on all Atlantic sailing charts used by mail packets between England and the colonies. Later, of course, he contributed to and was a signatory of the Declaration of Independence and spent the war years in France arranging for arms, credits, and ships, as well as cementing France's alliance with the new Republic. On his return home, at 80 years of age, he busied himself taking water temperatures so as to locate the boundaries of the Gulf Stream in passing.

One hundred and eighty-seven years later, in 1965, the Grumman Aircraft Company of Bethpage, New York approached Jacques Piccard to get his ideas on an undersea vehicle project. One idea that Piccard had which Grumman liked was for a submersible to make a month long drift in the Gulf Stream. In a single mission, extensive biological and physical information could be collected on the Gulf Stream and the sea floor beneath it. Grumman, of course, was not a public charity dedicated to studying the oceans. Long successful in aviation and more recently in aerospace, Grumman's corporate planners had noted *Trieste's* dive in 1960, the Deep Submergence Systems Project that followed *Thresher's* loss in 1963 and, following Kennedy's death, the many promises and predictions that Vice President Hubert Humphrey made to the effect that a "wet NASA" was next. Nor – as we have seen in *Deep Quest* and *Beaver* – was Grumman the only aviation/aerospace company to seek to position itself in a three-prong strategy to include undersea vehicles. Moreover, Grumman thought that the long term environmental aspects of a crew working in isolation might help them gain the experience and credentials needed for a contract proposal involving a long term manned space mission called Sky Lab, which they were planning to bid on at a later date — and eventually got.

In February 1966, Piccard visited the Grumman plant on Long Island, NY, where he met Walter Scott, the highly professional Director of Grumman's Ocean Systems Division and future manager of the Gulf Drift submersible project. Due to his previous work in Europe, Piccard asked to have the submersible built there with assembly in Switzerland. Grumman agreed on the condition that the American Bureau of Shipping (ABS) follow the construction and that each step met with ABS approval. Jacques's experience with his late father, the famous submersible designer, was considered as a possible asset to the project. The Auguste Piccard tourist sub had been sold recently at auction for a mere $100,000, so it was estimated that the younger Piccard might be anxious to reestablish his reputation. Walter Scott made another stipulation, that the submersible be named *Ben Franklin*.

The design function of the submersible was to drift, completely encapsulated the whole time, inside the Gulf Stream along the U.S. East Coast for 30 days at various depths. The scientific purpose was to make *in situ* measurements of all oceanographic physical functions (temperature, salinity, oxygen content, density, sound velocity, and so forth) and to observe, photograph, and in some cases, sample the biological material in the water near the submersible during the transit. Crew capacity was to be monitored with regard to both psychological and physical factors.

The Gulf Stream is made up of "two rivers" of tropically heated water. One pours through the Florida straights, south of Miami, from the Caribbean and then bends northward along the coast of Florida. The other flows northward to join it, off Florida, from the warm Atlantic equatorial waters. The two currents combine about midway up the coast of Florida, off shore from Cape Canaveral, at the outer edge of the Continental shelf. The heated current then flows about Northeast and parallel to the East Coast of the U.S., turning eastward at Cape Cod, Massachusetts toward Europe. A slower counter current between the Gulf Stream and the U.S. East Coast is made up of chilled Labrador waters creeping down from the north between the land and the western boundary of the Gulf Stream. The Gulf Stream varies in width between approximately 50 and 95 miles and runs on an average of about a mile deep. The general velocity is between 3 and 5 kts.; the volume is greater than that of several hundred Mississippis or that of all the land rivers in the world. In most places along the coast, the cold water and warm water can be distinguished from one another by their difference in color. The difference in temperature at the interface of the warm/cold waters is sometimes twenty degrees within a 200' distance.

The submersible was to have a pressure hull that was less compressible than sea water, so that once a desired depth and buoyancy status was attained, the submersible would not tend to drift vertically. Economy and the mission purpose set the maximum operating depth at 2,000' with a two ton payload capable of carrying six people submerged six weeks. The actual Gulf drift was planned for only 30 days, but redundancy was included in the design of all systems. Some additional features of the submersible were: The overall length was 49', beam 20', height 21', weight in air 15 tons, cruise speed 2.5 kts., and maximum speed 4 kts. The pressure hull was cylindrical with hemispherical ends, with an O.D. of 10'-4". The hemispheres and cylinders were fabricated of German steels similar to the Navy's H.Y. 80. Box-shaped internal ring stiffeners, instead of the usual "I" shaped stiffeners were approved, with a spacing of 27 1/2 inches apart.

The cylindrical hull, for practical reasons, had to be made in six segments, which looked like thick slices of baloney

***Name:*** *Ben Franklin* ***Builder:*** *J. Piccard* ***Length:*** *49'* ***Maximum Depth Rating:*** *2,000'*
***Capacity:*** *6 Man* ***Year Launched:*** *1968* Courtesy of Don Kazimar

prior to final assembly by welding. Four of the cylindrical segments welded together and to the forward hemisphere comprised one piece of the hull. Two of the steel segments welded together and to the other end hemisphere made up the second piece of the hull. The two assembled pieces were heat-treated individually and then bolted together by a connecting flange provided on each piece. The bisection of the hull and bolted flange assembly was necessary due to the limited size of the Swiss annealing oven. Twenty-nine conical shaped viewports were located throughout the hull. They all had an inside diameter of 6" and were 3" thick. Thirteen thickened penetrator areas were located on the upper and lower center line of the pressure hull. Seventeen to nineteen holes were in each penetrator area for various combinations of electrical, pneumatic, or hydraulic pass-throughs. Two spring loaded 30" hatches were located in the pressure hull. Both hatches are on the topside fore and aft centerline in the hemispherical end closures. The forward hatch provided access to and from the sail.

There were four externally mounted 25 H.P. reversible motors, with four-bladed propellers that could be pivoted 350 degrees for propulsion. Twin rudders, which could pivot 40 degrees to the left or right, augmented the motor thrust for turning, especially during cruise.

Power for the submersible was obtained from lead acid batteries totaling 26 tons in weight at 20% of the total weight of the submersible. The main batteries were located in the keel, while the emergency batteries were located adjacent to the head (toilet) inside the pressure hull. The hundreds of connections between the external batteries and the hull pass throughs, along with the hundreds of yards of cabling in sea water, were both potential and actual leak problems later. The life-support system for the submersible included 922 lbs. of liquid oxygen (LOX) in two cryogenic containers. Lithium hydroxide panels were used to scrub out the carbon dioxide in the air. Many 10 pound bags of silicate gel were used to remove moisture from the air. One project member said they were hung out everywhere in the submersible "like an old maid's laundry." Due to the long period of encapsulation, many contaminates could build up enough to require removal. Six different air contaminate removal systems were employed. The detection and measurement of contaminates was accomplished with a Draeger type gas detector, with 40 different reagent tubes for detecting different gases, many of which detected more than one contaminant.

A freshwater storage tank stored 4,600 gallons. Four huge thermos tanks stored a total of 1,600 lbs. of water, that was boiling when filled, to be used for reconstituted freeze-dried foods, hot tea, and the like. The main ballast system consisted of four fiberglass tanks outboard of the pressure hull which could provide 48,000 lbs. of buoyancy, when blown free of water with high pressure air for surface operations or emergency lift at any depth. Two pressure resistant steel tanks in the keel provided up to 6,800 lbs. of variable buoyancy for depth control. The emergency ballast of six tons of droppable iron shot, was controlled by electromagnetic valves.

After the hull was completed at Monthey, Switzerland, in the Giovanola plant, the submersible was assembled at Lausanne, Switzerland. On 4 April 1968, the project was shipped by rail to Antwerp, Belgium then by ship to Palm Beach, Florida. Thirteen of the original design and construction crew, along with their families, moved to Florida with the project to prepare the submersible for launching. Upon their arrival in Palm Beach, Walter Scott put his best engineers to work on several systems which he judged would not operate the projected thirty days without extensive maintenance or failure. The miles of electrical cable and hundreds of connectors and pass-throughs were their biggest headaches. In desperation, a silastic compound was coated on all the external wires, connectors and pass throughs. Drastic measures were necessary to obtain the required reliability. (Failure during the test stages alone would require $5,000 per lift out of the water for repairs.) The environmental system was improved on by adding numerous and varied contaminant removal systems and sensors. Recreational pastimes were added by the crew, some of which were used later in long duration space projects.

After a very busy two months of checking, rechecking, tuning up, and repairing, the submersible was launched for the first time on 25 July 1968. It was christened the *Ben Franklin* at ceremonies on 21 August 1968 near Riviera Beach, Florida. Rear Admiral O.D. Waters, the Oceanographer of the Navy, made an observation at the ceremonies when he told an audience "... for the first time in history a submersible will move in and be able to study the same cubic mile of water for its entire voyage, in addition to making point-to-point checks on the changing ocean bottom." After the christening, but before the thirty day drift, families, friends, engineers, and shop workers were able to make brief dives in *Ben Franklin*. The first dive was made to 40' of water in the harbor at West Palm Beach, Florida in November, 1968. Every imaginable means of rescue had been thought of and prepared for. Check lists were followed before the crew entered the submersible as well as afterward. It was the largest machine ever built by Grumman

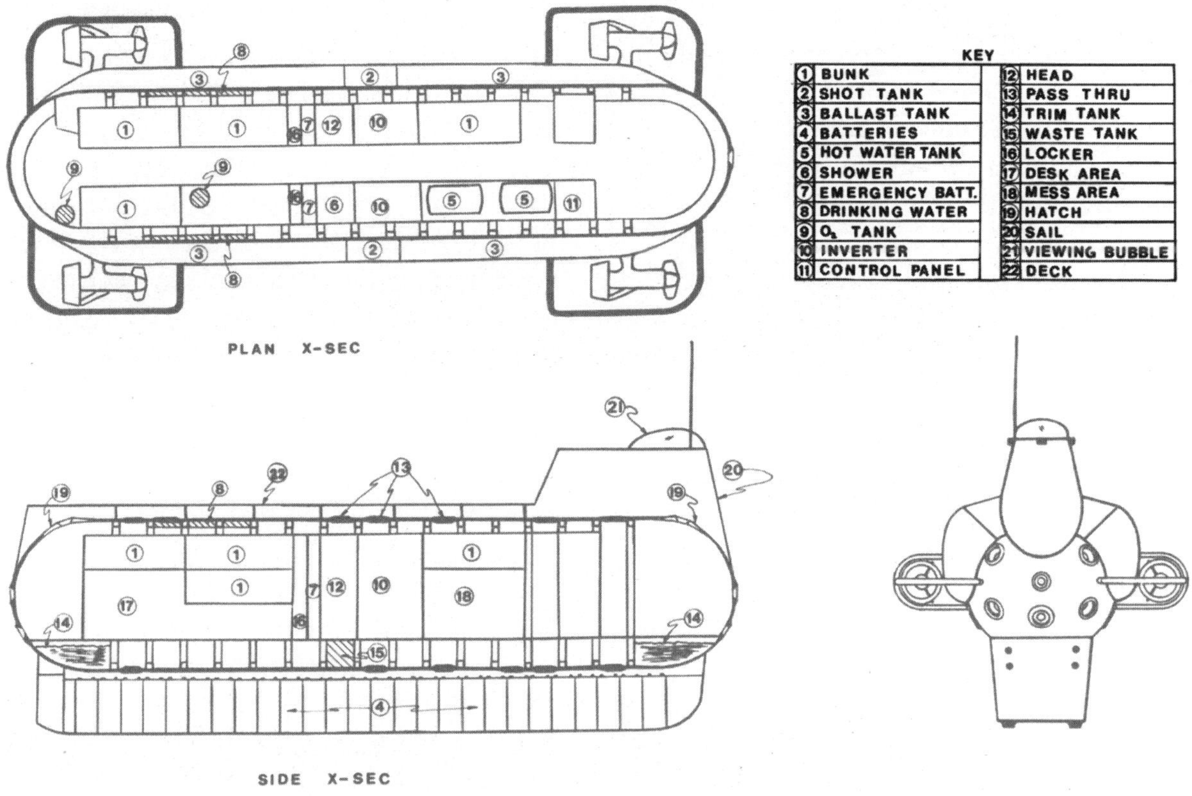


*Detailed Top, Side, and Rear View of Ben Franklin*

and every precaution was taken. For the first dive, two and a half hours were spent checking each system out after the hatch was closed prior to diving.

During the winter of 1968-69 and spring of 1969, many more sea trial dives were made to prove the systems and train the personnel. On 11 December, a three day dive on the harbor bottom was initiated, with each of the six man crew taking a two hour watch. The freeze-dried food was sampled and found boring and dull. Communications, life support, and crew compatibility were briefly sampled with no particular problems. About 40 dives for certification and testing of all systems were made before the Gulf drift expedition began. Certification dives included those for ABS and the U.S. Navy. In February 1969, an open ocean dive to 500' was made. A Thyristor inverter failed and the dive was aborted for repairs. In the next dive, all 1,650 lbs. of ballast in the shot hopper were dropped to check the emergency system and the resulting rapid ascent. The first strain gaged dive was made to 825', followed by incremental dives until a depth of 2008' was reached without abnormal strain data appearing. In early July 1969, Jacques Piccard suggested that a three day dive be made and, if successful, the 30 day dive be made by adding 27 more days to the mission. As the hurricane season was approaching, a few days of delay in starting the big dive might cost months of delay waiting for a break in the weather. Everyone wanted to get started on the Gulf Drift. Grumman agreed. After a 7 1/2 hour tow to the middle of the Gulf Stream, *Ben Franklin* was cut loose and on its own. At 8:34 P.M. on 14 July 1969, the hatch was closed.

The crew for the dive was senior, experienced, and multi-national:

- Don Kazimir, 35 years of age and the youngest on board, as Captain
- Erwin Aebersold, a Swiss, as Pilot and Assistant to the Captain
- Frank Busby, an oceanographer working for the Office of the Oceanographer of the Navy in Washington D.C.
- Ken Haigh, an oceanographer and member of the British Royal Navy assigned to a two year detachment service with the U.S. Navy
- Chet May, a NASA engineer and observer with interests toward future NASA long duration space projects like Skylab
- Jacques Piccard, the submersible designer with the title Team Leader

The escort vessel was the *Privateer*, which was hired by the Navy from the Reynolds Aluminum Company. Before entering the Gulf Stream, the crew and submersible spent the first night on the bottom at 1,640' moving at a mere 1/10 kt.

**Day 1 (15 July):** *Ben Franklin* was brought up to 1,300' to catch the current and start the trip. The sonar was a continuous transmitting frequency modulated (CTFM) type which scanned nearly a mile ahead of the submersible to warn of objects in the way. The average depth during the trip was to be about 650', so hazardous objects should not occur except during bottom excursions.

The off-duty crew had awakened to an inside temperature of 63°F while outside the ocean was only 44°F. A 1,500 mile, 30 day voyage lay ahead, so expendables, not time, were in short supply. To save expendable ballast, buoyancy, and compressed air, all ascents and descents had to be made in a very slow manner. After any change in depth, the submersible's hull required time to adapt to the temperature of the surrounding water. Since the water temperature could vary as much as 20 degrees within the 2,000' depth capability of the submersible, sudden changes in depth resulted in jockeying the ballast and buoyancy inefficiently while waiting for the hull to change temperature. Long stays in 44° deep water soon chilled the interior of the sub to very low temperatures, since no insulation existed on the hull and *Ben Franklin*, unlike Navy submarines, was severely power limited. At the surface, expendable bathythermographs were dropped over the side every few minutes of the day and night to ascertain the water temperature at all depths. This was an aid to the submersible pilot for selecting his operating depth and confirming that the submersible was in the Gulf Stream at all times.

**Day 2 (16 July):** communications were interrupted with the surface ship and a second underwater communication unit was employed. It turned out that the surface ship *Privateer* had lost the location of *Ben Franklin* and was too far away to communicate. On 16 July, the surface ship relayed the *Apollo 11* moon shot lift off information from Cape Canaveral to the submersible. It was becoming noticeable that the bow of the drifting *Ben Franklin* was almost always stern first or 180 degrees out of position. It swayed slowly from left to right, sometimes 90 degrees, before returning to its reversed orientation. Once the depth was selected and the buoyancy adjusted, the depth did not vary more than plus or minus six feet.

**Day 3 (17 July):** two of the four

thermos tanks had lost most of their heat, but all oceanographic measuring devices were working and recording correctly. Busby had assumed the night watches and Haigh worked the day shift. A motor in the life support system had failed and a request for instructions for its replacement with a spare on hand were called up to the surface for relay ashore. Since this was the third day and all was going reasonably well, approval for 27 more days was requested and received.

**Day 4 (18 July):** a deep dive was made to the bottom at 825'. While Busby was taking measurements and photos of the bottom, a 6' swordfish attacked the hull, just missing a viewport. Several of the crew came down with colds, probably due in part to the chilly, moist environment and changing temperatures of the inside of the submersible. Kazimir had a suitcase full of remedies that were administered to anyone willing to take them. When the floodlights near the viewports were turned on for about an hour, a myriad of plankton gathered in a dense mass near the lights. The biological mass usually included: "copepods; small crustaceans with oar feet; euphausid shrimp; krill-like crustacea; salpas; primitive chordates in chain formations; and arroworms, transparent rods that shoot like arrows a few inches at a time."

**Day 5 (19 July):** the sub was at 720', the water outside was 64°F; inside, it was a cool 61°F. The taped music from the only player on board became boring even when songs were playing. Exercising became popular, partly due to the room temperature. The submersible crew was told by topside that they were right in the middle of the Gulf Stream. The lithium hydroxide panels were changed since the $CO_2$ reached 1.5%. Internal microbe specimens were collected every day and cultured in broths. External plankton was also collected through an inlet device in the hull.

**Day 6 (20 July):** the 6th day was a Sunday. At 4:20 P.M., the *Apollo* people were reported walking on the moon. There was one Bible on board, but at least one atheist, so the sanctity of the day was ignored and the Bible made available to anyone interested in private reading. A measurement of the ocean chlorophyll and mineral content was taken as part of a routine sampling program. A summary of the vessel's drift path over the previous days was given to them by the surface ship. It indicated that they had been traveling about 12 miles east (seaward) of the 100 fathom depth line and in the flow of the Gulf Stream which is bounded on the shore side by the continental shelf and a counter current.

**Day 8 (22 July):** *Franklin* made a descent to drift along the bottom for a planned 24 hour stay in 46°F water, which was on the bottom at 1,200'. They were drifting slowly when the sonar showed a bottom object ahead, so they lifted the vessel 100' above the bottom to avoid it. Later deliberations suggested it was a fossil coral head. As the temperature outside was soon closely matched by the inside temperature of the submersible, the crew cut the 24 hour stay short. They slowly ascended to warmer water as their cruise depth. The freshwater supply was found to be contaminated and full of "wiggly bugs," so it was decided to cease the washing of clothes, use the contaminated water for showers, drink from the cooled down thermos tanks, and save the remaining hot thermos tank for tea and dehydrated food. It was further decided that future visits to the bottom for geological observations would be made in very brief excursions in order to minimize the misery that came with depth.

**Day 9 (23 July):** the off duty crew awakened to see the depth charts filled with great amplitudes of between 100' and 150' of depth variations, due to the effect of

*Individuals were tending to eat alone and read in their bunks more, instead of shooting the bull with others as much. Humor was less prevalent, and some were getting testy over insignificant events or topics.*

internal waves. These deep-water swells (not really waves like those at the beach) are the interface between the light surface waters and the dense deeper waters, which act a little like oil on water and often have much larger amplitudes than those interfaces between air and water.

**Day 10 (24 July):** *Franklin* crossed the Navy submarine training area. Seeing a submarine on sonar, they gave it a call since both underwater vehicles used the same frequency. The nuclear submarine *Lapon* responded and a brief conversation ended with a "bon voyage."

**Day 11 (25 July):** the surface ship *Privateer* notified *Franklin* that they had drifted out of the Gulf Stream 18 miles toward shore. After almost a day of trying to power back into the stream, *Franklin* surfaced and was towed with the hatches closed, 30 miles back into the middle of the current. The crew found the long day of being towed at the surface with sunlit portholes quite a pleasant change from the mole-like existence they had endured.

**Day 13 (27 July):** the day was without special events, minor changes in the crew were becoming more apparent. The need for privacy was beginning to be expressed in subtle ways. Individuals were tending to eat alone and read in their bunks more, instead of shooting the bull with others as much. Humor was less prevalent, and some were getting testy over insignificant events or topics.

**Day 14 (28 July):** *Franklin* reached a depth of 1,900' in 4,600' deep water. A battle against bacteria inside the submersible was being waged. Disinfectant had to be placed almost everywhere: walls, floors, tables, and the shower stalls. Things topside were different, but not any better, as the *Privateer* faced stormy weather and rain squalls with pitching seas.

**Day 16 (30 July):** a lengthy report was floated to the surface in one of the 6" diameter plexiglass spheres which were ejected from the SAS (a French acronym for the small lockout chamber on board). In it, reference was made to the thousands of pictures inside and outside the submersible that had been taken as well as literally millions of oceanographic samples and measurements made.

**Day 17 (31 July):** Don Kazimir, the skipper, decided that there was enough spare battery power on board to reheat the hot water in thermos tank #3 (the hottest one remaining) primarily for hot tea, but also to make the freeze dried stuff taste a little better if possible. Seeminly insignificant little things had become very important and hot tea had become one of the few amenities remaining. The crew was informed from the surface that they were passing Cape Hatteras some 42 miles away.

**Day 18 (1 August):** this was a Swiss holiday, celebrated all over Switzerland by the burning of huge bon fires. It was determined that the life-support system and environmental control management had sufficient leeway for the two Swiss members on board to light one match between them and to sing their anthem.

**Day 19 (2 August):** the crew discovered that during the night the submersible had averaged 5 kts. Hurricane Anna was reported 90 miles southwest of their position, just off Cape Hatteras.

**Day 20 (3 August):** the sea at the surface had turned rough with 40 knot winds while inside *Franklin,* all was calm and serene.

**Day 21 (4 August):** *Franklin* averaging 3.2 kts. and headed for Nova Scotia, so far from land that they were unable to estimate what port they would be taken to after their trip.

**Day 22 (5 August):** a school of large tuna swam by while they were at 600'.

**Day 23 (6 August):** the carbon monoxide (CO) reached 30 parts per million, exceeding the intended maximum of 25 ppm set before the dive. Someone on board had the then best selling fiction novel *Ice Station Zebra,* which describes life aboard a nuclear submarine. In the book, it is stated that 50 ppm was permissible, so the crew on *Ben Franklin* stopped worrying about their CO reading.

**Day 24 (7 August):** a surface aircraft supporting them reported that visually, they were exactly in the middle of the Gulf Stream.

**Day 25 (8 August):** Hurricane Anna had dissipated to a breeze. The seas were calm at the surface and *Franklin* averaged 3.3 knots per hour for twelve hours. The carbon dioxide level had reached 1 $^1/_2$ %, so the lithium hydroxide panels were once again changed.

**Day 26 (9 August):** another deep descent was made to check the current velocity at 1,650'. The four hour velocity average was 3.0 kts.

**Day 29 (12 August):** the surface weather had picked up with very high swells and waves into a choppy sea state. The submersible crew was told that the Coast Guard vessel *Cook Inlet* was on its way home from North Atlantic duty and would pass close-by on Day 30 to assist with disembarking and transportation ashore.

**Day 30 (13 August):** the final day was used to prepare for emergence at 8:34 P.M., which would be exactly 30 days since the hatch had been closed and dogged down. The carbon monoxide was at 40 ppm. *Atlantis 2* from Woods Hole was overhead to help celebrate the end of the trip. As the magic hour approached, everyone agreed to wait until 0800 the next morning to disembark since the late hour and faint light would, with all their scientific equipment, add hazard to transfer by rubber boat to the surface ship. The next morning, on schedule, the six man crew emerged like prisoners from low light dungeons to see a fleet of ships with crews cheering at them. Besides The *Cook Inlet*, the Navy-rented *Privateer* and the Woods Hole *Atlantis 2*, there was Grumman's scientific instrumented ship *Lynch* to welcome them. Later, the *Privateer* began the lengthy tow of the *Ben Franklin* to the port of New York. The crew arrived in Portland, Maine, to a waiting crowd of family and well wishers.

A report, given by Frank Busby to his Washington office, midway through the dive, gave a pretty good indication of some of the oceanographic data accumulated from the Gulf Drift expedition. In addition, interior cameras shot stills every two minutes for thirty days to record the crew's disposition and status. As one crew member put it, "the shrinks had a ball and asked questions so stupid and personal that I'm sure you have to be nuts to be a shrink." One of the disappointments expressed by Piccard was that

*Franklin* never ran into or through the Deep Scattering Layer (DSL), the layer of plankton and nekton that floats in layers in the sea at 900' to 1,200' during the day and migrates to near the surface at night.

Frank Busby's midway report to the Naval Oceanographer read:

- Measurements of salinity, temperature, and sound velocity at different times and depths; over 2.0 million
- Total accumulated time of measuring bottom currents; 6 hours
- Pairs of stereoscopic bottom photographs; 848
- Total accumulated time of measuring ambient light; 48 hours
- Accumulated time, measuring gravity anomalies;14 hours
- Miles of recordings of side scanning sonar; 3.5
- Experiments in acoustic reflections on the bottom; 425
- Hours of direct ocean observation at depths to 2,000'; 90

After the Gulf Stream expedition, the submersible was deployed to Atlantic City for a series of dives to study the sewage outfall on the slope of the Continental Shelf. This particular area was abundant with life. The observers interested in the outfall spent two days making observations, including a night on the bottom in the bunks, before returning to shore.

In Florida, a series of dives off Miami was made for geological and biological investigations. On board was Dr. Bob Ballard who had brought with him a borrowed manipulator from *Alvin* for collecting samples to go along with pictures taken. Another series was conducted off Cape Canaveral for Florida Atlantic University to study marine life. About 100 dives and 1.5 years after the Gulf Drift Expedition, a series of publicity dives was made in the Bahamas with the Explorer's Club members as passengers. On one of the nights during the series, the *Ben Franklin* was tied off as usual to a nearby surface ship at anchor in the harbor. The surface ship broke loose from its anchorage during the night and was blown ashore. The night watch on *Ben Franklin* was not on deck, so the submersible was blown ashore along with the surface ship, smashing the batteries in its keel.

Grumman's research budget for this purpose was exhausted, so they sold the submersible to the Horton Maritime Group. Horton cannibalized most of the equipment on board for use in the *Auguste Piccard*, which they had acquired previously. The exterior shell of the submersible *Ben Franklin* now sets in the public park in Vancouver, B.C. Canada.

Admiral Waters expectations were met by the *Ben Franklin* Gulf Drift. Grumman obtained data helpful toward obtaining a contract for Sky Lab. The Naval Oceanographic Office accumulated *in situ* oceanographic data unlike any other. The various long term human encapsulation projects obtained additional data to add to the meager amounts existing at the time. *Ben Franklin* was, in the final analysis, a valid and interesting milestone in the history of deep submersibles.

# *Sea Cliff*

***1968–1998***

# Sea Cliff

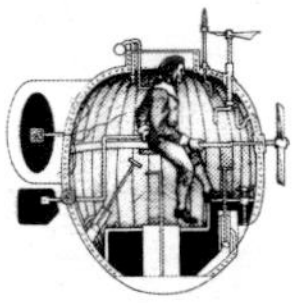

Soon after the original contract from the Office of Naval Research (ONR) was let to General Mills for the construction of the *Alvin* submersible, an amendment was made to it calling for the construction of three additional personnel spheres to be made of HY-100 (100,000 psi) steel. When spheres #1 and #2 were pressure tested, the strain gage data indicated that they both were much stronger than the minimum specified requirements for a 6,500' operating depth submersible. One of the tested hulls was fitted out for the *Alvin* project, the third hull was to have been pressurized to destruction but the test chamber failed at just below the equivalent pressure of 10,000' of sea water without harming the hull. It was evident that, without further testing, two very good spare submersible pressure hulls would be available to allow construction of two new submersibles for the operations and maintenance at the Navy's Atlantic Undersea Test and Evaluation Center (AUTEC) in the Bahamas. A contract was let for the construction of both submersibles based on *Alvin's* design, but with a few adaptations for range uses.

The Electric Boat Division of General Dynamics won the contract and *Autec 1* and *Autec 2* were launched on 11 December 1968 at Groton, Connecticut. About three dozen test dives were made by *Autec 1*, including one to its operating depth of 6,500', before it was assigned to Submarine Development Group 1 in San Diego in October, 1970. *Autec 1* and *Autec 2* became *Sea Cliff* and *Turtle*. Since they were identical when launched in 1968, *Sea Cliff's* original description, therefore, applies to *Turtle*.

*Sea Cliff's* spherical pressure hull was 1.33" thick with a 7' outside diameter, made of HY 100 steel in two hemispheres, welded together to make a sphere with an operational depth of 6,500'. Ballast/buoyancy was controlled by the flooding of the two main ballast tanks for submerged operations, and varying the oil content between four variable ballast tanks and two oil-filled bladders for minor adjustments in ballast/buoyancy. Propulsion is accomplished by a single stern propeller with shroud, powered by a 15 H.P. reversible hydraulic motor. The stern thruster can be pivoted 45 degrees left or right of center. Two four H.P. reversible electric motors power the shrouded propellers located midship, which pivot together in 360 degrees of rotation. The combined propulsion permits 1 kt. velocities over eight hour periods, or up to 2 1/2 kts.

***Name:*** *Sea Cliff* ***Builder:*** *Electric Boat* ***Length:*** *26'* ***Maximum Depth Rating:*** *6,500–20,000'*
***Capacity:*** *3 Man* ***Year Launched:*** *1968* Courtesy of Naval Photographic Center, Washington, D.C.

over a one hour period. Power was originally supplied by a 60 volt and 30 volt combination of lead acid batteries. Life support was originally supplied for 144 man-hours for a crew of three with an emergency supply for an additional 72 hours.

Five viewports — identical to *Alvin's* — provided viewing. One viewport was in the hatch with an inside diameter of 2 1/8". There were four main viewports in the personnel capsule: one dead ahead at the sphere's equator, one on the centerline at an 80 degree down angle from the horizontal, and one each on the port and starboard quadrants at 40 degree down angles. The four main viewing ports had a 5" inside diameter. The overall length of *Sea Cliff* was 26', its air weight was 24 tons, and its payload was 300 lbs. Jettisonable equipment included two manipulators, three banks of batteries at 3,400 lbs., and the personnel sphere at 1,800 lbs. of buoyancy, with emergency power inside.

*Sea Cliff* and crew completed their first operational tests successfully for Submarine Development Group One during deep water retrieval exercises in September 1971.

By December 1973, a total of 70 dives had been accomplished in the San Diego area in readiness for operational deployment.

From June through September 1974, *Sea Cliff* was deployed to the Hawaiian Islands for 40 dives to as deep as 6,432'. Fifteen of these were conducted at the Pacific Missile Range Facility at Barking Sands, Kauai, Hawaii. *Sea Cliff* located and recovered a MK-48 and a MK-45 torpedo, a current meter array, and a deep ocean transponder (DOT). It also inspected ten range hydrophones and conducted an intruder detection exercise.

Twenty-five dives were conducted for marine geology investigation for the Office of Naval Research, Woods Hole, Scripps, and the U.S. Geodetic Survey. The dives were for investigating submarine volcanic processes forming the Hawaiian Islands. Dives were made at several locations in the vicinity of Kauai and the big island of Hawaii. Submarine features investigated included lava tubes and pillows of various ages on slopes of 60-90 degrees, as well as previously unknown fault systems and columnar-jointed dikes. On the Puna Ridge near Hawaii Island, recent fissures of up to 30' wide were seen as well as vertical pillow walls as tall as 150' and as wide as 20'. The features appeared similar to many found in the mid-Atlantic Ridge area during project Famous. Detailed comparisons are planned.

A return to Hawaii was made in October 1974, with the dive series lasting until January 1975, in a continuance of the previous work there. Local San Diego dives were made until August when a dive series began for the search for an ambient noise and detector system array at Barking Sands, Hawaii. Additional dives were made to search, recover, or photograph lost torpedoes, current meters, and the like.

In June 1976, military hardware was photographed and its critical parts recovered off San Nicholas Island, California. In December 1976 and January 1977, dives were made to search for and photograph concrete chambers and spheres belonging to the Civil Engineering Lab at Port Hueneme, California. Thirteen of 18 spheres left on the sea bottom for time-related structural deterioration observance were located and photographed. Four out of five of the designated spheres were recovered. Underwater connectors were tested successfully by remotely making nine connection and disconnection tests.

After a dive series off Southern California in 1978, *Sea Cliff* returned to the Pacific Missile Range (PMR) at Kauai for hardware recovery and scientific operations lasting through the end of the year.

In February and March 1979, services for the Civil Engineering Lab at Port Hueneme were continued for inspection of a mock-up undersea habitat and concrete test spheres. A biological survey was made in Coronado Canyon and scientific dives were made for the Office of Naval Research and Scripps Institute of Oceanography in May. In September *Sea Cliff* recovered a Tomahawk missile near San Clemente Island in 6,500' of water for the Pacific Test Range. *Sea Cliff* returned to the missile range on Kauai in October and November for services to PMR and to perform volcanic origin research.

In March and April 1980, *Sea Cliff* was diving again for the Civil Engineering Lab at Port Hueneme for the recovery of concrete samples aging on the sea bottom and underwater connector testing. In April, *Sea Cliff* tested a submarine rescue reel system off Southern California. In July, a lengthy series of dives for biological and geological research was made for Scripps Institute of Oceanography, Michigan Institute of Technology, the University of California at Santa Barbara, and the Smithsonian Institute.

In January and February 1981, a search was conducted for a lost cruise missile off the Southern California coast. In March, the Civil Engineering Laboratory concrete spheres were again inspected and photographed and the underwater connector tests were continued. In May, a WWII Navy torpedo bomber, a TBM, was located and photographed for the Naval Oceanographic Office.

The year 1982 was spent in local dives and inspection of projects for the Civil Engineering Laboratory.

During January 1983, services were performed for the Office of Naval Research, sampling and surveying ecological communities off Santa Catalina Island. After dive #510 on 1 April 1983, *Sea Cliff* went to the Mare Island Shipyard for a complete overhaul and conversion to a 20,000' capability.

The completed conversion of *Sea Cliff* to a 20,000' operating submersible brought about many major changes to the vehicle. The new spherical shaped titanium hull is 7' in diameter with a nominal thickness of 2.9". The overall length of the submersible increased to 31' with the weight in air increased to 60,000 lbs., which includes 15,000 lbs. of syntactic foam. The number of viewports in the titanium sphere changed from five to four, one in the hatch, one forward and one on each side. The life support system was increased to 216 man hours. The electrical energy available was increased by over 50%, using silver zinc instead of lead acid batteries. The electrical cables are all oil filled. The main ballast system for surface buoyancy remained the same except that the old high pressure air tanks were replaced with titanium tanks and all piping is now titanium. A new seawater variable ballast system now consists of two titanium hard tanks with a high pressure sea water pump to fill or empty it. A new system of descent/ascent weights was added with 544 kg. (1,200 lbs.) available for dropping to descend before diving and 363 kg. droppable before ascending after a dive. An additional 30 drop weights of 9 kg. each are available as variable ballast. The stern and side pod thrusters are basically unchanged.

In April 1984, the first three dives after overhaul went to increasing depths finally reaching 15,000'.

In March 1985, *Sea Cliff* made its first 20,000' dive, during a dive in the Middle-America trench off the coast of Central America. During a 15 dive series in June, *Sea Cliff* recovered a very large Marine Corp CH-53E helicopter off San Clemente Island California.

In February, 1986, a mutual rescue agreement was signed between France and the U.S. for rescue of the French *Nautile* by *Sea Cliff* or visa-versa. The agreement cuts the response time for rescue assistance, if either one of the submersibles becomes disabled in deep water. During science dives in the summer, manganese nodules of about 6" diameter were retrieved at a 6,000' location ten miles south of San Clemente Island, near a hydrothermal vent, at the base of a large escarpment fault. During July and August, dives in the Gorda Ridge area were conducted off the northern coast of California. In January 1987, recovery and deployment of materials and structures were made for the Civil Engineering Laboratory off Port Hueneme.

In February, 1988, the 72 meter (236') long *Laney Chouest* was assigned as support ship for *Sea Cliff*. In March, *Sea Cliff* recovered the nose section of a MK-47 exercise torpedo. During July, *Sea Cliff* made a successful series of certification dives to 20,000' aboard *Laney Chouest* in an area northeast of Hawaii. The Gorda Ridge fracture zone was revisited in August and September, sponsored by the U.S. Geodetic Survey, the University of Washington, and the Department of the Interior. Unusually strong currents were encountered. The underwater vents in the area measured 500°F. There were many black smokers and the organisms near the vents and smokers were prolific beyond description.

*During July and August, 1992, Sea Cliff conducted operations in support of National Geographic's "Iron Bottom Sound" documentary, a photo documentary of the numerous Japanese and Allied war ships sunk in the battle of Iron Bottom Bay in the Solomon Islands in WWII.*

---

Near San Diego in February 1989, *Sea Cliff* recovered an SH3 helicopter and photographed a pre-WW II, F3F plane in April. *Sea Cliff* was deployed to Hawaii for classified range services to the Pacific Missile Range Facility.

In June and July 1990, the Monterey Bay Aquarium Research Institute (MABARI) brought *Sea Cliff* to the Monterey Bay area to recover lost scientific instrument packages, to survey a proposed current meter site and a slump area of the Monterey Canyon. Training dives were interspersed between the Institute's dives. On one of the training dives, the remains of a 1935 wreck of the dirigible Macon was found in 1,400' of water. Many photographs of the wreck and two of the four planes it carried were made for evaluation toward possible future recovery.

From September through October, Sea Cliff located and recovered pieces of a United Airlines cargo door at a depth of 14,000', blown off in flight 7 months earlier, 90 miles south, southwest of Oahu. In an 8 dive series, some lasting 18 hours, the missing door and other aircraft pieces were recovered for vital accident analysis.

In September 1991, *Sea Cliff* was deployed to the Farallon Islands near San Francisco to conduct a video survey of dumping sites there. Dives in the Monterey Canyon for a survey of the canyon walls were conducted later in the year.

During July and August, 1992, *Sea Cliff* conducted operations in support of National Geographic's "Iron Bottom Sound" documentary, a photo documentary of the numerous Japanese and Allied war ships sunk in the battle of Iron Bottom Bay in the Solomon Islands in WWII. From September through November, scientific dives off six Hawaiian Islands were conducted for the National Ocean and Atmospheric Administration (NOAA). Two weeks were spent on the Pacific Missile Range Facility (PMRF) in support of the PMRF command requirements for hardware recovery on the range including a MK-50 torpedo.

During 1993, *Sea Cliff* was on restricted availability status for major maintenance work and made no dives.

*Sea Cliff* was back diving on February 1994. The first two dives were sea trial dives to 10,000' after overhaul. There were, in addition, 18 training dives at various depths and locations near San Diego. Two dives were made to 10,500' for recertification after the major maintenance overhaul. *Sea Cliff* departed San Diego in July for inspection of the nuclear waste site off the Faralon Islands. The next dives were made near Pioneer Canyon in the Pacific Northwest, where 47,000 drums of paint scrapings and similar debris had been dumped in 3,700' of water following the tests at Bikini shortly after WW II. *Sea Cliff* completed 26 dives during the year.

After a "complete and extensive overhaul" *Sea Cliff* made a series of local dives in the Southern California area for test and training purposes in August and September 1995. An "entirely new crew" was in training for deployment to the Pacific Northwest later in the year.

*Sea Cliff* was named after the small town on Long Island, New York. After 1985 it represented the deepest diving American submersible and performed the tasks too deep for any others in the U.S. at the time.

# *Turtle (1968)*

***1968–1997***

# Turtle (1968)

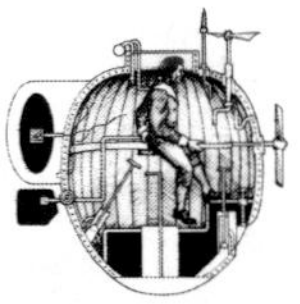

*Turtle*, named after Turtle Town Tennessee, was launched at the same time as *Sea Cliff* on 11 December 1968 at Groton, Connecticut as *Autec 2*. Originally destined to be an Atlantic Underwater Test and Evaluation Center (AUTEC) submersible, it bore that acronym at launch time. The specifications, construction details and operational characteristics of *Turtle*, as originally constructed, were identical to those of *Sea Cliff*.

Basic characteristics that would later be modified included a 7' diameter HY-100 steel hull, lead acid batteries and an air weight of 24 tons. About a dozen test dives were made on the East Coast of the U.S. before *Autec 2* ended up with its reptilian name and the Navy title *DSV-3* at the Submarine Development Group One in San Diego, California in September 1970. During dive #21 on 4 April 1971, the submersible reached its design operating depth of 6,500' for the first time.

In May 1972, the *Turtle* made close up inspections of some large concrete spheres placed on the sea floor in 1971 by the Navy Civil Engineering Laboratory (NCEL) of Port Hueneme, California. The spheres were part of an on-going, long term, "weathering" test of underwater structures and materials by NCEL.

During the fall of 1973, a 35 dive series was completed on a classified project at 6,206' prior to going in for overhaul.

When diving resumed, following overhaul, approximately 50 dives were completed to as deep as 6,200' by the end of 1974.

In the spring of 1975, *Turtle* was deployed to Central and South America on a 39 dive series sponsored by the Office of Naval Research (ONR) with observers from seven major universities. Many of the dives were at or near the maximum operating depth of the submersible at 6,500'. Objectives during the cruise were:

- Determining the the existence of volcanic and non-volcanic seamounts near Puerto Vallarta, Mexico
- Dynamic processes operating in various canyons along the coast of Central America
- Morphology (biological make up) of sediment filled valleys of the Coco Ridge area

- Geological and biological studies and sampling of the slopes near Malpelo Island, Mexico
- Examining and sampling deformations in the Coco Ridge subduction (plate collision) zone

During 1976, the submersible was out of service for six months, but completed 19 dives during the winter at the Pacific Test Range Facility in Hawaii.

The work in Hawaii continued in 1977 as *Turtle* performed maintenance and survey work on hydrophone arrays at Barking Sands, Hawaii. In June, a search was conducted for an electronic communications aircraft EC-130Q that crashed near Wake Island. The plane was located 1,600' off the end of the runway in a debris field filled with many other aircraft parts and pieces, as well as coral heads reaching 100 meters above the sea floor. Recovery of the most sensitive and critical hardware was completed in August.

Mostly local Southern California dives were made in 1978, with a brief series of dives made off Guadalajara, Mexico aboard Lockheed's *Transquest* in April to as deep as 6032'.

In 1979, *Turtle* made 23 dives were made. In May, June, and July, *Turtle* worked at the Pacific Missile Range Facility on Kauai, making hardware recovery and servicing hydrophone installations. A successful search for a lost Tomahawk missile was made in September, in the Southern California area. *Turtle* was in major overhaul at Mare Island from October 1979 to September 1980.

During overhaul, the old *Alvin* HY 100 hull (that had been lost at sea for 10 1/2 months in 1968 into 1969) was installed in *Turtle*. The hull was the best (strongest) of the four hulls originally made for *Alvin*. It became available in 1973 when the original Alvin hull was replaced with one made of Titanium. Since the original *Alvin* hull was thicker than *Turtle's* first hull, the submersible was upgraded for 10,000' operations after coming out of overhaul in September 1980. Life support time was extended to 240 man-hours. The air weight was increased to 25 tons. The propulsion system remained the same. After two equipment check-out dives, the first dive to 10,000' was made in October 1980 in the San Diego area.

After 20 training dives in 1981, *Turtle* attached two lines to the 10 1/2 ton practice submarine hatch used by the DSRVs for training purposes. *ASR-21 Pigeon* made the lift for DSRVs practice mock escape hatch, *Deep Seat*. Another two dozen training dives in the Southern California area were made later in the year.

In August 1982, *Turtle* was deployed to the Hawaiian Islands to provide services to the State of Hawaii, NOAA, ONR, and the Barking Sands Underwater Range Facility. The first dive was made for project OTEC (Ocean Thermal Energy Conversion). The project involved energy generation by the transfer of deep cold water to the surface through large pipes. A total of 23 dives on OTEC included *Turtle's* assistance in the lift of a 100 ton cold water pipe.

During January 1983, *Turtle* recovered hydrophones from the San Clemente Island Range. Operations in the Gulf of Mexico took place in May and June before being stormed out by hurricane weather. Dives off the Hawaiian Islands followed. The first ones were made off the Kona Coast to ascertain the lava flow limits of the 1877, 1919, and 1950 eruptions in the area. The next dives off Oahu Island were for determining the optimum locations for OTEC pipes.

During a dive in August, 1984, an electrical failure occurred and power to the port motor ceased, causing the dive to be aborted.

**Name:** *Turtle (1968)* **Builder:** *Electric Boat* **Length:** *26'*
**Maximum Depth Rating:** *6,500–10,000'* **Capacity:** *3 Man*
**Year Launched:** *1968* Courtesy of Peter Juhos

While being lifted from the water on the Lockheed support ship *Transquest's* elevator, the oil-filled port distribution box, housing the electrical breakers, was seen to be smoking. Oil from the distribution box ignited soon after the submersible left the water. Available fire fighting efforts were ineffective, until the vehicle was lowered back into the water to put out the fire. Damage to electrical wiring and equipment, as well as the syntactic foam was extensive. During overhaul, the fire damage was repaired and many upgrades were made as well. The lead acid batteries were replaced with silver zinc batteries. Propulsion was increased to a maximum speed for 8 hours at 2 1/2 kts. or 20 hours at 1 1/2 kts.

In July 1987 *Turtle* made the first dive after overhaul. On the third dive after the overhaul, *Turtle* was taken to 9,650' as a test. The remaining 39 dives made during the year were mostly for training operators, plus several "deep seat" hardware recoveries.

The first dive in January 1988, involved searching for and recovering a MK-27 mobile target near San Clemente Island (SCI). In March, one MK-27 target and four Deep Ocean Transponders (DOTS) were recovered for the Naval Undersea Warfare Engineering Station near SCI. In April, two DOTS and a string of current meters were recovered. In July, a survey of hydrophones on the MK-48 range was made. In August, *Turtle* conducted a survey of the *Deep Seat* training fixture used by the DSRVs for practice submarine rescue. The twentieth anniversary of *Turtle's* launching occurred in December.

After eleven months stand down for standard maintenance, as well as repairs caused by damage from a handling crane, *Turtle* regained operational status in August, 1989. Dives to search and photograph a lost F-14 in a debris field in 4,000' of water were made in September at San Clemente Island. Search for and video coverage dives of a UH-1N helicopter were made in 2,800' water in late September. In October, dives were made on Fieberling Guyot and Jasper Seamount at depths of 3-4,000'. On Tuesday, 7 November *Turtle* was trapped by two cables as it was attempting to recover a hydrophone off San Clemente Island at a depth of 1,277'. The skipper of the sister sub *Sea Cliff*, Reed Popovitch, viewing the entanglement from above while using a remotely operated vehicle with camera, was able to advise the *Turtle* pilot how to maneuver to clear the lines of entanglement. After a 14 hour, 12 minute dive, the *Turtle* and crew surfaced tired but safe.

In March 1990, *Turtle* was back in the Hawaiian Islands. While at the Pacific Missile Range Facility, the submersible recovered classified objects, torpedoes, and two Tomahawk test shapes. Four dives were made near Hilo on the big island prior to finding an SH-2 helicopter and two dives before finding a CH-46 helicopter. In November, a dozen dives were made off the tip of Baja California to investigate a hydrothermal vent field filled with black smokers, sheet and pillow lava beds, and marine life that thrived on the hydrogen sulphide in the water. One area is so thick with large clams that it has been nicknamed "clam acres." Many physical measurements and samples were taken for chemical and biological content.

In February 1991, *Turtle* recovered a Scorpio remotely operated vehicle from the sea floor. In mid-April, the first dive was made in a series to search for two P-3 Orion aircraft that crashed in a mid-air collision. In late April, *Turtle* conducted many video survey dives of Monterey Canyon, California. In May, *Turtle* recovered a MK-50 exercise torpedo, two Tomahawk capsules, made surveys of the range

*...the oil-filled port distribution box, housing the electrical breakers, was seen to be smoking. Oil from the distribution box ignited soon after the submersible left the water. Available fire fighting efforts were ineffective...*

---

hydrophones, and searched for a SH-2F, a HSL-37 and a CH-46E helicopter at or near Barking Sands. In November *Turtle* was again deployed off Mazatlan, Mexico for a ten dive series involving hydrothermal vent research.

In April and June 1992, *Turtle* was involved in resetting the U.S. Geodetic Survey's laser instruments for measuring ocean floor movement, a hydrothermal vent survey for the University of Washington, and the recovery of seismic receivers and seismometers. In July and August, local area training and recovery of an AQS-14 minehunting hydrophone was accomplished. A total of 56 dives were made in 1992 to as deep as 9,680' for as long as 9.5 hours.

Dive #700 was the first dive made in January 1993. Fifteen dives of eight to nine hours duration were made in the Sea of Cortez (Gulf of California) for video surveys of hydrovents, retrieving cores, sampling vent water, collecting animal samples, and collecting oil samples. It also deployed Jason, a Remote Operated Vehicle, as a part of a program to show school children all over the U.S. live pictures from undersea research programs. Most of the rest of 1993 was spent in training dives to complete a total of 67 for the year.

The first dive of 1994 was made in June, followed by a series of local dives in the Southern California area. A series of dives in August and September were made in the eastern extremity of the Mendocino Fracture zone off the Oregon/California border. Typical dives were 7,300' to 7,400' in hydrothermic vent areas.

Early in 1995, a swimmer delivery vehicle (SDV) operated by a Seal Team was lost near Hawaii. After a lengthy search, using a deep side scan sonar sled towed from a surface ship, the general location of the SDV was determined. *Turtle* located and recovered the SDV during three lengthy dives in May and June and, then, went on Restricted Availability Status until February 1996. In spite of a history of productive schedules and a successful dive record, budget cuts reduced *Turtle*'s future operations to a bare minimum with scheduled retirement in 1998.

# *NR-1*

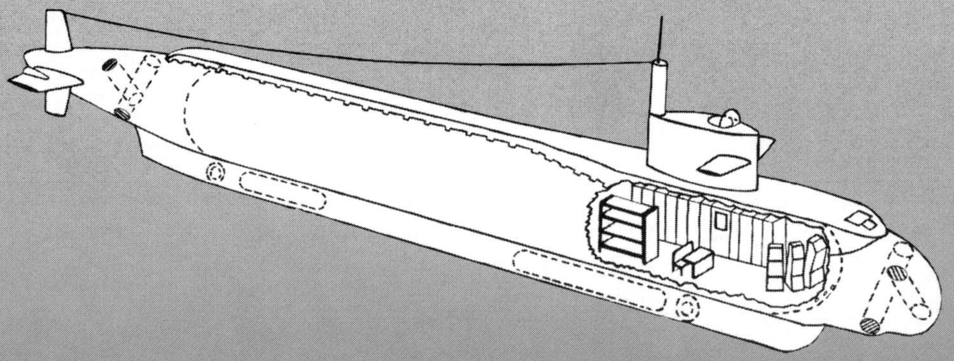

**1969–(          )**

# NR-1

The Deep Submergence Systems Project (DSSP) under Dr. John Craven was established after the loss of the nuclear submarine *Thresher*. One of the many segments of DSSP that Craven managed was the Advanced Seabed Deterrent Program which included a special deep running, small, nuclear submarine study. Craven assigned the study to the Bureau of Ships that designed nuclear submarines with power plants furnished by Admiral Rickover. When Rickover heard of the project, he ran to Craven's boss, Admiral Red Raborn, claiming anything nuclear was in his bailiwick. Admiral Raborn, handpicked for his job by the Chief of Naval Operations, gave little heed to the academic Rickover whose primary job was only to furnish the power plants. As time passed, the frustrated Rickover continued to monitor the program, allegedly claiming the study "incompetent."

One day Captain Anderson, the first skipper of *Nautilus*, America's first nuclear submarine, called Craven and asked him to be the project manager of a "crazy idea for a small, deep, nuclear submarine," if funding could be found for it. Unspoken, but obvious, was the fact that the submarine was the one Craven and Bureau of Ships had been studying for some time, so Craven said "yes." A meeting was set up with Rickover, the Assistant Secretary of the Navy for R&D, the head of the Polaris program and Craven. Rickover started the meeting by announcing that the submarine was to be called the *NR-1*, which implies the first in a series. Rickover then asked the group if there were any existing programs that could fund the project. The Admiral in charge of the Polaris program indicated that about $22 million from the ballistic submarine programs could be reallocated to *NR-1*. Craven had $10 million from the Advanced Seabed study, which could be released. Rickover is alleged to have then announced to those present "O.K. then, the *NR-1* will cost $32 million." The others at the meeting all looked at each other in disbelief over the low cost estimate based on the money available, instead of a realistic projected cost estimate. Rickover assured the group that additional funding was his problem. (At launch time four years later *NR-1* was estimated to have cost $135 million.)

Two days after the *NR-1* meeting in April 1965, one of Rickover's people called Craven and warned him that

President Johnson had been told of the *NR-1* by Rickover and that the president was announcing *NR-1* to the press in the afternoon. It was further recommended that the Secretary of Defense and Secretary of the Navy be told about the *NR-1*, before they heard about it in the newspapers or by rumor. The next day, President Johnson made the announcement about the *NR-1* in a speech from the Texas White House. Congress read about it in the papers and, needless to say, were less than pleased, since all ships must be approved by them prior to construction or public announcement of their approved construction. The powerful Armed Forces Committee scheduled a meeting with those involved for the next day. Craven and company, armed with slides, charts, and various forms of justification went before the committee with a presentation. Heated arguments went on and on between the committee and Rick-over, but finally the committee approved the building of *NR-1*, establishing a record for approval of a ship's construction in less than two weeks. Indeed, *NR-1* probably would not have been approved by Congress had it gone through the usual budget reviews, since at the time, the Vietnam war was consuming almost all of the defense budget.

Rickover and Craven then charged ahead with finalizing design and construction plans. The steel was selected from the best available steel (HY-80) that was left over from defense contracts for nuclear submarine construction. Due to *NR-1*'s smaller diameter, it could go much deeper than the larger nuclear submarines made of the same material. The Integrated Control and Display Systems (ICAD), a combination of several functions into one or more displays developed for *Trieste 2,* were used on *NR-1* and provided a significant savings in development costs.

***Name:*** *NR-1* ***Builder:*** *Electric Boat* ***Length:*** *146'*
***Maximum Depth Rating:*** *3,000'* ***Capacity:*** *16 Man* ***Year Launched:*** *1969*
***Description:*** *NR-1 at Christening* Courtesy of Department of Defense Still Media Depository

The $32 million initially provided was expended instantly and additional money had to be either obtained from Congress or from other programs. Somehow, the money was obtained and on 25 January 1969, the *NR-1* was launched at Electric Boat in Groton, Connecticut. Sea trials were completed four years after the "premature" announcement – in August, 1969. For all the good Rickover did to get nuclear power into the Navy, his flawed estimate on *NR-1* put him in conflict with many in congress, limiting some of his future plans.

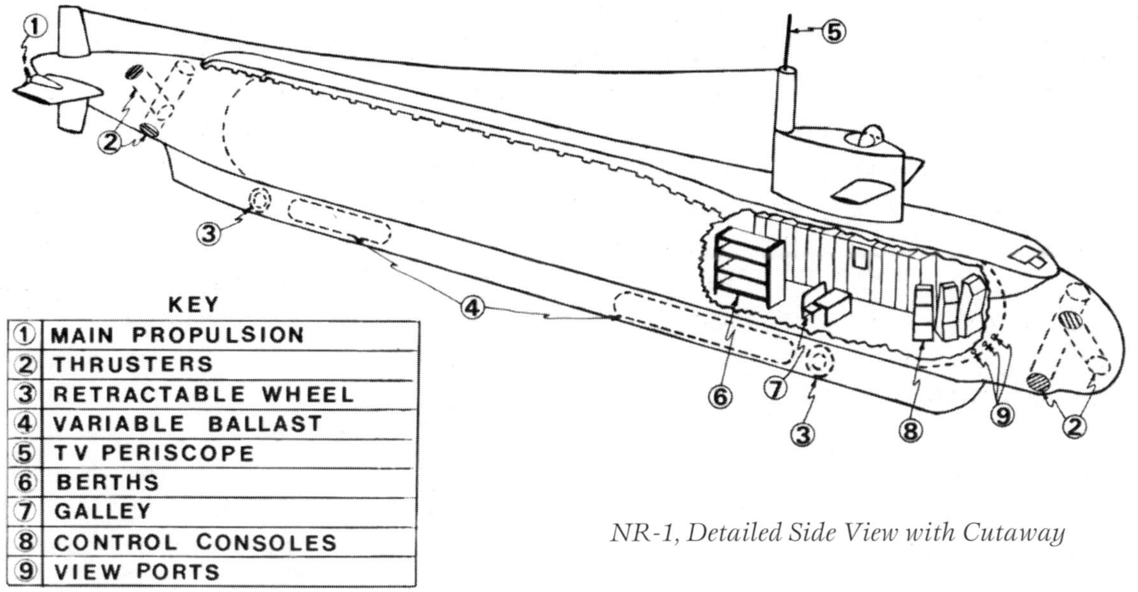


*NR-1, Detailed Side View with Cutaway*

The overall configuration of *NR-1* is quite similar to a scaled-down version of a fleet ballistic submarine (SSBN) since scaling down existing drawings was an obvious cost-saving technique and even the materials were the same. Like an SSBN, the streamlined sail on top of the main hull has an access hatch to an upper landing for conning the submarine during surface operations. An acrylic "windshield" protects the watch standers from most of the whitewater during rough weather. The dive planes are located in the sail and are hydraulically operated from the control room. The pressure hull located under the sail was originally 96'-1" long and 12'-6" in diameter. Recent elongation of the pressure hull has increased the overall length of the submarine from 137' to 146'. Displacement became 400 tons.

The operating depth is advertised as 2,375', but several unofficial published sources claim it can operate deeper, and a philatelic cover sent from the Straits of Sicily showed a dive to 3,000' on 20 September 1995. In any case, its depth capability is ideal for investigating the continental shelves. A four foot high box keel is located below the pressure hull. Retractable 2' diameter by 12" wide liquid-filled wheels are located at the fore and aft ends of the keel for bottom travel – as originated by Simon Lake.

Possibly, the most significant feature of *NR-1* is its distinction as the only nuclear powered research submarine. An overhaul in 1990 included, in addition to the added length to the pressure hull, a new nuclear core giving the submarine about another twenty years of operation before needing "refueling." Power is available for scientific equipment in all of the normal laboratory voltages, frequencies, and currents, with hull pass-throughs and connectors for customers requiring special oceanographic instrumentation not already on board *NR-1.* The power plant itself is fairly small with a nuclear core that isn't much bigger than a coffee pot. Shielding, steam generating equipment, a steam turbine, the motor generator, and so forth take up most of the classified machinery space aft of the operations area.

*Unlimited time submerged is possible from the power/propulsion system, but the submerged duration of NR-1 is, in fact, limited by the food on board to about 33 days — although dives lasting 40 days have been made.*




The forward operations area takes up about one-third of the pressure hull. The generated electricity powers the two 150 H.P. reversible propulsion motors with propellers mounted at the rear of the fixed horizontal stabilizers. The rudder control surfaces are located in the vertical stabilizers and are hydraulically operated for steering. At 4 kts., the sub's maximum underwater speed, the turning radius is 900'. During operations under 2 kts. the maneuvering thrusters can be used for slow or hovering operations. Four thrusters are located in two "X" configured tubes. One "X" is at the bow and one is near the stern. Each tube has a 7 H.P. reversible motor and propeller for providing 300 lbs. of thrust in any of the 24" diameter tubes. These thrusters permit crabbing traverses on the bottom for inspecting cables, pipes, and the like, and can compensate for a 90 degree side current of a 1/4 kt. Unlimited time submerged is possible from the power/propulsion system, but the submerged duration of *NR-1* is, in fact, limited by the food on board to about 33 days — although dives lasting 40 days have been made. Typically, four expeditions a year are made, lasting about a month each with an additional eight training dives of shorter duration.

The surfaced speed of *NR-1* is little better than its 4 kts. submerged capability. This requires a surface support ship for transits of any significant distances. Usually an ASR (Submarine Rescue Vessel) with a satellite global positioning system aboard, giving very accurate navigation information in any weather, is employed as a tow. The surface towing technique allows a speed of 12-13 kts. with the submarine manned and submerged for a smooth ride in any sea state.

Buoyancy/ballast control can be accomplished by several different systems. A variable water ballast system provides the normal method of changing the submarine's density. One thousand lbs. of water can be added through flooding or by being pumped overboard, using a 4,500 psi compressed air system. A freshwater internal trim system provides trim for pitch and list. This system uses two tanks forward and two tanks aft, with displacement pumps for transferring the water in any combination between the four tanks. Lead shot totaling 22,000 lbs. can be dropped to provide emergency buoyancy. Most of the external appendages (wheels, manipulator, and work packages) can be dropped in an emergency.

Several methods for viewing outside of the hull are possible. Three, four inch diameter acrylic viewports are located in the lower forward end of the pressure hull. The periscope in the mast of the sub houses a 360 degree viewing TV camera, instead of the older optical system used in WW II and before. Ten underwater TV cameras are strategically located about the sub, including one at the rear of the submarine.

Three monitors select the viewing cameras and areas of observation. One 70 mm and two 35 mm cameras are provided for still photography. Eight 1000 watt and nineteen 250 watt lights are provided for illumination. A manipulator with an 8' reach that can lift up to 240 lbs. is installed at the bow of the sub. It has wrist, elbow, shoulder, and hand functions. A work module can be extended below the hull in front of the viewing area and manipulator. It has a recovery claw and storage bins for stowing up to 500 lbs. of recovered material.

The normal crew compliment consists of three officers, 11 enlisted men, and two scientists. All of the crew are volunteers with previous nuclear submarine experience. There is a long waiting list for all billets. The average floor area available for each person is ten square feet. Only four bunks are available for everyone on board, so "hot bunking" is required, but sleeping in the viewing area and passageways is a common procedure. The galley table seats three people, and everyone fixes their own meals, which are primarily TV dinners stored in a freezer and microwaved before consumption. Normally, everyone on board stands six hour watches with six hours on duty and six hours off.

The navigation on board is done by dead reckoning, using Doppler sonar to determine speed made good over the bottom and the sub's headings from gyro and compass readouts computerized together to determine courses and positions. Forward looking obstacle avoidance and side scanning sonar give a high degree of electronic visibility that is far ranging with high resolution both for the area ahead and around the sub.

Most of the operations to date have been in the Atlantic Ocean from Iceland to the Gulf of Mexico and the Caribbean. If required, the submarine can operate in any ocean in the world. It is possibly the only research vessel capable of working under polar ice for lengthy periods of time.

The history of *NR-1* operations is somewhat sketchy due to the classified nature of much of the work done by the submarine. Beginning in 1972, *NR-1,* under Submarine Development Group One, conducted geological, geophysical, and photographic studies of the submarine canyons Lydonia, Oceanographer, Walker, Gilbert, and Hydrographer, all off the coast of New England. It also deployed to operate on the Blake Plateau near Florida. Test dives were made in Narragansett Bay near New London, Connecticut. Upon deployment to Roosevelt Roads, Puerto Rico, a complete underwater circumnavigation of the island was made.

In the Fall of 1974, lengthy operations were conducted in the Puerto Rican Trench. In September of 1976, an F-14 slid into the water off the aircraft carrier *U.S.S. Kennedy* (CV-67) in 1,890' of water near Scotland. The pilot ejected but the plane went in with a Phoenix missile on board. Soviet Cruisers and intelligence ships were in the area and security compromise was greatly feared. Heavy seas and equipment failure prevented an ROV making a search and recovery.

On 19 October, *NR-1* arrived on the scene and dove on a transponder dropped near the plane. The plane was found with a lot of fishing net around it, but the Phoenix was missing. The *NR-1* picked up a heavy cable dropped from the surface support ship and put it around one of the landing gear wheels of the F-14. While being hoisted near the surface the hoisting cable broke and the F-14 dropped back to a depth of 1,500'. The lift drama was repeated with a second line sent down. This time it parted near the surface. Heavy weather was continuous throughout the operations and, after the second try, most of the surface ships headed for port to wait out the storm. Unaffected by the weather, *NR-1* searched and found the Phoenix on 30 October, surfacing with the 15" diameter missile in its manipulator. At the surface, divers put many straps and lines on the missile in 10' seas. The half million dollars worth of hardware was then hoisted aboard the surface support ship without compromising security. In November, the F-14 was again lassoed by *NR-1* and this time successfully recovered.

During June and July 1977, operations near Iceland were conducted on the Reykjanes Ridge, an extension of the Mid Atlantic Ridge (MAR). While doing Geological studies on board, Dr. Bruce Heezen, the National Geographic Society cartographer, died of a heart attack.

Operations recorded in 1978 included sea trials off the Connecticut Coast during February and March. Florida Coast operation took place in March and May with more sea trials in August. Operations in the East Atlantic occurred in October and November.

In 1979, *NR-1* operated in the Baltimore undersea canyon in May and the Florida Straits during August.

In 1980, the submarine returned to the Blake Plateau, near Florida, to conduct investigations of the distribution and thickness of phosphorate-manganese pavements and nodules prevalent in the area. On one dive, a break in the concrete-like floor was "traced for many kilometers." Sea trials in the North Atlantic were also conducted.

Maximum depth, classified operations took place in March 1981. In September, a Groton, Connecticut to Scotland transit was made. From October through December, classified operations took place in the North Atlantic.

In April 1982, the submarine operated in the Florida straits at maximum depths. *NR-1* had its first overhaul in Portsmouth beginning 4 November.

The Reykjanes Ridge, about 770 miles southwest of Iceland, was revisited in 1984 after overhaul. The expedition was highlighted by the April 1985 *National Geographic Magazine* in an article by Bob Ballard. This area of the Mid-Atlantic Ridge (MAR) is important to geologists for making comparisons of it to the more southerly MAR near the Azores and to complete Dr. Heezen's studies begun in 1977. The search for active volcanism related geothermal vents, common in some areas of the MAR near the Azores, proved fruitless during this dive series, even though volcanism had created an island near Iceland in 1963. Ballard, having spent 20 days on board *NR-1*, compared it in many ways to the NASA Space Shuttle, in particular, due to its close quarter isolation and extensive capabilities.

In 1985, a Continental Shelf survey was made off New Jersey.

Following the loss of the Space Shuttle *Challenger* in 1986, the *NR-1* was brought in with several other undersea vehicles to search for, identify, and recover critical parts of the spacecraft. The submarine's ability to remain on the sea floor, without the need to frequently surface, resulted in it being used as the major tool for searching deep waters to locate and identify objects for recovery. Because of its independence from the surface, *NR-1* was able to remain submerged and on station, even when heavy weather and rough seas forced the other search and recovery ships into port for many days. Most of the *Challenger* material recovery was made by *NR-1*, due to its extended bottom time being measured in days instead of hours as was the case for the other undersea vehicles.

During 1987, 1988, and 1989, in the Gulf of Mexico, natural petroleum seeps were investigated in *NR-1* by Texas A&M scientists. Organisms similar to the ones found in deep thermal hydrovents (limpets, mussels, clams, and 3' tubeworms) were found to be thriving on the hydrocarbon chemicals instead of sunlight. Brinepool surveys were also made in the Gulf over the same period and hypersaline pools were identified and recorded.

In 1990, extensive research in the Gulf of Maine was conducted under an agreement with the National Undersea Research Center. Shrimp, scallop, lobster, herring egg beds, and wolf fish were observed, and their quantities estimated. Gilbert and Oceanographer's submarine canyons were surveyed by scientists for lobster content. The *NR-1* went into overhaul and drydock late in the year. At the same time, a new support ship was under construction.

After a two year repair period, the *NR-1* returned to service. The first unclassified work reported was made for marine life and coral mound surveys off the coast of Florida in November 1993.

A Mediterranean series of dives began in late summer of 1995 to conduct archeological surveys of ancient Roman trade routes between Carthage and Rome, with Bob Ballard as chief investigator.

The literature describing scientific uses and results of *NR-1* may not add up to Alvin's accomplishments but, considering the non-scientific demands – mostly classified – on *NR-1*, its history is impressive and stands as a growing monument to the merit of occasionally rushing a good idea through Congress's budget process. It is unfortunate that *NR-2* was never funded or built, especially since *NR-1* has fully justified its existence by so many lengthy deep dives for both science and defense purposes.

# *Kumukahi*

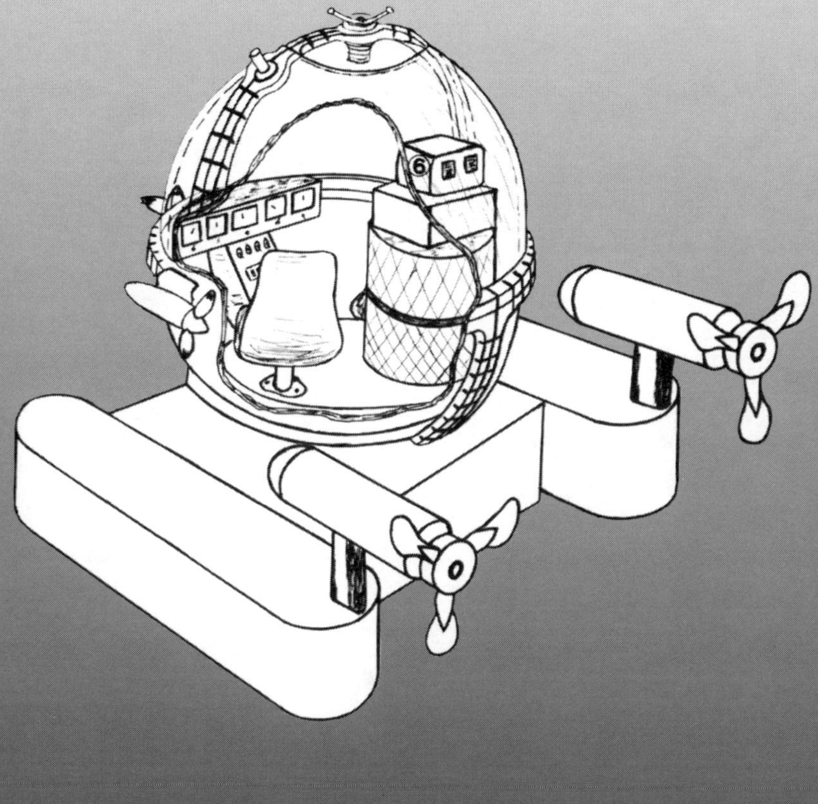

## *1969–1970*

# Kumukahi

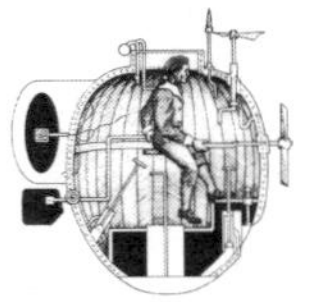

Observing the undersea world through viewports or through optical systems does provide minimal viewing, but what if one could cruise helicopter-like, with no observational limitations whatsoever? Two materials that would make this possible were under serious consideration at the Naval Ordnance Test Station at China Lake, California in the early 60's. These were acrylic and glass. The choice of acrylic would limit the operational depth and require some form of segmenting and/or laminated construction, due to limitations on the size of manufactured sheets available. The technology of casting large heavy sections had not been developed at the time. Initially, the glass option seemed barely feasible in view of the manufacturing risk, the very high cost, and the lengthy developmental time.

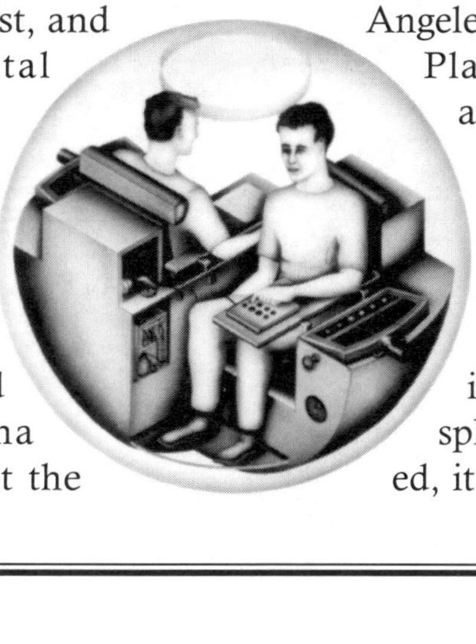

Dr. Tom Lang, a research fellow and specialist on hydrodynamics, was the first one to come up with a reasonable proposal for an acrylic hulled deep submersible at China Lake. He called his project the Utility Submarine. It consisted of an acrylic, spherical pressure hull, large enough to hold two people, with a set of internal batteries for propulsion and lighting.

When asked by Lang to find fabricators and run model tests, I approached various acrylic manufacturers in the Los Angeles area. Chet Fortin of Fortin Plastics in Saugus, California agreed to "take a crack at it" and make a 16" diameter sphere in two halves using five plys of acrylic. The results were optically outstanding and when the one-inch-thick-walled assembled sphere was hydrostatically tested, it withstood pressures equal to

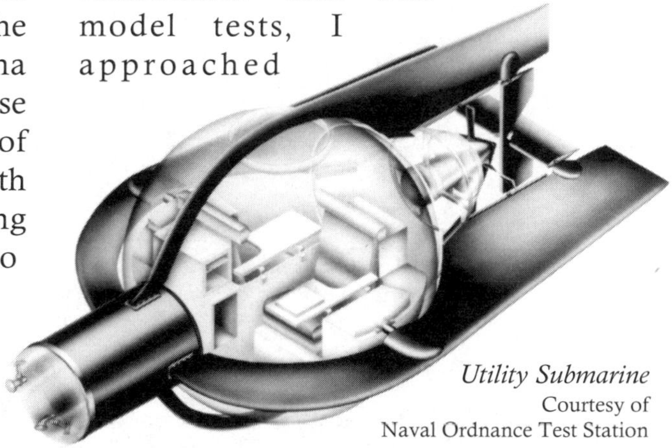

*Utility Submarine*
Courtesy of
Naval Ordnance Test Station

those at a mile depth of water. The model tests suggested that a 5" thick hull, 5' in diameter would satisfy the project's 2,000' depth requirements and could be fabricated economically. Unfortunately, the use of batteries inside the sphere was considered unsafe by the reviewing officials.

A technique we used on several projects as a low cost simulator was the operation of remote-controlled submersible models in a large tank full of water. Fine wires from the models to a control box provided the power and controls for the scaled down submersibles. One of the several model submersibles used in the tank was an all-acrylic-hulled submersible called the "Hot Rod." The $1/10$ scale model represented the simplest most maneuverable version of a transparent hulled submersible conceivable at the time.

Once a year, China Lake's top management invited an off-station group of experts, called the Advisory Board, to comment and evaluate the different projects either in progress or under consideration, at the base. Tap Pryor, who was one of the younger members of the board, had originated, owned, and operated the oceanarium Sea Life Park in Hawaii, as well as its sister institutions the Oceanic Institute and Makai Range, which were chartered for and active in marine biology and ocean engineering, respectively.

Dr. McLean, the Director at NOTS and my mentor, was aware of Pryor's research team's desire for a small submersible. He was also aware that while we at the Navy lab could easily design and build an all-acrylic sub, the necessity of taking it though the research, material testing, data collection, and numerous committee reviews required for Navy certification for the first such use of acrylics for a pressure hull would make such a project too costly. During one of the Advisory Board's annual visits to China Lake, McLean suggested to Pryor that he might want to visit our submersible model tank, having a good idea ahead of time what might develop.

Pryor showed up, introduced himself and was placed in the control chair to operate the Hot Rod. Running the little acrylic model called "Hot Rod," he was as enthusiastic as a kid on Christmas morning. After about an hour of "oos," "ahs," and "oopses," he finally stopped and asked if I would design a Hot Rod for his Oceanic Institute scientists and Makai Range Engineers. I said yes even before we discussed the details — and there were many.

The main problem was obtaining a good acrylic hull at an affordable price. The laminate technique could be used, but it involved skill and an art form bordering on witchcraft. Chet Fortin wasn't as excited about making a full scale submersible hull as he had been to make the 16" models. This may have been because the models alone had been a demanding exercise with several throw-aways before attaining success.

Dr. McLean and I brain-stormed the problem and came up with a four piece sphere, or quadrant concept. A flat sheet of acrylic when heated with the edges clamped and then blown with air (like bubble-gum) into a hemisphere, thins at the pole to one-third of the original thickness. By comparison, a sheet of heated acrylic when forced over a mold shaped like a quarter of a sphere thins only six or seven percent. The technique works quite well for thin, as well as thick sheets of acrylics and for spherical submersible hulls operating at depths down to about 1,000 feet. Twelve piece spherical segmented spheres, as used for *Nemo* could go deeper, but are much more difficult and expensive to form.

**Name:** *Kumukahi* **Builder:** *Oceanic Institute* **Length:** *5.9'* **Maximum Depth Rating:** *300'*
**Capacity:** *2 Man* **Year Launched:** *1969* **Description:** *Kumukahi on LRT*
Courtesy of Time Life Syndication

Sixteen-inch-diameter models were made by this quadrant method, heated in McLean's kitchen oven and formed over a home-made mold. Hydrostatic testing of the models indicated a close comparison to the theoretical strains and stresses predicted. After viewing the results of our fabrication and hydrostatic testing, Fortin agreed to make the full scale hull.

The project was initiated in June 1967. For the next two and a half years, I spent every other weekend commuting to Hawaii working on what was to be named *Kumukahi* (Hawaiian for first in a series) and *Kumu* for short. Bill Walker, who worked for Tap Pryor as a multi-discipline engineer at the Oceanic Institute, accepted the job of Hawaiian project manager for Kumu. Phone calls between Walker and myself, during lunch-time or after work, averaged about two or three a day the first year. A month after the project was authorized, I submitted a preliminary, forty-page design package to the Oceanic Institute which included cost estimates and manufacturer's sources of materials for approval. The package, after acceptance, became the construction manual.

Fortin was given a contract and proceeded to develop the forms and hardware to make the 1 1/2" thick 5' diameter hull to provide a 300' operating depth hull with a large safety factor.

One-third H.P. Sears Roebuck electric trolling units were oil filled to provide the thrust for propulsion and vertical control. The small Sears units were installed as the main propulsion thrusters only for the initial tank trials and demonstrations. Funding was to be from philanthropic sources and like most other funding sources, required some "show and tell" as early in the program as possible. Environmental equipment and tests included a manned encapsulation test and a manned test tank trial of the assembled propulsion system and controls.

A remote-controlled model was made for use as a simulator and as a demonstration device. A full scale mock up of thin-walled acrylic hemispheres was built for demonstration purposes as well as for manned, encapsulated, environmental tests. The recommended hardware was purchased by Bill Walker via phone calls and purchase orders and installed as received.

After Fortin delivered the four-piece sphere to Hawaii, assembly began. Hydrostatic model tests had indicated that the connecting flanges which protruded out from the nominal shape of the sphere, added a stiffness which increased the stress in other parts of the sphere. To relieve the excess stress, the flanges were sawed part way through, leaving a configuration not unlike the bent back of a mammal's vertebrae.

On 7 December 1967, using the thin mock-up hull, the first manned encapsulation test was conducted with 82°F water pumped from the nearby ocean over the outside hull's surface. The moisture scrubbers couldn't keep up with the respiration and perspiration given off. Within fifteen minutes, Bill Walker and I were in a 100°F and 100% humidity environment, sweating like we were in a steam room. The oxygen supply and carbon dioxide removal systems worked fine, but the scrubbers were inadequate for moisture removal. Hindsight analysis showed that we were sitting in the sun surrounded by 82°F water, using a scrubber system that worked well in 65°F water, but not in the tropics in a greenhouse. Later, a much larger scrubber worked well enough, but it was always necessary to shade the hull and blow-dry to

*The moisture scrubbers couldn't keep up with the respiration and perspiration given off. Within fifteen minutes, Bill Walker and I were in a 100°F and 100 % humidity environment, sweating like we were in a steam room.*

cool it well before diving. Blocks of ice were often taken along to pre-cool the sphere, and condense internal moisture.

During an after-work bull session, I made the suggestion that we needed a "wet aircraft carrier." In other words an underwater platform to launch and recover from. Pryor said that there was a catamaran platform which had been used as a prototype for Makai Range's AEGIR manned undersea habitat program. Makai manager, Hank Horn, a Naval architect who had designed the surface support ship LULU for *Alvin* at Woods Hole, readily added variable ballast chambers in the two foot diameter pontoons, added synthetic foam buoyancy above the deck for stability, and installed a platform on the 17'x 33' long catamaran. The vessel was called *LRT* for "Launch, Recovery, and Transport. When *Kumu* was ready for operations, it was towed on *LRT* to the dive site, the catamaran was flooded down with *Kumu* on board to a depth of 20'-60', where the water is calm even in the roughest sea. *Kumu* would launch itself after release from the catamaran by the divers. The process would be reversed for recovery and transported back to port. The system worked well and was later used for 100 dives by *Star 2* as well as launch/recoveries of *Nekton* and *Deep Rover*. An improved second generation version of the *LRT* is now in daily use by the Hawaiian Undersea Research Lab at the Makai pier near Waimanalo for *Pisces 5*.

After the pressure hull was completely assembled, a strain gaged hydrostatic test was needed in order to see if there was a favorable comparison between the model tests and the full scale hull. No facilities for hydrostatic testing existed in Hawaii at the time, so an at-sea, 450' deep, proof test was performed with strain gages placed on the submersible while the strain indicators were read on aboard a surface support ship, connected by 500' of lead wire to the submersible. All lead wires were of equal length and resistance balanced in a vernier resistance balancing box, prior to submerging the pressure hull. Temperature-compensated gages were used with frequent read-out periods. The highest stress indicated, as predicted, was found in the hatch reinforced area (a fiberglass built up section). All strains were reproduced in near equality in all ten cycles of pressurization to 450' (except for two strain gages that flooded). At the operating depth of 300', the highest stress was well within the existing standards of practice and all read-outs were proportional to the depth.

The first manned operations in August 1969 were made with the small, temporary Sears motors. In spite of their minimal thrust, control, and maneuvering were nearly as good as had been experienced with the model "Hot Rod." Additional propulsion power was indicated, as had been anticipated, for operations in strong currents. Two 1 1/2 H.P. motors with propellers were added at the rear of the battery pods for additional thrust. Shallow tests resulted in fast speeds and good maneuverability.

In January 1970, an unmanned proof test of the submersible to 450' for six hours was being conducted. During the operations in a rough sea, the submersible was smashed against the fantail of the surface support ship *Holokai*. A thick 12" diameter circular acrylic plate was substituted for the damaged area and bonded in place.

The *LRT* was used for the next *Kumu* operations. A suitable location for the undersea habitat AEGIR was sought, using *Kumu* for the underwater search. Several areas at different depths along a 600' slope were investigated and reported on. During one of the area searches, an electrical short inside the pressure hull occurred that failed to blow a fuse immersed in oil in the battery compartment. Next, the electrical pass-through in the hull heated up enough to warp the adjacent acrylic area, causing damage to the hull. In later testing of the fuse, which was rated at 100 amps in air, it took 350 amps to blow it while immersed in oil.

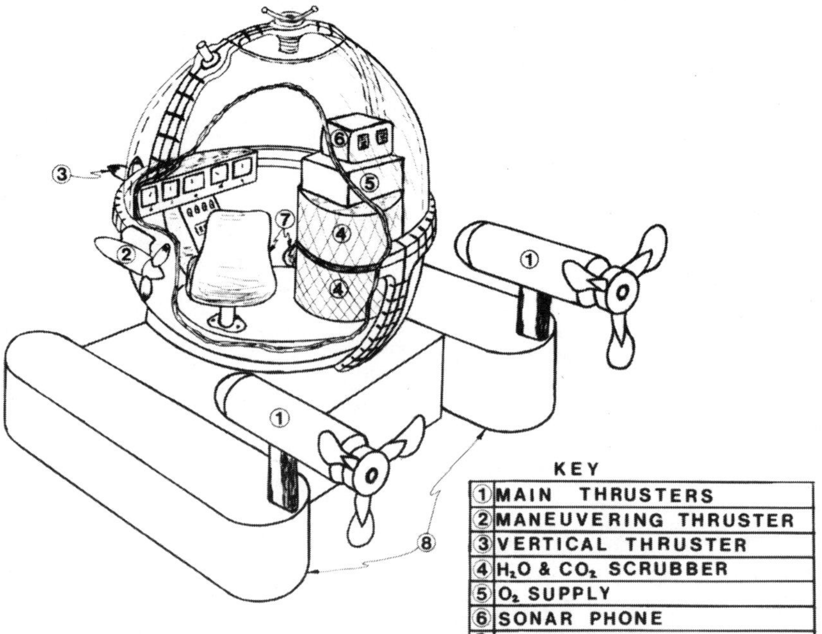


*Kumukahi, Detailed Side View with Cutaway*

One last dive was made without power to get a picture in the March 1970 issue of *Life Magazine*, with *Kumu* sitting on the *LRT* at sea. Although its life and uses were limited, *Kumu* served as proof of concept, was catalytic in the development of the *LRT,* and finally, served out its life as an exhibit in Sea Life Park. The design and development process proved pivotal for later transparent hulled projects.

# DSRV-1 Mystic

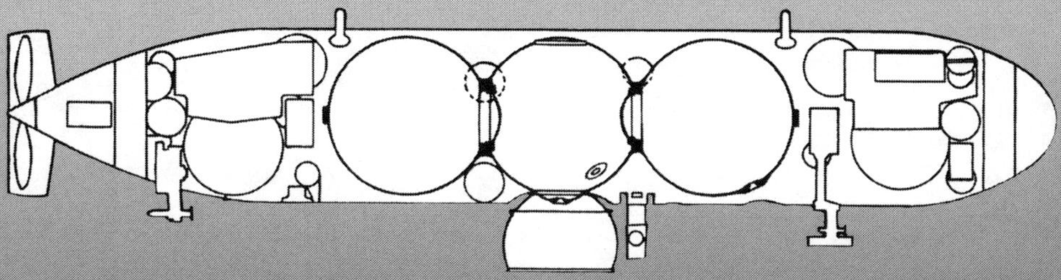

1970–(    )

# DSRV-1 Mystic

One of the major programs specified by the Chief of Naval Operations after the loss of the nuclear submarine *U.S.S. Thresher* was Deep Submersible Rescue Vehicles (DSRVs).

Lockheed Missile and Space Company was awarded a contract in June 1966 for *Mystic*, the first DSRV construction. The award was based in great part on Lockheed's success with building and operating *Deep Quest*. The Charles S. Draper Laboratory of the Michigan Institute of Technology (MIT) which had the guidance system for Polaris with Lockheed was awarded the navigation and controls contract for *Mystic* as well.

The HY-140 (140,000 psi strength) steel specified by the Navy contract for the fabrication of three pressure hull spheres for a 5,000' operating depth submersible had never been produced and welded in the sizes and shapes required. Sun Ship, under contract to Lockheed, first welded hot-spun formed hemispheres to the two intermediate ring forgings located between the three spheres. After the hatch bosses, trunion plates, pass-through plates, and viewport bosses were installed, the six hemispheres were then welded together to form the three spheres. The transfer skirt was the last piece to be welded to the center sphere to complete the pressure hull assembly.

Some important operational requirements of the DSRVs included the capability to "mate" their transfer skirts onto a submarine hatch at angles of up to 45 degrees in pitch or roll, rescue 24 crew members at a time, and transfer them to another submarine or surface ship. A maximum speed of 4 kts. with a cruise speed of 3 kts. for 12 hours was also required. The initial plan was to load the DSRV into a C141 and fly it to the nearest seaport, where a submarine or ASR (Submarine Rescue Vessel) would piggyback it to the site of a disabled submarine. There, within 24 hours of startup, the rescue process would begin.

The completed vehicles weigh 38 tons and have a length of 49.3'. The three spheres in the pressure hull are 7'-6" in diameter with a nominal thickness of .738". The transfer skirt, which is a pressure vessel during rescue transfers, has a 53" diameter opening, where it connects to submarine hatch areas. The hatch in the control sphere is 28" in diameter, while the other three hatches are 25" in diameter. Reinforced areas for the trunnions are located at the forward and

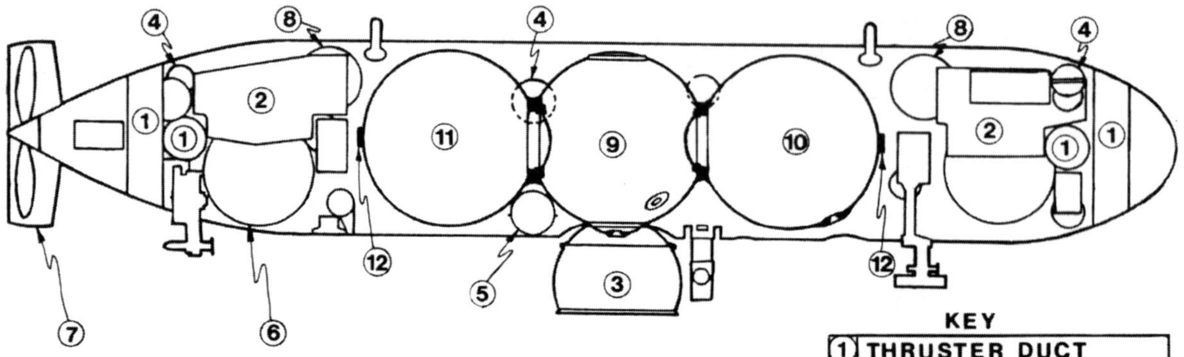

**Names:** *Mystic and Avalon (DSRV-1 and 2)*
**Builder:** *Lockheed Missile and Space Co.*
**Length:** *49'3"* **Maximum Depth Rating:** *5,000'*
**Capacity:** *3 Crew/24 Rescues*
**Year Launched:** *1970*
**Description:** *Detailed Cross Section*

KEY

| | |
|---|---|
| 1 | THRUSTER DUCT |
| 2 | MAIN BALLAST TANK |
| 3 | TRANSFER SKIRT |
| 4 | MERCURY TRIM TANK |
| 5 | MERCURY RESERVOIR |
| 6 | TRANSFER SKIRT $H_2O$ |
| 7 | TILTING SHROUD |
| 8 | VARIABLE BALLAST |
| 9 | MID SPHERE |
| 10 | CONTROL SPHERE |
| 11 | RESCUE SPHERE |
| 12 | TRUNNION |

aft extremities of the assembled pressure hull system.

The main ballast system consists of four saddle tanks which displace 6,400 lbs. of sea water to provide buoyancy at the surface. Variable buoyancy control is provided by floodable fore and aft tanks. Four water-filled, collapsible bags in each sphere can be emptied to compensate for the weight of 24 people when brought aboard. Fore and aft tanks are provided to store water when it is pumped out of the rescue skirt during transfer.

Fore and aft propulsion is provided by the stern-mounted reversible 4' diameter, 3 bladed propeller, powered by a 15 H.P. electric motor. The shroud around the propeller can be pivoted in two axes for both depth control and steering. Vertical tubes forward and aft contain ducted thrusters for pitch control, especially during hovering. Horizontal tubes fore and aft provide turning control, particularly during hovering. The four ducted thrusters are each driven by 7 1/2 H.P. electric motors.

The pitch and list trim of the submersible is provided by altering weight and balance in three ways. One method is pitch by transfer of up to 1,428 lbs. of mercury carried in tanks located forward and aft. Another is list by the transfer of up to 440 lbs. of mercury between port and starboard trim tanks. And finally, the submersible's stabilizing moment arm, which is the distance between the center of gravity and the center of buoyancy, can be altered by shifting mercury between the storage reservoir, low in the hull, to the above centerline pitch trim tanks.

Main power for the submersible is provided by externally located silver-zinc batteries. Emergency power is available from silver-zinc batteries inside the pressure hull.

Separate environmental systems are provided for the forward (control) sphere and the two aft (rescue) spheres. Air conditioning is provided to control heat and humidity. Emergency breathing systems are provided for all compartments. There is a total of 384 man-hours of life support for the 3-person operating crew and 24 members of the rescued group.

There are five 4" diameter viewports in the pressure hull system. Two in the forward sphere that includes one forward that is down-angled at 30 degrees and one viewport that is 140 degrees from dead ahead on the starboard side, which is also 30 degrees down-angled. Two viewports are in the central sphere, both are 70 degrees from dead ahead on the port and starboard sides and 50 degrees down-angled from the horizontal. One viewport is located in the center of the bottom hatch in the mid-sphere for viewing the transfer trunk.

In conjunction with the DSRV Inertial Reference Unit (IRU), the on-board electronic system can operate as an integrated, transponder, doppler, or effector navigation system using doppler and/or pulsed reply sonar. Horizontal and vertical obstacle avoidance sonar, as well as downward looking and altitude/depth sonars, aid in both navigation and general operations of the submersible. A manipulator is carried to clear hatches on disabled submarines, as well as for use in scientific research. Video and still cameras, with appropriate lighting, provide the recording and observing capabilities.

*Mystic* (DSRV-1) was launched at San Diego on 24 January 1970, and made its first dive in October 1970.

Its first deep dive to its early limited operating depth was made in January 1971 to 3,520'. The design depth of 5,000' was not realized initially, as operations were limited to 3,500' until 1973, when analysis and tests proved that the pressure hull was safe to 5,000'. In February, the first underwater mating with a simulated distressed submarine was made. In October, LCDR Raetzman mated *Mystic* to the *U.S.S. Hawkbill* (SSN-666) while submerged for the first transfer of personnel between a submarine and a rescue submersible. In December, nine mating procedures were completed at San Clemente Island, California.

On 29 May 1973, *Mystic* practiced being loaded and unloaded aboard a Navy LKA (transport ship) as a sample of "transport of opportunity." During dives in August, 1973 Lockheed's *Transquest* was used for the first time as a launch/recovery platform. The first 45 degree pitch and roll mating procedures were completed on the practice submarine hatch *(Deep Seat)* during a dive in September.

The first joint DSRV/ASR launch and recovery was made in April 1974, aboard *Pigeon* (ASR-21). In May, at 2,000', *Mystic* simulated a mating on the practice hatch fixture demonstrating the ability to transfer

*DSRV-1 Mystic on HMS Repulse*
Courtesy Naval Photographic Center, Washington, d.C.

*During February and March 1982, Mystic completed the most successful Mosub operation in three years with U.S.S. Pintado (SSN 672).*

---

personnel at greater depth. In late May, off the California coast, *Mystic* conducted mother submarine (Mosub) operations with *U.S.S. Seawolf* (SSN-575) to test high speed transit effects on the submersible while attached to a Mosub.

In January, 1975 *Mystic* was launched from *Transquest* and recovered on *Pigeon* (ASR-21). Eight dives in May were made on the practice submarine platform as well as a deep dive to 5,000'.

An F-4 aircraft was searched for, located, and photographed during five dives in January 1976. After six months of overhaul, *Mystic* made sea trials, practiced mating procedures, and made two test dives to 5,000' in August.

Following training dives and practice mating dives, *Mystic* was launched and recovered eight times on *U.S.S. Pintado* (SSN 672) in March 1977, to as deep as 2,036'.

After training dives on the mock up submarine hatch was performed, true mating maneuvers and latch ups were made on the *U.S.S. Pintado* (SSN 672) in April and May 1978. During August and September, *Mystic* was mobilized and made three dives aboard *U.S.S. Bergall* (SSN 667) on the East Coast. In September, the citizens of Mystic, Connecticut, near Groton, had a chance to observe their namesake in port.

In August 1979, multiple dives to 5,000' were made for training and pilot familiarization. Hatch mating training on the mock-up submarine continued during February and March 1980. After 16 routine training dives, *Mystic* conducted mother submarine training operations off the Southern California coast aboard *Pintado* (SSN-672). In April and June, an anchor that had been missing from *U.S.S. Proteus* (AS-19) was located. In November, a lost Swimmer Delivery Vehicle (SDV) was located off San Clemente Island and subsequently recovered. An eight month major maintenance overhaul began in November.

After overhaul, refresher training was conducted in August 1981, followed by assignment to standby status as a rescue submersible, while the DSRV-2 named *Avalon* was on an overseas training trip.

During February and March 1982, *Mystic* completed the most successful Mosub operation in three years with *U.S.S. Pintado* (SSN 672). Five dives were performed in five days in the Southern California area for submarine/submersible mating and personnel transfer training. Operations with *U.S.S. Pigeon* (ASR-21) were conducted to certify *Pigeon* as a DSRV support platform. In December, a fly away (transport by aircraft) was made to Charleston, South Carolina in an Air Force C-5a by *Mystic* for operations, including two dives, to certify *U.S.S. Ortolan* (ASR-22) as a DSRV support platform.

Training dives and dives to test new equipment were made in early 1983. An F-18 crash site in 2,000' of water off Dana Point, California, was searched for, surveyed, video-taped, and the first DSRV/aircraft recovery was conducted.

In June 1984, *Mystic* performed a CH-53E helicopter debris field survey and object recovery. In July, *Mystic* participated in the first Mosub/ASR/DSRV operation and fly away. The fly away was conducted to Cape Canaveral, for operations with *Seahorse* (SSN 669) and *Ortalan* (ASR-22). In December 1984, *Mystic* and *Avalon* (DSRV-2) performed the first MOSUB operations with two DSRVs, an ASR, and *Transquest*.

During February 1985, *Mystic* was involved with classified object recovery. During March and April, *Mystic* participated in the recovery of the Amatec-Straza Remote Operated Vehicle (ROV). In May and June, services for the Naval Civil Engineering Laboratory and the Naval Ocean Science Center were conducted. An anchor from the *Arcadia* was located which *Pigeon* (ASR-21) recovered. In December, services for the Pacific Missile Test Center were performed.

During May 1986, *Mystic* made four dives with *U.S.S. Blueback* (SS-581) making six seals on the after hatch and one on the forward. During September, 1986 a NATO exercise called Sorbet Royal 86 was conducted in Stavanger, Norway. The *U.S.S. Billfish* (SSN-676) was the mother submarine. The simulated disabled submarines were *HNOMS Utsira* (Norwegian) and *FS Agosta* (French). Both a C-141 and a C-5A were used for the fly away. Two dives on each submarine were made from the Mosub averaging eight hours per round trip. Crews of the two foreign submarines boarded *Mystic* and visa-versa for tours. After the fly away and return to San Diego, a lengthy overhaul was begun.

The first dive of 1987, after overhaul, was made in September for sea trials. An equipment problem caused an abort, and repairs and additional overhaul were required.

The overhaul was completed in August 1988 with sea trials conducted in September. Ten testing and 13 training dives followed in September and October. *Mystic* conducted Mosub operations with

*DSRV-1 Mystic Hovering over Test Pool*
Courtesy Naval Photographic Center, Washington, D.C.

*U.S.S. Pintado* (SSN-672) in November and December.

During June 1989, *Mystic* conducted a search for the tugboat *Barcona*. In late June, *Mystic* conducted search and recovery operations for missile components for the Naval Ocean Science Center. In August and September, *Mystic* conducted operation Sedgemoor 89 with the Royal Navy near Glasgow, Scotland. After transit in an Air Force C-5 aircraft, *Mystic* was embarked in Mosub, *HMS Repulse* and conducted simulated rescue "flights" to "disabled" submarine *Opposum*. During the course of seven days underway on *HMS Repulse*, *Mystic* completed six sorties for 60 hours of operational time which included 18 complete mating cycles. She also provided submerged, underway DSRV indoctrination and transport for 165 VIP passengers from six NATO countries. After return to California, *Mystic* conducted a mine shaped target recovery for the Pacific Missile Test Center.

The first dives in January 1990 were for personnel training purposes. In May, the upgraded Integrated Control and Display (ICAD) system was tested and validated. Dive #700 was made in June during a *Deep Seat* practice. In July, a search for an Air Force short range missile was attempted. The remainder of the year's dives were for *Deep Seat* practice and ICAD test and training.

In June 1991, four dives were made for the Naval Oceanographer's Office, which located, identified, and video/photo-surveyed 18 targets in an aircraft debris field, with recovery of key items. The *Deep Seat* frame was mated on for demonstrations for the Naval Oceanographic Office representatives in June and again in September. Training dives were conducted in November and December.

*Mystic* was on standby during January and February 1992, with a disassembly and refurbishment period from March through September. The Submersible and crew were recertified to 5,000' in October via dock and sea trials, followed by training dives in November and December.

*Mystic* made 71 dives in 1993. Four were for training on Mosub *Tunny* (SSN 682); the rest were for training. *Mystic* made only one dive in 1994 for training and then went into an extended overhaul lasting through 1995.

Both *Mystic* and *Avalon* have fully met their original design requirements. They have repeatedly demonstrated (both nationally and internationally) their capacity to perform submarine rescues to depths of 5,000'. They are capable of being transported from wherever they are stationed to any potential rescue site. Their well-trained crews are ready at all times.

# *NEMO*

*1970–1973*

# Nemo

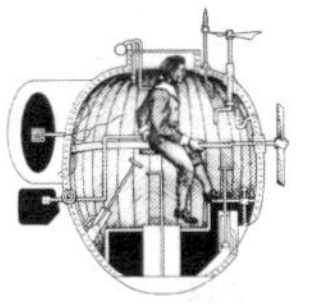

After graduating from Pennsylvania State University, Jerry Stachiw worked at the Naval Civil Engineering Laboratory (NCEL), Port Hueneme, California, where he assisted with the development of some underwater concrete test chambers called SEACON. The unmanned chambers were being constructed for lengthy offshore submerged material evaluation purposes and required recovery and reinstallation every year or two for inspection. Stachiw proposed an all acrylic tethered submersible for interim inspections to be called the Naval Experimental Manned Observatory or NEMO.

Auguste Piccard, the father of Deep Submergence, had written about such a transparent hulled submersible in his 1956 publication *Earth, Sky, and Sea*. Calling such a middle depth ship a Mesoscaphe (in Greek: mesos scaphos) : "... as for the cabin, it could be made of steel like the bathyscaphes ... but it would be very much better to build it entirely in transparent Plexiglas ... it would present a wonderful panoramic view to the observers ... to calculate the strength of a Plexiglas cabin would not be easy ... it would be necessary before having a real cabin built to proceed to numerous tests with models subjected to high pressures. It would probably be impossible to run it off in two hemispheres ... it would be possible, for example, to replace the two hemispheres by twelve-twelfths of a sphere, all exactly equal and glued or welded together. Their contours would be

**Name:** *Nemo*
**Builder:** *Naval Civil Engineering Lab*
**Length:** *7.5'*
**Maximum Depth Rating:** *600'*
**Capacity:** *2 Man*
**Year Launched:** *1970*
**Description:** *Artist's Rendition*
Courtesy of Naval Ocean Science Center
San Diego, California

the central projection on the sphere of the twelve pentagons of a regular dodecahedron. The thirty joints would be flat surfaces ... If circumstances, financial or otherwise, do not permit me to construct this new apparatus myself, I hope that someday someone will be found to take my project in hand and bring it to a happy conclusion."

Auguste Piccard would have been glad to know that, two years after his death in 1962, the twelve piece, spherical-shaped, acrylic hulled NEMO project was initiated at NCEL. Twenty-two 15" diameter scale models were made from 1/2" sheet acrylic. Twelve pentagon-shaped spherical segments were glued together. Vacuum-forming of the spherical segments was done against an interior mold of fiberglass. In addition, two of the full-size 66" diameter hulls with 2 1/2" thick walls were fabricated as prototypes for testing and evaluation. The first full-size hull was completed in December 1967. Steel inserts were located at the top for a hatch and at the bottom for pass-throughs. For the full-scale hulls, gluing was modified from the close fit method used on the models to a 1/8" wide controlled width joint. A commercially available glue (PS-18) was used with moderate optical success and more than adequate structural strength. Reaction of the models and prototype spherical hulls to short-term pressurization, implosion, sustained loadings, and to cyclic loading was determined experimentally. The calculated collapse depth was confirmed by destruction under pressure of one of the full-scale prototypes at 4,150 psi.

Following the in-house development work at NCEL, a contract for three of the 66" diameter *Nemo* hulls with 2 1/2" walls was let to Swedlow Inc. of Los Angeles. Swedlow was able to provide strong, optically clear joints, using a proprietary modified version of the PC-18 glue. The successful delivery of the three Nemo hulls with 2 1/2" walls soon led to the fabrication of three *Nemo*-type hulls with 4" thick walls for the Link Industries' submersible, *Johnson-Sea-Link* (see Chapter 23). The design and fabrication for the *JSL* hull was virtually identical to the *Nemo*, except that 6061-T6 aluminum was substituted for the metal areas.

The design of *Nemo* as a tethered underwater vehicle is unique, in that its travel is primarily vertical, about a 380 lb. anchor. Basically, the spherical pressure hull sits on a cylindrical housing for a winch and a battery compartment. The main ballast tank is located between the sphere while the cylindrical housing and provides a geometric transition between the two. Two 1 1/2 H.P. hydraulic motors with 8" diameter propellers, outboard of the ballast tank, provide minor excursions to the limits of its tether when neutrally buoyant, as well as rotation about the anchor line. The main ballast tank has an 8 cubic foot capacity, which can be recycled twice at the 600' operating depth, using cylinders of compressed air. A 5 H.P. pump can deballast at 2.8 gpm at the operational depth.

The main electrical power for the winch, motors, and environmental systems is provided by lead acid batteries. Emergency silver zinc batteries are located inside the sphere. Life support is supplied for eight hours of normal usage with a 24 hour emergency capacity. The greenhouse effect, due to the transparent hull, requires ice to be carried for temperature and humidity control in near-surface operations and/or tropical waters. The 2 H.P. winch can raise or lower the vessel at 60-100 feet/min. An underwater phone provides communications while submerged, with a radio for use at the surface. The pressure hull design

*Nemo on Display at Disneyland*

partment via the ballast tank. Three curved vertical structural beams provide a cage around the pressure hull. Pad eyes at the top of the beams provide for lifting the entire weight of the vessel.

After assembly, the complete vehicle was manned and pressure-tank tested at Southwest Research Institute (SwRI), San Antonio, Texas, for strain gage analysis and operation of equipment under hydrostatic pressure.

In the May of 1970, sea trials were initiated at the Bahama Islands. A total of about 90 dives were made at the operating depth of 600' or less, with one dive to 720'. After leaving the Bahama Islands, the vessel was operated during early 1972 in the Santa Barbara Channel for about twelve dives for certification and inspection of SEACON. *Nemo* was transferred to the Naval Undersea Center in San Diego in mid-1972. Operations at San Clemente Island for shark research, and recertification occurred until its final dive, #119, in October 1973.

After NUC completed its work with the submersible, *Nemo* was transferred to Southwest Research Institute in San Antonio for storage. In 1993, it was displayed in front of the submarine ride at Disneyland, California. It can be said that Stachiw took the acrylic-hull concept to the "happy conclusion" that Piccard envisioned, for

provides a safety factor of seven for short term pressurization. If pressurization occurs over a 50 hour period or longer, a safety factor of only two exists, due to the flow characteristics of the acrylic. When out of the water, the weight of the spherical pressure hull rests on the bottom cylindrical com-after *Nemo* other *Nemo*-like hulls went into projects such as *Johnson-Sea-Link* and these eventually led to the casting of solid acrylic hulls. While not the first acrylic submersible, *Nemo's* 12 piece construction technique bridged a gap until it was possible to cast thick hemispheres in one piece. The vast amount of quality research accomplished by Jerry Stachiw on the *Nemo* project made him the world's most respected expert on acrylics. He is consultant for most, if not all, of the official specifications on acrylics including Lloyds of London and the American Bureau of Shipping.

# *Johnson-Sea-Links & Clelia*

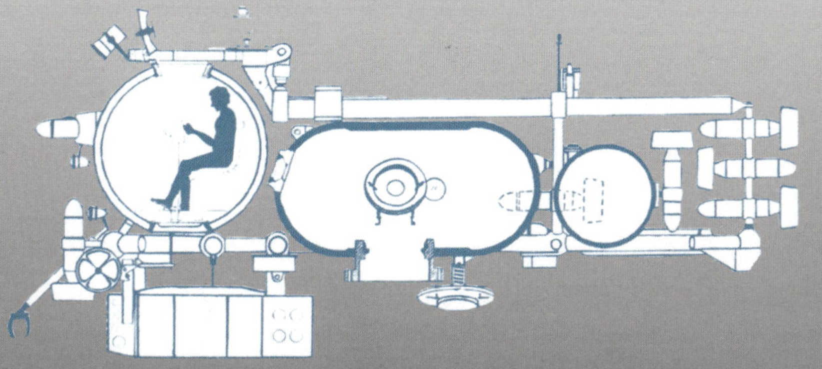

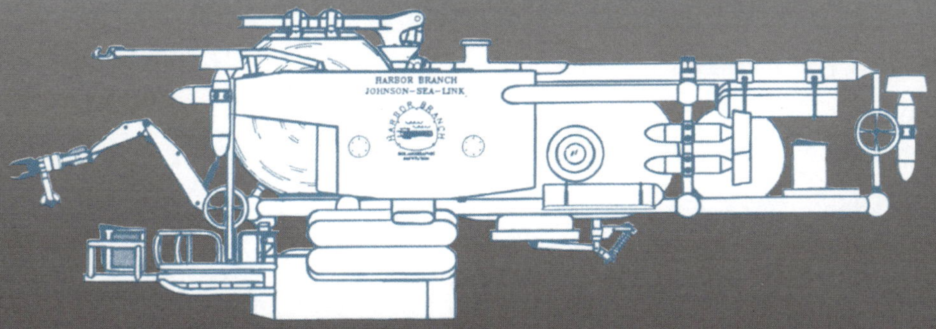


## *1971–(     )*

# Johnson-Sea-Links & Clelia

Ed Link developed the famous flight simulator called the Link Trainer, used in WW II by almost every U.S. aviator in the armed forces. After selling his business interests in the trainer, he became interested and pursued underwater projects which lead to his joint effort with John Perry to build the Perry/Link submersible, *PL-4*, which launched in March 1967. After the *PL-4,* Link shared his dreams and plans for an acrylic submersible with Dr. William McLean and myself during brainstorming sessions while on a half dozen business trips to San Diego. Years later, at Link's invitation, Dr. McLean and I attended the christening ceremonies of the *Johnson-Sea-Link (JSL)*, as his acrylic submersible was called, in January 1971. In a lucky photo shot, we got the only picture taken of Seward Johnson Sr., Chairman and co-founder of Johnson & Johnson with Ed Link of the Harbour Branch Foundation in the *JSL* during the launching. This was the only dive that these two visionaries and pioneers made together.

***Name:*** *JSL-1* ***Builders:*** *S. Johnson and E. Link*
***Length:*** *23'* ***Maximum Depth Rating:*** *1,000–3,000'*
***Capacity:*** *5 Man* ***Year Launched:*** *1971*
***Description:*** *Johnson and Link at Christening of JSL-1*

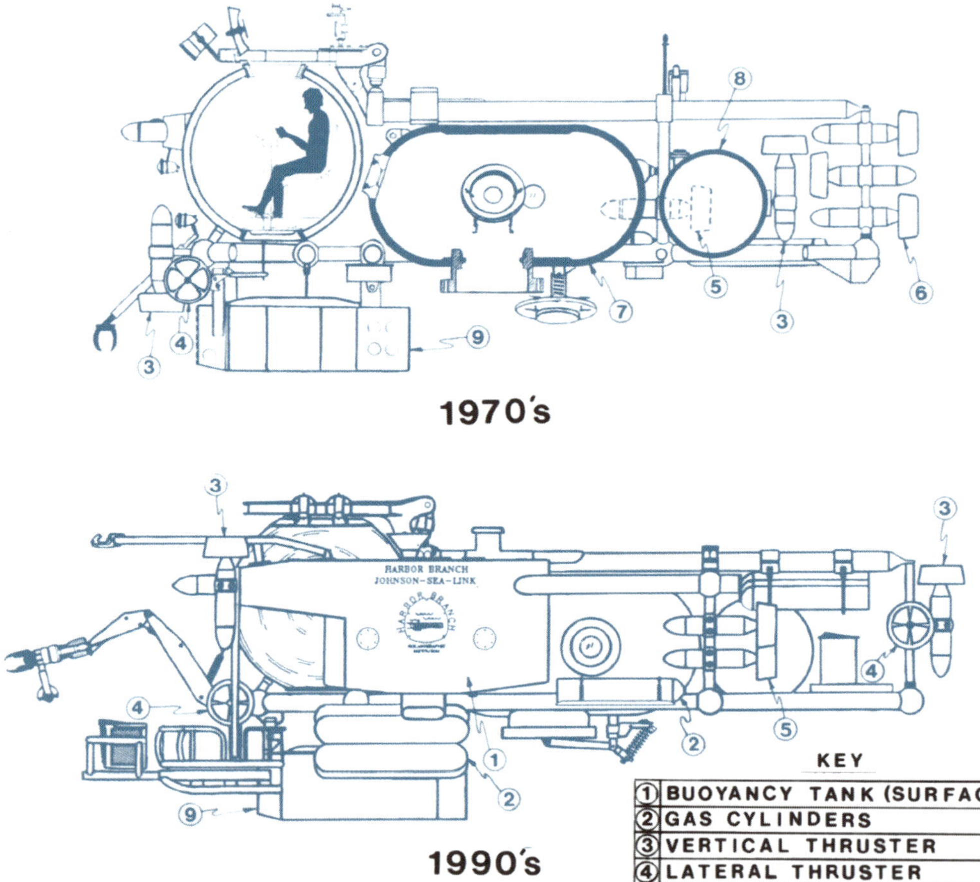


KEY

| | |
|---|---|
| 1 | BUOYANCY TANK (SURFACE) |
| 2 | GAS CYLINDERS |
| 3 | VERTICAL THRUSTER |
| 4 | LATERAL THRUSTER |
| 5 | SIDE THRUSTER |
| 6 | PIVOTAL THRUSTERS |
| 7 | DIVERS COMPARTMENT |
| 8 | HELIUM SPHERE |
| 9 | BATTERY COMPARTMENT |

*Comparison of JSL-1 Before and After Modifications made in 1990*

The *Johnson-Sea-Link* hasn't changed in any major aspect in the two plus decades since it was christened. A welded aluminum tubular frame with spherical joints at all intersections, supports all the vital components. The pilot's acrylic sphere, an aluminum cylindrical-shaped diver's chamber, batteries, buoyancy/ballast tanks, and thrusters are all mounted to the frame for easy accessibility for maintenance or replacement. Link thought that a "smoothly styled exterior would only add greatly to the cost and have little effect on streamlining at slow speeds and that higher speeds were unnecessary." Instead of one large and expensive motor and speed controller, he substituted six propulsion pods, each having a motor with propeller on each end. Link said, "it would be a simple procedure in case of trouble to quickly substitute a fresh motor unit for the defective one without delaying use of the submarine."

Three of the units were pivotable at the sub's stern, acting somewhat like a rudder. Two units were for vertical thrust, one at the bow and one at the stern to provide pitch and/or depth control. Two of the pods were located at the sides of the vehicle to assist with fore and aft thrust. A single pod at the bow provided lateral thrust to augment turning. By the 1980s, all of the rotating pods had been altered to fixed positions. They included:

- A vertical thruster at the bow and one at the stern
- A lateral thruster at the bow and one at the stern
- Two thrusters pods on each side of the vehicle for primary fore and aft propulsion

Each of the 1980s thruster pods consisted of 1 1/2 H.P, electric motors with four-bladed 14" diameter propellers providing a forward velocity of 1 1/4 kts.

The 8' long by 50" inside diameter diver's chamber consists of thick walled aluminum instead of ring stiffened steel, to gain inside room and corrosion resistance. It holds three divers and has been used for underwater lock-out, as well as for personnel transfer to submerged chambers. The overall length of the submersible is 23' with a weight of 9 1/2 tons. In 1971, the total cost of the submersible of $450,000 was furnished by Ed Link.

The life support system was originally 72 man-hours but, following improvements made in the 1990s, it is now good for 480 man-hours. Oxygen tanks and the diver's helium sphere are all located external of the two personnel capsules. When the submersible is surfaced in tropical sunlight, it is air-conditioned by external blowers and ducts. Otherwise, the greenhouse effect would be intolerable.

Shortly after the first 66" diameter by 2 1/2" thick *Nemo* hulls were delivered to the Nemo project at the Naval Civil Engineering Laboratory (NCEL) by Swedlow Plastics, Link ordered three spheres of the same diameter, but with 4" thick walls for the *JSL*. Swedlow was able to use the same molds and fabrication equipment for *JSL* as had been used on *Nemo*, so Link got a good price for his spheres that had a deeper depth capability. A 24 strain gage test was conducted in the 72" diameter pressure test tank at NCEL which established that the thick spheres could be used safely to 1,000' and this established the operating depth of *JSL*.

The power is provided by the lead acid battery pack which, in an extreme emergency, can be jettisoned for 2,000 lb. added buoyancy. Surface buoyancy is provided by tanks on either side of the submersible outboard of the acrylic sphere, when blown with compressed air. In an emergency, they can also be blown at depth since they are flooded before each dive. Smaller tanks for fine trimming the buoyancy status are located within the large surface buoyancy tanks. Compressed air tanks provide the air for blowing the trim and surface buoyancy tanks.

A six function manipulator with rotating grip is located at the bow. It can be extended 6' to obtain and place objects in a collection tray located forward of the battery compartment. An underwater phone, surface radio, doppler, and sonar navigation systems aided by pingers and echosounders, provide the operations/scientific equipment.

In 1971, almost all the dives were made in the Bahamas for equipment testing and/or pilot/diver training. In May, Ed Link orchestrated a 190' lock-out dive. During the later part of the year, *JSL* was refitted with new motors and electronic equipment.

Part of the early 1972 dives were for testing new equipment and collecting data on reef life down to 1,000'. The usual bad weather, loss of communication, or failure of seals on motors and other pieces of

*A later investigation indicated that "Stover and Clayton were not adequately equipped for long term survival under water because this was intended to be a short dive in which no lock-outs were planned."*

---

equipment occurred. A thumb rule of one hour of maintenance for each hour of dive time was established. The first of several dives to the operating depth was made in January to 1,020' near Fort Pierce, Florida. Approximately 75 dives were made during the year in the Florida Keys, the Bahamas, and British West Indies. Over a half dozen lock-out dives were made for fish and bottom sample collections. Early in May, Clayton Link, Ed and Marion Link's son, was made Director of Diving for the Smithsonian Insitute which at the time, was making extensive use of *JSL*. In late May, Ed and Clayton made a lock-out dive together.

The first few months of 1973 were spent in lock-out training and coral reef surveying. In early June, *JSL* surveyed an artificial reef consisting of two WW II Navy vessels, old tires, and other junk, located about fifteen miles southwest of Key West. Return trips and dives were made with ichthyologists to study the marine life, fish traps, and ecosystems that were there.

On Fathers Day, 17 June 1973, Ed and Marion were aboard the submersible's support ship, *Sea Diver*, when *JSL* glided toward the man-made reef. Quoting from Marion's book *From Sky to Sea*, "the sub, at a depth of 362', was heading for a fish trap that had been left on the *U.S.S. Berry* by researchers who had begun reef population studies the previous day, when suddenly the pilot reported, at 10:13 A.M., that he was snagged in what was later described by rescue divers as 'a lot of spaghetti, consisting of cables, wires and antennae.' "In the forward sphere were pilot Menzies and marine biologist Meek. In the diver's chamber were Al Stover and Clayton Link. Menzies had made about 40 dives in *JSL*. Stover, at 51, was an expert diver and submersible pilot. Clayton Link had just left the Navy as a Diving Officer. A later investigation indicated that "Stover and Clayton were not adequately equipped for long term survival under water because this was intended to be a short dive in which no lock-outs were planned." The water at the entanglement depth was 40°F. The pilot Menzies had reported to the surface ship, prior to the entanglement, that the Gulf Stream current at that depth was stronger than any they had encountered before.

Link called the nearby Key West Naval Station for help "with complete faith in the results." Because of the Sunday crew liberty, it took nearly four hours for the *U.S.S. Tringa*, a submarine rescue ship, to get underway. Since they knew that help was on the way, all occupants aboard the submersible decided against any desperate attempt at making a free ascent from 362'.

*Tringa* arrived at the dive site at 4 P.M., some six hours after being called. It took *Tringa* four more hours to complete a four point mooring, which had not been expected, nor had the occupants of *JSL* been forewarned. A four point moor consists of four anchors, two aft and two forward, spread apart to form the corners of an imaginary rectangle with the ship in the middle. The heavy currents had made the four point moor necessary for the safety of the rescue divers. The first two sets of divers arrived on *JSL* at midnight, eight hours after the entanglement and call for help. The 3 kt. current and clutter prevented the divers from helping to free the submersible. Two hours later, a second team had the same result. At 1:30 A.M. communications ceased from the divers chamber. At 7:30 A.M. breathing sounds ceased from the diving chamber. At dawn a Navy diving bell arrived from San Diego, but they were unable to get through the debris to *JSL*. At 2 P.M., now Monday, John Perry's minisub arrived but could not penetrate the debris or fight the current. The next to arrive was skipper Hugh Bowlin, a civilian, aboard the *A.B. Wood 2*, some 23 hours after getting underway. He managed to free the sub within one and a half hours after arrival, using a tethered tow sled, an underwater camera, a grappling hook, and a lifting crane on his boat. The accident board later reported, "Menzies and Meek were alive in the forward sphere, but after 32 hours in the sub it was too late for Clayton and Al Stover."

Some months later, when Ed Link began to recover from the loss of his son and friend Al Stover, he began the development of the Cable Observation and Rescue Device (CORD).

No more dives were made until April 1974, when a three-week cruise in the Bahamas was conducted. In May, another series of dives was made in the Bahamas for training purposes and to photograph the *Hydro-Lab*, an undersea habitat located there. Five two-week operations took place during the rest of the year for the Smithsonian Institute as part of an Indian River study of the reefs off Florida. A two-week cruise in the Bahamas ended the year's dive programs. During the year, nine lock-outs had taken place, two of them from depths of 500'. Using new decompression tables developed by Duke University, the standard Navy decompression table time of seventeen hours was reduced to twelve hours without adverse affect. The 120' surface ship R/V *Johnson* was added to the Harbor Branch fleet as a self-contained rescue system for submersibles during the year.

*Johnson-Sea-Link 2* was christened in January 1975, and was operational in March. The new submersible had the #2 Swedlow sphere as the pilot's compartment with Lexan (polycarbonate plastic) as an interface or gasket material between metal and acrylic parts. Extensive testing of the 4" thick *Nemo* hull using the buffer material provided a 2,000' operating depth capability. The duration of the life-support system was extended from 24 hours to five days to ensure that no repetition of the *JSL 1* tragedy of 1973 could occur. *JSL 2* was partly intended as a backup and rescue capability for *JSL 1* and received ABS certification in November.

April 1975 was a busy month for *JSL 1*. Project SCORE (Scientific Cooperative Operational Research Expedition) was conducted in four segments during the month. Using the undersea habitat Hydro-Lab as an underwater home at 60', four teams of divers made excursions to as deep as 200' to the nearby habitat, or using *JSL 1,* went to deeper, more remote locations, locking out upon arrival to return the same way. Final decompression was done aboard the surface-support ship R/V *Johnson* with a backup chamber at the Hydro-Lab base. *JSL-1* spent May and June making inspection transits of the shore-line from 100' to 600' between Melbourne and West Palm Beach.

Project CORD (Cable Observation and Rescue Device) was made operational in July as a submersible rescue tool using video and manipulator equipment on a specially designed ROV for the purpose.

Many of the dives in 1976 included diver lock-out operations. The new capability of having one of the *JSLs* standing by as the other did diver lock-out was used to good effect. The deepest lock-out was made from *JSL 2* at 600' for 15 minutes. A decompression period of 18 hours was required as *JSL 1* stood by under water as back-up if needed. All dives were local or in the nearby Bahamas.

In 1977, a total of seven dives to 1,000' were made in *JSL 2* in the Bahamas in February and March. Each dive consisted of very slow ascents and descents for 3-4 hours. Most dives were at night with all lights, including a new xenon, turned on to observe the contents of the deep scattering layer. In July, The National Oceanographic and Atmospheric Agency (NOAA) enlisted the use of *JSL 2* for a survey of the Civil War ironclad *Monitor*, sunk off Cape Hatteras, North Carolina during a storm in December 1862, in 220' of water. Underwater sleds with cameras had made a composite collection of pictures into one complete image in 1974, but "hands on" close-up inspections were needed to see if recovery, in whole or part, was indicated. The results of dives made by lock-out diver inspection was that *Monitor* was in too delicate a condition for recovery of anything more than minor artifacts either from on board or nearby.

In the fall NOAA, University of Washington, Western Washington University, and the Harbor Branch Foundation Inc. coordinated a project for a deep water platform for deep water cosmic ray tracking. The platform permitted *JSL* to land on a lock-out compartment for the transfer of equipment and people to a dry underwater chamber containing photographic emulsion plates. The plates had to be poured, dried, and assembled at the work site, using 1,000' of water as a shield against background radiation. The first cosmic ray detector was deployed in October. The mission was repeated 3 days later with the installation of the second detector in the chamber and completed on Day 4. In March 1978, the recovery of the first photographs of cosmic ray tracks from the deep ocean occurred as divers transferred dry from *JSL 2* into the cosmic tracking chamber located near the Bahamas, at 1,000' beneath the surface. The photograph plates were developed in the divers' compartment and were sent ashore for scientific analysis by computer scan. In October, the remaining plates, left one year before were transferred and developed in *JSL 2* during a three dive series.

During the year CORD 2, the second generation submersible rescue ROV, was completed and tested. In addition to CORD's various special tools other tools were developed to be used on the *JSLs*. These included; a suction device for biological sampling called a "critter gitter," water quality analyzers, and biological culturing trays dropped by surface ships, later to be floated to the surface by bags attached to them from the *JSLs*.

During a dive in 1979, near the Bahamas, *JSL 2* was testing a recovery reel of line at a depth of 1,040' when they encountered a school of 50-100 sharks ranging in length from 6-12'. One of the sharks was seen to circle the submersible several times just before a loud bump was felt and heard on the right side of the submersible, at the same time as the underwater telephone gave out a loud metallic ringing sound. The pilot stopped all power to the only propulsion unit running on that side of the submersible and continued the mission without it. Later, upon recovery, large teeth marks were found on the cover around the propeller blade. The consensus among the crew and biologists was that the sharks were schooling for the mating season and that the motivation was either territorial protection or jealousy. A total of 49 dives were made on the *Monitor* wreck during the summer. Only a signal lantern and a 65 lb. piece of the iron plating were recovered.

In May 1980, *JSL 1* and *JSL 2* dived to 2,020' at a point six miles west of West End, Grand Bahamas Islands, for ABS certification of both submersibles to 2,000'. Until that dive, both submersibles had an operational depth limit of 1,000'. The depth capability was increased by redoing the pressure tests and changing the pass-throughs to lower the stresses in the hull areas near them. During May, a one atm. personnel transfer was made between *JSL 2* and a sea bottom chamber, while mated at a depth of 1,980'. During the summer, 31 dives were made for the National Marine Fisheries Service as a baseline study prior to the area being drilled for oil. The operations were 1,000' or less and located 140 miles East of Woods Hole on George's Bank. Later that summer, nineteen dives were made for NOAA and Texas A&M in the Gulf of Mexico at the East Flower Garden Reef, 110 miles SE of Galveston. The beautiful area had been nominated as a Marine Sanctuary and support data was gathered to help justify the nomination.

*Beasley in JSL Hull*
Courtesy of
Bruce Beasley

Late in the year, 18 glass-filament-wound aluminum gas storage tanks were installed in *JSL 1*, after exhaustive testing in pressure tanks, to replace the existing heavy steel tanks. A layer of insulation foam was installed around the divers' compartment for safety and comfort. During the end of a season refit period, the first cast-acrylic hull was installed on *JSL 1* to replace the 12 piece, nine-year-old, *Nemo*-type sphere. It was the culmination of a long process and the forerunner of things to come.

At a 1969 underwater convention in San Diego, an artist named Bruce Beasley carried about the exhibit hall an optically clear 3" thick 18" diameter acrylic hemisphere for all to see. Upon inquiry, he told Bill McLean and myself of how he had cast it by a new process. Since commercial manufacturers of the day could not accomplish thick castings without bubbles, his model definitely got our attention. Beasley told of how he had won a $50,000 prize in a contest put on by the state of California for his 13,000 lb. clear acrylic sculpture displayed on the grounds of the Capitol at Sacramento. We put him in touch with Jerry Stachiw, who then had him make spherical then models for testing, which proved to be as strong or stronger than the 12-piece Nemo hulls. The model tests led to the fabrication and test of a 66" diameter cast sphere 4 1/4" thick that destructed at the equivalent depth of 9,000'. And so, the first submersible to have a cast acrylic hull was *JSL 1* in 1980 because a sculpturer had done the research, acquired the fabrication equipment, and extended the state of the art in acrylic manufacturing.

In March 1981, psychology research assistant, Penny Hall made a 20 minute lock-out at 282' on a 10% Heliox (helium and 10% oxygen) gas mixture from *JSL 1.* The purpose of the deep dive was for algae collection off St. Lucie Inlet, Florida. During July, *JSL 1* was dived in the George's Bank from Lydonia south to Oceanographer's Canyon to recover pingers and do photographic reconnaissance as well as to collect anemones, lobsters, and samples of any and all biological life to establish a pre-oil drilling baseline of the area before pollution might occur.

On Labor Day, 7 September 1981, Ed Link died after his health had rapidly failed, following a trip to Europe to celebrate the 50th anniversary of the Link Trainer. Marion Link reported in her book that two weeks prior to his death, he was redesigning his wheel chair and that "his attitude toward work had never changed."

*JSL 1* was used for commercial fishing gear evaluation in September. Long bottom lines anchored at either end were interspersed with buoyant vertical poles bearing hooks and bait. The system keeps the gear from tangling and prevents unwanted bottom scavengers from eating the bait. During a two week multi-objective mission in the Bahamas, they did five days of filming for a television special.

The first two months of 1982 were consigned to the overhaul of both *JSL 1* and *JSL 2.* During the year, *JSL 1* had a complete change in syntactic flotation to increase its payload capability. The sphere was removed to replace the bottom penatrators with a new design fabricated in the Harbor Branch shop. The old Strazza CTFM sonar was replaced with a newer and smaller unit, which improved the visibility from the pilot's sphere.

In 1983, *JSL 1* completed a 73 day cruise in mid-September after being joined by 30 officials from NOAA and the Virginia Historic Landmarks Commission. The 73 day mission was funded for $700,000 and was intended, in part, to help commercial fishermen find new fishing grounds, as well as to help government officials determine new fishing areas to be protected.

A five year project was initiated to bring up the turret from the *U.S.S. Monitor.* Of special interest was the 1,500 lb. anchor lying near the *U.S.S. Monitor.* In August, a lift bag attached to the four-pronged anchor by *JSL 1* broke water 16 miles SE of Cape Hatteras, North Carolina. Lock-out divers had used an underwater oxycetelene cutting torch system to sever the anchor chain. The anchor was transported to the East Carolina University campus at Greenville, North Carolina for preservation.

In April 1984, both *JSL 1* and *JSL 2* were certified to 2,640' (an upgrade from their previously certified depth of 2,000') in a dive 10 miles SE of Freeport, Grand Bahamas Island. Chief Pilot Tim Askew took Otis Barton on a 50th anniversary dive of the Beebe/Barton Bermuda dives in 1934 while operating off Bermuda. A four month dive series began in June in the Gulf of Maine, extending along the East Coast of the U.S. to the Gulf of Mexico. Part of the mission was to observe old abandoned gill nets still catching fish and estimating the damage occurring. Estimating the population of marine life such as lobster, tilefish, and grouper in potential oil drilling areas was another. Evaluating the effectiveness of commercial fishing gear in the different areas was yet another. Besides Harbor Branch, nine universities and government agencies were involved. *JSL 2* made a series of dives collecting zooplankton in the Bahamas in November, off Chub Cay. An investigator in *JSL 1* while off San Salvador Island, in the Bahamas, discovered new coralline algae plants growing abundantly, deeper — at 884' — than had been previously recorded. *Science Magazine* indicated that the deepest growth recorded previously were found at 693'. Specimens retrieved were found to be photosynthesis dependent at light levels 1/100 less than available to shallower water forms.

The first year that Harbor Branch operated both submersibles and their support vessels simultaneously was 1985. To obtain neutral buoyancy in freshwater, 950 lbs. of syntactic foam were added to *JSL 2*. A four week expedition on Lake Superior was conducted by *JSL 2* during July and August. The $550,000 expedition that was sponsored by NOAA included:

- Geochemists from Michigan State
- Biologists from the University of Wisconsin
- A curator from the Lake Michigan Maritime Museum
- Biologists from the University of Michigan and the University of Connecticut

Lake Superior has a maximum depth of 1,325', which was reached in July. At a depth of 1,300' primitive hydras (freshwater polyps with tentacles) were recovered where none were expected. Lake trout, sunken ships, pollutant chemicals (pcbs, etc.), and various zooplankton were under observation and sampled for later examination. Abandoned gill nets, bottles, and beer cans were photographed as part of the 700-1,000 pictures taken per dive in the 34°F water.

In January 1986, the Space Shuttle *Challenger* exploded over the Atlantic, 73 seconds after liftoff from the Kennedy Space Center. For nearly seven months an area of 500 square miles was systematically inspected and mapped from depths of 10' to 1,200'. The area was in the midst of the Gulf Stream so that surface currents ran as high as 5 kts. The time of year presented frequent storm fronts to limit the number of dive days available to surface launched submersibles. The nuclear powered submersible *NR-1* spent more time on the bottom than any of the other submersibles, but *JSL 1* and *JSL 2* are credited with bringing back the first photographs, video recordings, and small pieces of debris thought to be from the right-hand solid rocket booster. Early reruns of the explosion had made the right hand booster suspect as seen from recording cameras. Existing bottom debris and the mixture of right- and left-hand solid rocket motor parts made the jigsaw puzzle difficult. *JSL 2* first spotted shuttle motor pieces in March and proceeded to recover parts at the crash site some 35 miles NE of Cape Canaveral, as directed, after the videos and still camera photographs were reviewed by rocket experts. Both *JSL 1* and *JSL 2* recovered motor parts and otherwise assisted in operation "Big Lift" for the major part in April.

Using a $500,000 grant from NOAA, 28 scientists made a 25 day research expedition into Lake Superior and Lake Huron beginning in July. *JSL 1* took members from Michigan State, University of Michigan, and seven other universities and government agencies on dives for five days into Lake Huron and 16 days into Lake Superior. The Lake Superior dives were similar to the ones made a year before. The group was better prepared for observing the abundance of fish and other life forms at depths below 1,000' that had surprised scientists during the 1985 dives. Baited fish traps and shockers were used to gather specimens during the series. A comparison of the Lake Huron lake trout was made to those in the less polluted Lake Superior. Geologists made dives in the Caribou Trench near the Canadian border of Lake Superior to sample the huge trench left by glaciers during the last ice age. The Great Lakes fishing industry, which harvests between $2-$4 million annually in commercial fishing, was primarily interested in the lake trout, survival, the lamprey population that kills the trout and the results of new stocking procedures.

Both *JSL1* and *JSL 2* spent a good part of the fall months in search of pharmacy medicines from sea life, such as soft coral sponges and algae. For this work, *JSL 1* was sent to the Galapagos Islands and *JSL 2* went to the Bahamas. The project for pharmaceutical research from sea life was a joint venture between Sea Pharm Inc. of Princeton, New Jersey and Harbor Branch Oceanographic Institute. Sea Pharm was partly funded by the National Cancer Institute of Washington D.C.

During the dive year of 1987, both *NR-1* and *JSL 1* made dives in the Gulf of Mexico about 150 miles off the Louisiana coast to investigate natural oil seepage areas. The areas were expected to be developed into new oil fields, and the biological life was recorded prior to any drilling. During the investigation, a Texas A&M scientist found that many deep biological varieties seemed unaffected by the natural seepage. *JSL 2* was active on the East Coast from the Carolinas to the Bahamas.

In June 1988, Lake Michigan was added to the freshwater lakes that were investigated by *JSL 2.* Lake Superior was revisited in July in a continuing investigation of the lakes for fish studies, pollution effects, geological research, etc. *JSL 1* received 3,000' certification in August, off Freeport, Grand Bahamas Island prior to a series of biological survey dives made in August through October.

*JSL 1* began the 1989 dives in the U.S. Virgin Islands, returning to the offshore area of Florida, sampling biological respiration research and returning to the Gulf of Mexico for most of September. Bioluminescence studies were then conducted in the Bahamas at depths to 3,000'. *JSL 2* made surveys in the Mobile Oil lease blocks for the biological status and as a base-line to measure later pollution effects. Three separate missions were conducted by *JSL 1* to locate and collect pieces of a failed Trident missile off Cape Canaveral for Lockheed Missile and Space Company.

In 1990, the second Beasley cast acrylic hull was installed in *JSL 2.* (*JSL 1* had the first Beasley hull installed in 1980.) The 66" outside diameter size was maintained with a 4 1/4" hull thickness. The first of four wreckside monuments was placed in June by *JSL 2* on the *U.S.S. Monitor* in the National Marine Sanctuary off North Carolina. During June, *JSL 1* revisited a WW II Navy Torpedo Bomber (TBM) seen on the bottom at 390' during the Challenger search. The missing "Avenger" was thought to be part of the six plane flight carrying fourteen men, lost in the Bermuda Triangle in 1945 on a flight out of Fort Lauderdale. The wreckage was videotaped and photographed for possible identification by the Navy. In July, *JSL 1* installed the last three monument markers on the *U.S.S. Monitor* and on 18 August, made the first dive by the Harbor Branch Submersibles in Mexican waters, near Isla, Cozumel, for algae collection.

Beginning in April 1991, *JSL 1* began a multi-disciplinary expedition to the Mediterranean for bioluminescence studies in the Alboran Island region between Spain and Morocco, and mediator ecology studies in the Ligurian Sea off Genoa, Italy, and Monaco (seaward from the French Riviera). Biochemical research, which began in May in the Eastern Atlantic and included the Canary Islands, continued, extending to the Southeast Coast of the U.S.A. from North Carolina to the Gulf of Mexico for NOAA, doing biogeochemistry research. After transiting aboard R/V *Ed Link*, *JSL 1* made mediator studies and acoustical research in the Caribbean Sea. During August 1991, the TBM Avenger aircraft, missing since WW II, was rigged with lifting slings by *JSL 2* and recovered, while nationally televised throughout the United States. A disabled submersible recovery method called the "Pull Down Retriever" system was successfully demonstrated by the recovery of *JSL 2* from a depth of 2,400', using surface ship hardware. Afterwards, the system was carried as standard equipment on all Harbor Branch surface support ships. Late in the year, the first 5 1/4" thick cast-acrylic hull was received from Stanley of England and installed in *JSL 1*.

*JSL 1* was in major overhaul from January through May 1992, which included the installation of the new Stanley sphere. Operations for the University of North Carolina at Hattaras and Onslow Bay, N.C. took place in June and July. The National Undersea Research Center sponsored biochemical research dives in the Gulf of Mexico in August. Dives in the Dry Tortugas and in the Bahamas were sponsored by the University of North Carolina in September. Biochemical research dives were made off the coast of Peru in October. *JSL 2* underwent routine maintenance from December 1991 through April 1992. Operations for the year included dives for biochemical research in the Great Lakes Huron and Michigan. The National Undersea Research Center sponsored dives for bioluminescence studies from the Gulf of Maine to North Carolina and from Florida to the Bahamas for acoustic research.

Harbor Branch Oceanographic Institution acquired a Perry submersible, *PC-1204*, from Subsea Workboats of Seattle, Washington in January and renamed it *Clelia* for Clelia Johnson. After acquisition, it went into overhaul. After sea trials and certification dives in November, it was certified by ABS in December following 12 dives.

In 1993, *JSL 1* conducted dives for the National Science Foundation in the Bahamas in February, May, and June. Dives in late June and July were made for the University of North Carolina and Texas A&M in the Gulf of Mexico. During August and September, *JSL 1* alternated dives between the Bahamas and Jamaica. The Marine Sanctuary was monitored by *JSL 2* in July and August off Cape Hatteras. During August and September, the University of North Carolina and University of Connecticut sponsored dives off New England. *Clelia* made training dives in March and April in the Bahamas and off Florida. The University of North Carolina sponsored dives in the Florida Keys in June, while the University of Connecticut sponsored dives off New England in July and August. Rutgers University sponsored dives off New Jersey in August.

In 1994, *JSL 1* conducted dives off Florida, the Bahamas, the Turks and Caicos Islands, St. Croix, the Gulf of Mexico, and Cape Hatteras, North Carolina. *JSL 2* saw no operational service in 1994 due to complete disassembly and overhaul while installing the Stanley sphere. *Clelia* operations in 1994 included; dives off New England, North Carolina, Florida, the Bahamas, as well as the Great Lakes;

Superior, Michigan, Erie, Ontario, and the St. Lawrence Seaway. The high point of the Great Lake dives was the expedition to photograph the sunken vessel *Edmond Fitzgerald* in Lake Superior.

During the first part of 1995, *JSL 1* operated in the Bahamas for research accomplished by the Biomedical Marine Research and Marine Sciences Divisions of Harbor Branch Oceanographic Institution. The Bahama dives were followed by research in the Gulf of Maine. *JSL 1* underwent maintenance from mid-July through early October. During November, a month-long expedition off the Galapagos Islands was conducted aboard R/V *Seward Johnson* for biomedical marine research and filming of a TV documentary. On 15 December 1995, *JSL 1* completed dive #4000 off Isla Jicarita, Panama. Early in the year, a complete refit of *JSL 2* was made, which included a new 5 1/4" thick acrylic sphere, new fiberglass-wound aluminum gas storage bottles, and new high pressure stainless steel pressure tubing. A shakedown cruise in May resulted in an ABS recertification for a 3,000' operating depth. During the cruise, the Harbor Branch Oceanographic Institution's rescue ROV simulated several rescues of *JSL 2* on the bottom at 2,300' by cutting strategically located steel cables and conducting video reconnaissance of the entire sub to determine other simulated hazards. The R/V *Edwin Link* made three successful simulated rescues of *Clelia* using the pull down retriever system that has been developed for use with *JSL 1* and *JSL 2*.

*JSL 2* completed dive #2,600 in December 1995, after a year long dive program that included projects for the National Oceanic and Atmospheric Administration's National Undersea Research Programs (NOAA/NURP). *Clelia* completed dive #390 in December 1995, while conducting dives from the R/V *Sea Diver* in the Gulf of Maine also in support of the NOAA/NURP. The latter included the High School Aquanaut Program sponsored by the National Undersea Research Center at the University of Connecticut-Avery Point.

It is difficult to overestimate the professional contributions made to marine science and submersible development that began with the remarkable partnership of Johnson and Link, carried on by Seward Johnson Jr. The quality work continues to this day, supported and guided by the Johnson family's Harbour Branch Foundation. Continued safety improvements, greater depth capabilities with acrylics, a commitment to careful engineering innovation, and a wide range of applications in cooperation with many private and public institutions is remarkable enough. That these are granted the support of a fine surface fleet and a solid, well-staffed research insitiution, lends further credibility to the dream that two pioneers shared.

# *DSRV-2 Avalon*

***1971–(     )***

# DSRV-2 Avalon

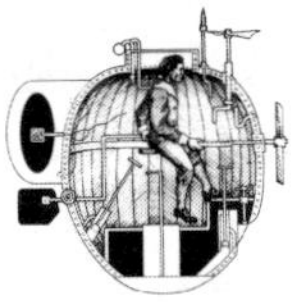

A contract for *DSRV-2* was let to the Lockheed Missile and Space Company and Draper Laboratory in October 1967, about sixteen months after the *DSRV-1* contract had been awarded. As intended, the design specifications and operational requirements were the same for both submersibles. The hull was hot-rolled into hemispheres at Hahn and Clay of Houston, Texas and assembled in Silver Creek New York by Excel-Co. The hemispheres, transfer chamber, and transition rings were made of HY-140 steel furnished by Standard Steel. The Hitco Company formed the fiberglass exoskeletal hull. A third DSRV pressure hull was pressure tested to destruction in the new high pressure test tank at the David Taylor Model Basin in Annapolis, which gave reassurance to the safety of the DSRV's operating at a depth of 5,000'.

Named the *Avalon* when launched on 1 May 1971 at the Lockheed Ocean Laboratory in San Diego, the first dive was made in September 1971. The first dive to its operational depth of 5,000' was made in December.

The first underwater mating practice took place in February 1972. Late in the summer of 1973, *Avalon* went into an extended overhaul and maintenance period.

Launch and recovery operations utilizing submarine rescue vessels (ASR's) as optional support ships began in August 1974, on the *U.S.S. Pigeon* (ASR-21).

***Name:*** *DSRV-2 Avalon*
***Builder:*** *Lockheed Missile and Space Co.*
***Length:*** *49'3"*
***Maximum Depth Rating:*** *5,000'*
***Capacity:*** *3 Crew/24 Rescuees*
***Year Launched:*** *1971*
***Description:*** *Avalon on Deck of HMS Repulse off Clyde, Scotland*
Courtesy of Naval Photographic Center Washington, D.C.

In July 1975, a 12 dive series was made during a fly away, using three U.S. Air Force C-141's for transport to Norfolk, Virginia, with the *U.S.S. Ortolan* (ASR-22) as the support ship. In October *Avalon* made a dive on an F-1 submarine hull that had sunk in 1917. The F-1 had been rammed by her sister ship, the F-3, in a fog bank during a surface run from Los Angeles to San Diego.

During early 1976, training and test dives were completed. A five dive series in mid-year was made for practicing mock submarine hatch mating techniques at different angles of the distressed submarine hatch. In July a series of matings and dematings was made aboard the *U.S.S. Gurnard* (SSN-662), which was followed by recovery aboard an ASR surface ship. Two practice mock-submarine hatch apparatus existed. One was the Simulated Distress Submarine (SDS), located near San Clemente Island, the other was the *Deep Seat*, located near La Jolla.

By November 1977, *Avalon* had passed all mating, maneuvering, and performance tests for acceptance and certification as a U.S. and NATO available DSRV.

In early 1978, a series of dives for rescue protocol practice and *Deep Seat* training at 1950' off La Jolla was made. An official visit to the city of Avalon, the DSRV's namesake, on Catalina Island, was made in April, followed by a five month overhaul period.

Emergency tow tests of *Avalon* were conducted in March 1979, in preparation for the possible event that the DSRV could not land on its mother sub or an ASR and had to be towed to a recovery location. During Operation Latent in April, the British diesel-electric submarine *Odin* reported, as a drill, that it sank in 400' of water off Scotland. The Air Force immediately sent a C-5 and a C-141 to North Island NAS, loaded *Avalon* aboard and took it to Glasgow, Scotland. From there, the submersible, with all its support equipment and the 16 man crew, were taken to the British Clyde naval base and loaded piggyback onto the British nuclear submarine *Repulse*. *Repulse* sailed to the underwater area where *Odin* was on the bottom and then released *Avalon*. The DSRV located *Odin* by sonar in murky water with only 5' of visibility. *Avalon* made eight trips between *Odin* and *Repulse* to affect personnel transfers. British technicians, not officially part of the exercise, rode the dead head return trips, just for the fun of it. The fly-away from San Diego, the rescue, and the return to San Diego, lasted two weeks. It was the first time that men had transferred from one submarine to another while both were under water. British, as well as most NATO countries, from that time on have been equipped for DSRV rescue. During the next five months, *Avalon* was assigned to the Naval Reserve unit at the 32nd St. Naval Station in San Diego for training purposes.

In April 1980, a dive on *Deep Seat* was made off La Jolla in 1950' of water. A dive made in October was made for a test to 5,000'.

January through August 1981, were spent in periodic training dives to maintain competence and availability status. During early September, a fly away to Pearl Harbor was made for additional transport on mothersub *U.S.S. Cavalla* (SSN-684) for operations in the mid-Pacific. After return to Pearl Harbor on the submarine, *Avalon* was flown back to San Diego to complete a busy nine day deployment. A nine day Mosub operation was made during November and December, which was initiated aboard the *U.S.S. Pintado* (SSN-672) and terminated aboard the R/V *Transquest*, after a submerged launching from the submarine.

Three sea operations in April 1982, included dives on *Deep Seat* and a deep dive to 5,000' for test purposes. During July and August, five dives were conducted for pilot training and side looking sonar indoctrination.

During January 1983, *Avalon* completed a 16 day operation in conjunction with and to certify *U.S.S. Pigeon* (ASR-21) as a DSRV support ship. A 5,000' dive in April recertified the submersible for deep operations. A three dive Mosub operation with *U.S.S. William Bates* (SSN-680) took place in early June. In August, a second joint UK/US fly away of *DSRV-2* to Edinburgh, Scotland, took place using a C-5A and two C141's from the Air Force for transport and *HMS Revenge* as Mosub. *Avalon* conducted four dives in the Firth of Clyde, three of which were to *HMS Opossum*. In November, a six month overhaul was initiated.

In August and September 1984, *Avalon* conducted ten search and recovery dives at the crash sight of a USMC C-53E helicopter, west of San Clemente Island. As much debris as possible was collected to determine the cause and circumstances of the accident. From October to December, several training dives from the *U.S.S. Pigeon* were conducted, a new Westinghouse side scan sonar was evaluated, *Deep Throne* was moved to a 300' depth, four miles west of Mission Beach, California, and an operation was conducted with *U.S.S. Cavalla* (SSN-684), operating from *U.S.S. Pigeon*.

During January and February 1985, *Avalon* conducted Mosub operations with *U.S.S. Pintado* (SSN-672), while operating from *U.S.S. Pigeon*. Nine dives in April and May were conducted in the Southern California area recovering a Mk-46 torpedo and a classified object. Five science dives to 3,400' were made at the Farallon Islands near San Francisco, for the Environmental Protection Agency, to investigate the waste site there. During July to November, 18 dives were made for crew training and to conduct post-maintenance tests. In December a Mosub operation from the *U.S.S. William Bates* (SSN-680) was conducted from M/V *Transquest*.

*Avalon* conducted ASR training in January 1986. In May a fly away to Alameda, CA, was conducted using an Air Force C-141 for *Avalon* and three flatbed trailers for transporting the support equipment overland from San Diego. A dive from the Mosub *Richard B. Russell* (SSN-687) was made for recovery on *U.S.S. Pigeon* at the surface. Training on *U.S.S. Pigeon* continued in November.

In early January 1987, MOSUB launch and recovery training dives took place aboard the *U.S.S. Pintado*. Nine dives were conducted in February and March aboard M/V *Transquest* simulating ASR operations. A fly away exercise was conducted to Port Canaveral in late March, using a C-5A and a C141 for air transport. Initially *Avalon* was loaded aboard the *U.S.S. Mendel Rivers* (SSN-686). Three dives were made from the *U.S.S. Mendel Rivers* with final recovery aboard the *U.S.S. Ortolan* (ASR-22). In late April, *Avalon* was embarked on *U.S.S. Pigeon* and conducted photo and video survey of the *Deep Seat* installation. Two weeks of pilot training on the *Deep Seat* and *Deep Throne* were made in June and July from M/V *Transquest*. Mosub operations from *U.S.S. Pogy* (SSN-647) were conducted in late June. Operations were conducted on Mosub *U.S.S. Gurnard* (SSN-662) between November and December.

From February through May 1988, training dives, hydraphone surveys, and transponder recovery were made on the MK-48 torpedo test range off the R/V *Transquest*. A brief Mosub operation on the

*William H. Bates* (SSN-680) was made in early June. Training dives in late June on M/V *Transquest* were followed by in-port maintenance and restricted availability through December.

In mid-1989, a post maintenance deep dive was made to 5,000'. Many training dives in the Southern California area occurred in the later part of the year, including photo video surveys and inspections of *Deep Throne*.

Several dives in 1990 were made for state officials in the San Diego area, searching by side scan sonar as well as photographing, and videotaping the resting place on the bottom of a civilian boat involved in a homicide case. The M/V *Dolores Chouest* became a support ship for *Avalon* in April. No dives were made in 1991.

In June 1992, *Avalon* conducted Mosub operations with *U.S.S. Russell* (SSN 687). In July *Avalon* conducted the first mating of a U.S. DSRV on a Japanese Maritime Self Defense Force (JMSDF) submarine off La Jolla, with *JMSDF Mochishio*. In September, Project Sorbet Royal was part of a fly away by *Avalon* to Malaga, Spain. The *U.S.S. Billfish* (SSN-676) was Mosub with Dutch, Spanish and Italian diesel boats as the simulated disabled submarines. The foreign submarines were *HNLMS Zwaardvis* (Dutch), *SPS Galerna* (Spain), and *ITS di Cassato* (Italy).

In February 1993, *Avalon* was relieved as standby DSRV by *Mystic*. In April, *Avalon* commenced scheduled restricted availability (RAV-15). After the 1993 RAV-15 period five training dives were completed.

In February 1994, *Avalon* made training and equipment check out dives culminating in a dive to 5,000'. A Mosub operation with *U.S.S. Gurnard* (SSN-662) from M/V *Dolores Chouest* as support ship was conducted in June with excursion exercises to the *Deep Seat* Platform. A fly away to Brest, France called operation Pilou 94 began with transport by a C-5A from San Diego to France. The Mosub was the French *Le Tonnant*. Four dives were made to the "disabled" submarine *Le Beveziers* in 490' of water, each lasting 10 or 12 hours. After returning to San Diego in September, an upkeep period followed.

During 1995, four major mother sub (Mosub) operations took place. The Mosub operations which included the mating of the rescue sub to the mother sub and the transfer of personnel, were with the *U.S.S. Pogy* (SSN-647), *U.S.S. Sea Horse* (SSN-669), *U.S.S. La Jolla* (SSN-701), and the British missile sub *Victorious*. Mosub operations typically consist of a series of three to six dives during a five to ten day period. The Mosub operations with the Royal Navy Ballistic Missile Sub *Victorious* took place in November off the coast of Scotland. A 13 hour flight from San Diego to Scotland by Air Force C-5A began operation Sedgemoor 95. It was the third Mosub operation by U.S. DSRV's in England and the first to operate with the newest class of British missile subs. In addition to the Mosub operations made by *Avalon*, other training and operational dives were made in the San Diego area, including a sub of opportunity mock rescue from the surface to the fast attack sub *U.S.S. Chicago* (SSN-721). *Avalon* made approximately 70 dives in 1995, which may be a record for both the numbers and types of dives made. The year's work earned the *Avalon* crew the Deep Submergence award for the second year in a row. It was the first time the award has been given to the same sub group two years in succession. *Avalon* completed dive #800 in 1995.

During the Cold War that ended when the Berlin wall fell in 1989, the two DSRV's operated "piggy-back" and covertly from U.S. nuclear submarines in the Sea of Okhotsk north of Japan. So as not to arouse suspicion, they would make infrequent visits to the bottom-planted communications cable between the USSR and their Pacific fleet. Using the DSRV's helicopter-like maneuverability and its very efficient manipulators, listening and recording devices were implanted and periodically recovered and replaced. The operation was of tremendous value until its end when a National Security Agency employee sold information about the program to the Russians and they recovered the spy devices to put in their KGB museum in Moscow.

As stated in the chapter on *Mystic*, the success of the deep submergence rescue program's original goals is undeniable, even exceeding original goals because of how functional the fly away and submarine delivery options have proven to be. Fortunately, there has been no reoccurrence of the earlier submarine accidents requiring actual missions. There seems little doubt that *Mystic* and *Avalon* with their various delivery systems will perform admirably if called upon to do so in a real emergency.

# *Deep View*

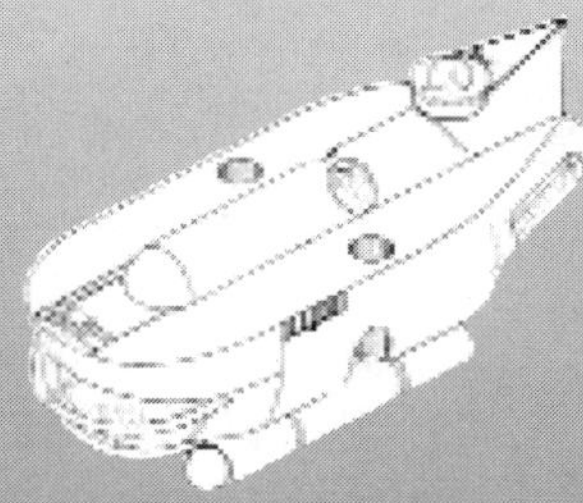

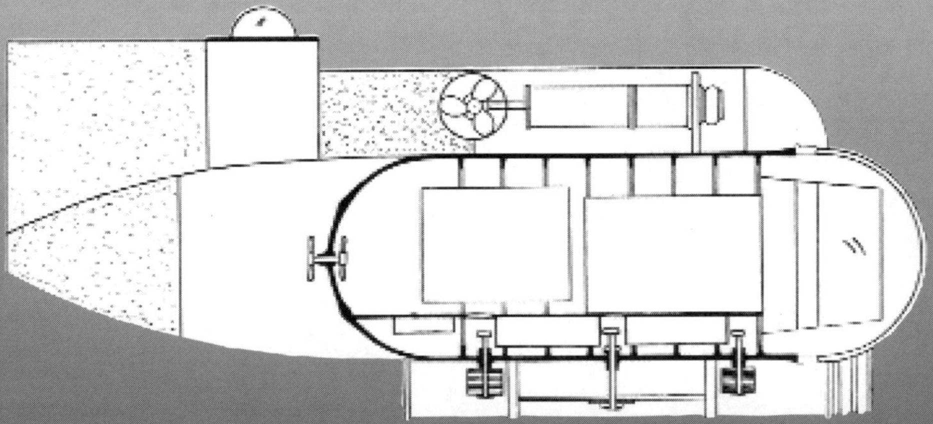

***1972–1975***

# Deep View

In mid-summer of 1966, the China Lake Moray Group was making a presentation to the Director at the Naval Ordnance Test Station (NOTS) China Lake, California. They wanted to obtain funding from the station's research account to initiate the construction of a submersible with a spherical glass pressure hull. At China Lake, we realized that glass could be the ideal material for use as pressure hulls in the deep sea. Under extreme pressures, the relatively high compressive strength of glass compensates for any brittleness. As a result of these characteristics, both high strength to weight and optimum visibility are attainable. The data from testing glass spheres gives the appearance that they grow stronger as they go deeper. Thus, it would seem to be the ideal material for very deep submersibles seeking maximum observation, an improvement over acrylics if extreme depth is the goal.

These concepts were well known in 1966, but design and testing towards actual use was almost nonexistent. However, the Moray Group had done their homework. Their presentation was commendable, but there was a major flaw in the program they proposed. Although our two groups were in competition for the station money, building submarines on a base in the middle of the Mojave dessert that mainly tested explosive weapons, made for a certain compatability: we often helped one another in a pinch. With this in mind, I kept my mouth shut about the obvious flaw during the question and answer period.

After the meeting Dr. McLean, the Director, asked me to "please stay a minute." He then wanted to know why I had been uncharacteristically quiet. I explained that the worst thing one group could do to another was to muddy up their water when they were pitching for the limited station money available. He nodded knowingly, and then said in a tone fathers reserve for wayward sons, "What's wrong with their proposal?" I paused before telling him that I thought their idea of a 6' or 8' diameter spherical glass-hulled submersible was a great idea, but that the condition of waiting two to three years to see if Corning Glass Works could produce the glass hemispheres was impractical. Even if Corning eventually agreed to fabricate the glass hulls, it meant that they had to interrupt or slow their production line of TV tubes and the like. I didn't think they would do that for a mere million dollars, as estimated in the proposal, counting the production interruptions, start up time, mold and oven preparations, and turn off time. Let alone the fact they would have to make at least ten hemispheres to get the five good ones needed. McLean nodded knowingly and asked what could be done. Off of the top of my head, I suggested that if someone really wanted a glass submersible in a hurry for a reasonable price, why not buy some of the available ready-made 44" diameter Corning hemispheres that were left over from the AT&T transatlantic cable crossing project to buoy the cables up where they crossed over canyons. It should be a simple matter to take a hemisphere and stick a steel cylinder on the back end. The glass bow would provide a wide angle view from a prone position. The Director said, "Why don't you do that?" I was a little surprised, but agreed to get him cost estimates. He asked if I thought the glass might be a problem as a structural material. With the confidence of someone who is totally naive about a subject, I replied somewhat casually that it was just another material and that, once its physical properties were determined, the structural application would be feasible. In retrospect, that was about like Columbus saying to Queen Isabella in 1491 that, to get to America after leaving Spain, just turn right at the Canary Islands and keep on going.

As I left the administration building and walked down the outside stairs to ground level, it dawned on me that I had just picked up a $500,000 project that would last at least two years. The event took place so unexpectedly and rapidly that some of the Moray project people were still walking toward Michelson Laboratory, where we all worked, only a block away.

After considering the design parameters, it was obvious that the only significant unknown was the physical properties of glass. A general design of a ring-stiffened steel cylinder, connected to a glass hemisphere on one end and a steel hemisphere containing a hatch on the other end, would be the pressure hull. Testing of available 10" diameter glass hemispheres against metal rings, representing the cylindrical hull, was started immediately. One thing was a given: the full scale glass hemispheres, since they were already fabricated and sat in Corning's warehouse in Corning, New York. Contracts for the steel hull were initiated as soon as my design drawings could be completed. No matter what the interface between the glass and the steel hull might eventually look like, the centerline of the glass rim had to line up with the centerline of the steel cylinder's rim, so the steel hull was designed immediately. The interface between the glass and the steel would take some development time and would have to wait.

The brittle properties of glass are well known. Most glass compositions can withstand extremely high compression forces as long as tensile, or pulling apart, forces are not induced to cause failure. Glass has a particular sensitivity to failure relative to the surface finish, if stressed in tension. The as-cast surface of glass is much stronger than a surface that has been polished. The best theory for the stronger as-cast or fire finish surface, is that the ground surface contains minuscule surface cracks, which under even minor amounts of induced tension from whatever cause, propagate into cracks that fail. Glass spheres, such as the Japanese floats, can withstand tremendous hydrostatic pressures even if badly scratched because under pressure the spheres are normally only subject to compression forces. Numerous tests with the available 10" diameter glass hemispheres set into metal rings were made. Gaskets and other interfaces were experimented with in attempts to improve the joint strengths. The glass-to-metal ring tests indicated that depths of several miles in the ocean could be obtained. A 10" diameter model of the complete *Deep View* hull was made. An elaborate strain gaged test indicated that the glass/metal interface design was a success and that all areas of the steel hull were also well within structural limits.

The next test was full scale with a 44" diameter glass hemisphere having a wall thickness of 1 1/4" against a steel ring, duplicating the earliest model tests. The test was run at Southwest Research Institute in San Antonio. The glass hemisphere was inserted into a large test tank with the convex glass surface in the tank and the inside or concave surface exposed toward the ceiling of the facility. In other words, the glass formed one end of the test chamber. An unknown mistake by a local contractor in fabricating the steel ring had resulted in the use of a weaker steel than specified. Without warning during the test, the glass exploded toward the ceiling with a detonation equivalent to about three sticks of dynamite. Several of us were standing 12-15 feet away reading the gages when the explosion occurred. No one was even scratched, but we were all petrified to say the least. There were no pieces of glass bigger than a cubic inch left and a lot of glass dust filled the room. Since a large concrete slab had been installed between the test chamber and the ceiling for the test, only minor damage to the roof occurred. (George Horn's laconic comment after a few moments of stunned silence was, "Does this mean we can take the early flight home today?")

The test had to be repeated, so we installed electronically amplified listening devices on the pressure test tank since model tests had demonstrated that we could hear the glass crack long before catastrophic failure occurred. The tank pressures in the next test had not reached the equivalent of a few hundred feet of depth in the ocean when the cracks began to sound off. The tank was shut down and the glass examined; it was cracked and chipped throughout the entire area of the glass/metal interface. A worse failure couldn't have been designed.

In an effort to figure out what was wrong with our joint, one man was sent to a photoelastic school for a few days, a laboratory compression instrument (a calibrated hydraulic press) was obtained from government surplus and photoelastic analysis equipment was obtained and set up at the press. It was obvious that the large glass was not behaving like the models, so testing was initiated on peripheral segmented samples (cut chunks of the dome edges) against segments of the steel rings. Under polarized light, transparent materials display the stress induced upon them by "distorted rainbows."

Knowing this, when the *Deep View* is 44" diameter hemispheres were selected from the many available at Corning, New York, I had improvised a crude, photoelastic inspection technique to get the glass with the least amount of residual stress in them left over from the cooling process. Using a broken pair of Polaroid sun glasses, I rotated one of the lenses in front of my Polaroid sun glasses, to discern the "rainbows" of stress left in the different hemispheres and picked out the best ones for purchase.

A trained technician can count the "number of rainbow changes" in a specimen under pressure, calculate the internal

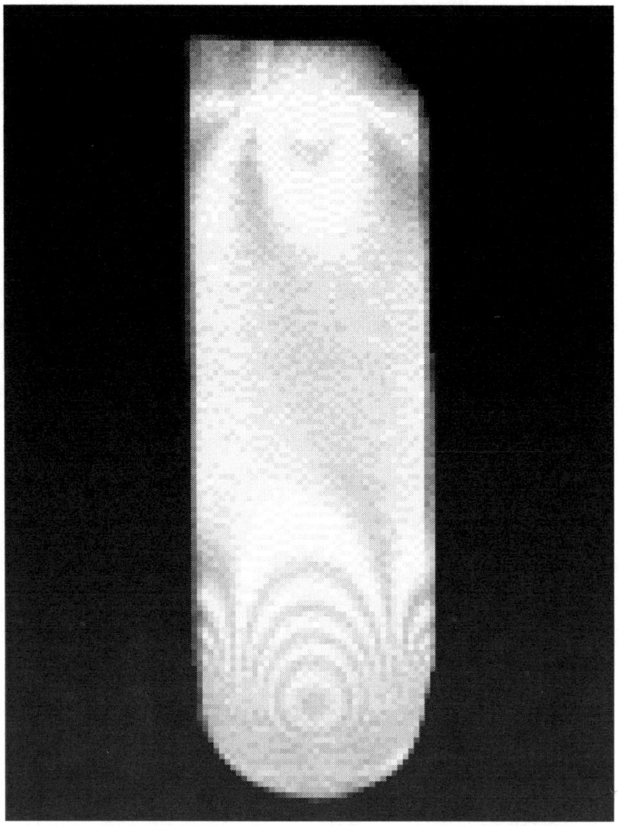

*Deep View's Glass/Metal Joint Under Polarized Light*
Courtesy of NCCOSC-RDTE Sandiego, California

stress in the glass, and photograph its distribution pattern. Analysis showed that the physical distortion of the steel under pressure was pulling the glass apart at the interface. In an effort to prevent this condition, the glass samples were rounded into a half circle at the edge while the steel was concave machined to match. Thereafter, we were able to induce extremely high pressures on the samples without failure of the material. There was some tension indicated in the interior of the samples, far from the sensitive edges and outer surfaces, but the adage at the time among experts was that "glass always fails at the surface."

The segmental samples looked so good that full scale tests were then conducted, using a new steel ring with a groove to fit the rounded glass edge of a 44" diameter dome. The glass hemisphere did not crack at any point on the surface or ground edge, but — contrary to the adage — cracked internally away from all of the previous problem areas. We had solved one problem only to invent another. Experts from Corning and Pittsburgh Plate Glass came to our shop to see for themselves the first known failure of glass occurring internally only. To further challenge the surface crack theory, not one of the internal cracks went anywhere near any edge or surface.

A technique to minimize the internal stress was to give the rounded edge of the glass a longer radius. Instead of a half circle roundness, in which the edge has a radius of half the thickness of the glass ($r = 1/2\ t$) an edge radius equal to twice the thickness ($r = 2\ t$) was used. To balance the radial deformation or change in the diameters of the two dissimilar materials under pressure, a short (6") transition ring of titanium was used between the two shells. The complete pressure hull was then tested to 2,500 psi (5,600 ft. of seawater) and cycled without failure.

Two years of research and development taught us, among many other things, that glass model test data did not scale up proportionately to full scale test results, unlike most other structural materials.

At the same time as the pressure hull development was being carried out, a full scale mock-up of *Deep View* was developed. Inside the mock up, the instrument panels, environmental systems, communication systems, and emergency batteries were placed on a trial and error basis for optimum effect. Since the mock-up duplicated the completed outer shell of the submersible, it was used to determine the hydrodynamic drag characteristics of the vessel. A large, open reservoir in a vacant section of the test station at China Lake was used to measure the drag, thrust requirements, and general hydrodynamic characteristics of the hull to better determine

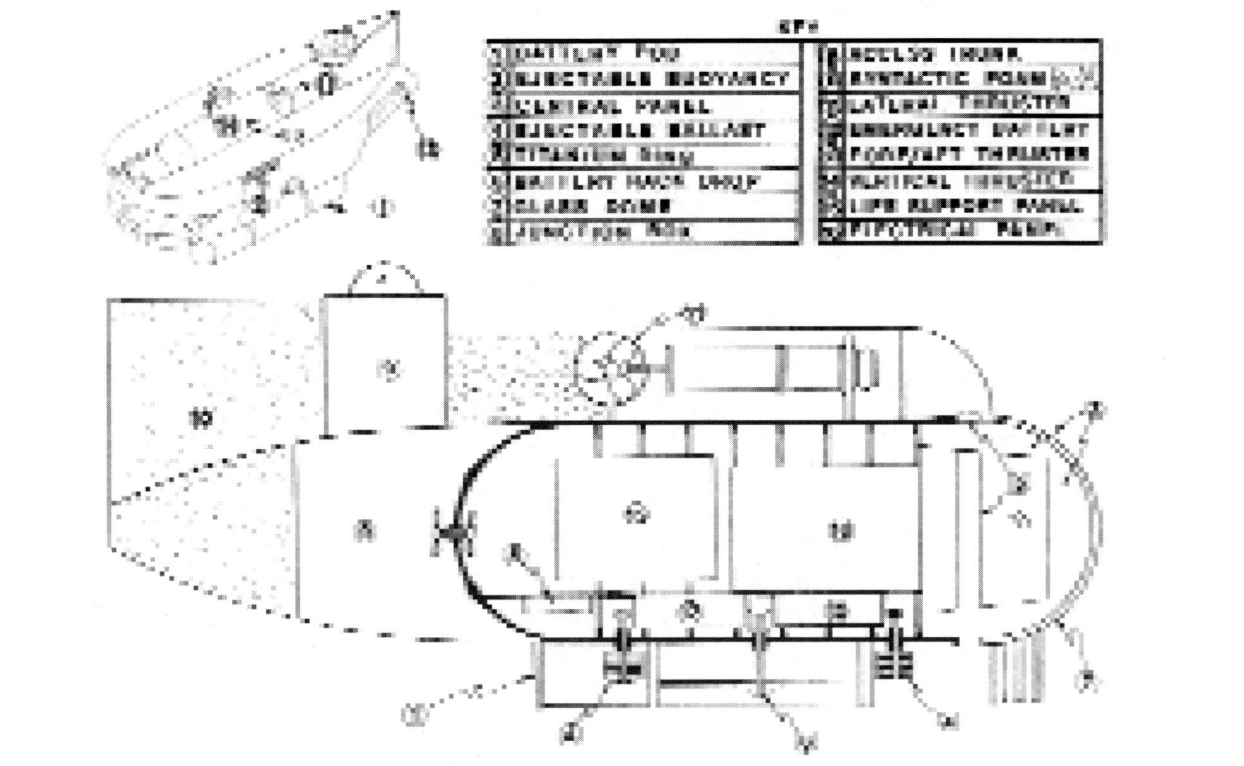

***Name:*** *Deep View* ***Builder:*** *Naval Undersea Center*
***Length:*** *16'6"* ***Maximum Depth Rating:*** *1,500'* ***Capacity:*** *2 Man*
***Year Launched:*** *1972* ***Description:*** *Detailed Cross Section of Deep View*

the propulsion and power requirements. The full scale mock-up was pulled through the water by winch, and the pulling force was then measured for several different speeds. Once it was determined that 16 golf cart batteries would be required for the vehicle's propulsion and ancillary equipment, the external battery pods were designed.

Mark 34 electric torpedo motors from Navy surplus were encapsulated in housings, filled with oil, and then tested with propellers in a pool for thrust measurements and power consumption. The 200 lbs. of static thrust from each of the motors rewound to provide 5 H.P. of continuous thrust at 48 volts was established. A contract was let to provide electronic variable speed control for the d.c. motors. Later tests of the contract fabricated control system proved it to be unreliable, so a three-speed forward and three-speed aft system was substituted for each individual motor/propeller thruster unit. The three-speed system was so satisfactory under the most difficult maneuvering and hovering situations that the variable control system was not missed and better reliability resulted. Once all the systems of the submersible were assembled, they were tested and adjusted until they worked properly. A 12 hour, two man, closed hatch encapsulation was made in the shop to test and evaluate the environmental system. During the test every other operable feature on the submersible was also tested. In the final configuration, *Deep View* weighed 6 tons was 16 1/2' long and had a payload of 800 lbs.

The H.Y. 100 ring-stiffened steel cylinder portion of the pressure hull was 6' long,

44 1/2" in diameter with a shell thickness of .65 inches. The 300 lb. glass hemisphere section of the pressure hull was made up of a composition classified as borosilicate, the same material used in making Pyrex kitchen glassware. Viewing from the forward glass hemisphere could be made in all forward directions, as well as in all directions behind the steel hull for as much as 45 degrees aft of the beam. The hatch in the H.Y. 100 steel hemisphere was 20" in diameter and opened into the transfer trunk of the sail. The trunk, when blown or pumped free of water, provided a surface buoyancy of 900 lbs. Access to the submersible pressure hull was feasible while afloat at sea through the trunk in the sail. An indicator inside the pressure hull showed that, if the trunk had been adequately emptied at the surface prior to opening, it was safe to open the pressure hull hatch from inside.

The fore and aft propulsion thrusters consist of two, 5 H.P. electric motors and propellers located at the stern of the tapered afterbody (ogive). The 10 degree angle of the motors and propellers with the axis of the submersible caused the thrust to act as if on a long lever arm extending outward from the vertical pivoting axis of the submersible to facilitate turning. Model tests showed no loss in forward propulsion efficiency during straight line courses since the heading corrections, caused by variable cross currents and drift, were more efficiently corrected for by this method. The power available was such that when divers were in the water, the 10 H.P. thrusters made it possible to flatten them against the glass bow and propel them, pinned for a short ways, as a lark. The two 5 H.P. thrusters used for vertical movement made it possible to rapidly make depth corrections. The full power in the vertical mode was seldom used except for short bursts of acceleration, instead of sustained speeds. The lateral 5 H.P. thruster was a joy to have at times, especially during a hovering mode when straight side motion was desired. The hovering capability and control was best tested and demonstrated on one occasion when *Deep View* was hovering under water with the glass nose a few inches away from a spherical acrylic chamber 15' under a boat *(Sea See)*, while the observers in each hull made gestures at one another.

The main electrical system consists of eight, 6 volt golf cart batteries in each of two droppable pods external of and below the pressure hull. A non-gassing battery was located inside the pressure hull for life support and communications during emergencies, if and when the main batteries were jettisoned or failed. An oil-filled junction box outside of the pressure hull housed the many electrical circuit breakers. All circuit breakers were tested and rated for their designed amperage in oil at sea water temperatures at the surface or under pressure.

The batteries on a rack in each pod could be rolled out on a track for maintenance or replacement, but were normally flooded in oil with gassing vents for the release of oxygen and hydrogen during discharge or while being charged. Either of the two battery pods could be operated independently of the other, in case of a power outage. The two battery pods were connected mechanically by a frame which could be released and jettisoned from inside the hull, in an emergency, for a gain of 1,100 lbs. of buoyancy.

Life support was provided to the crew of two for a normal dive of five hours plus 24 hours of emergency time. Oxygen from pressurized storage tanks and carbon dioxide removal by lithium hydroxide panels made up the primary portion of the environmental system. Scuba air tanks and breathing systems were provided in case smoke or other contaminates occur inside the pressure hull.

Ballast/buoyancy control was provided by 40 droppable 5 lb. steel plates and 20 ejectable buoyancy blocks of 4 lb. buoyancy each. The ballast plates were located under the hull in both forward and aft locations. Pitch trim was done by internally movable ballast and, in some temporary cases, by slightly shifting crew positions. Launch and recovery was made by marine railway, pier side crane, and ships of opportunity. Communications were made by underwater phones, VHF radio, and sound phones when surfaced.

*Deep View* was christened pier-side at the Naval Undersea Center (NUC) in Point Loma, California by 18-year-old Susan Forman on 15 December 1972, while about 100 observers looked on. Following the christening, the first manned underwater tests were conducted pier side and the battery pods were dropped during a simulated emergency.

A few more dives were made from the old marine railway at NUC at Point Loma to check the propulsion and navigation systems. Next, *Deep View*, a 1 1/2 ton truck, a mobile crane, and all the ancillary equipment for the submersible were loaded aboard a Navy LST landing craft and shipped to the Isthmus at Santa Catalina Island. Upon arrival at Catalina the LST let down its bow ramp on the marine railway at the water's edge, and the entire cargo load was transferred ashore. *Deep View* was loaded onto the tram car of the railway and pulled up and into the hanger built for, and formerly used by, *Beaver 4* a few months earlier. At the Isthmus, *Deep View* made two weeks of extensive dives resulting in the certification of the submersible and two pilots by the Navy before being sent to nearby San Clemente Island, California. Although the submersible had undergone documented pressure and cyclic tests equivalent to more than a mile deep in the ocean, the initial certification was limited to 100' pending an unmanned immersion to 1,500' at a later date. Lack of funding for any kind of operations later on made the deeper certification unnecessary.

After a week of dives at San Clemente, the submersible was shipped to the Naval Undersea Laboratory facility in Hawaii. A month of dives were made there in conjunction with a classified porpoise program related to the Vietnam War. About 25-30 dives were made there until the program was canceled. The submersible was transferred back to San Clemente for a week of demonstration dives for representatives of prospective funding sources.

*DSRV-1 Mystic and Deep View*
Courtesy of NCCOSC-RDTE
San Diego, California

Due to a lack of funded dive programs, *Deep View*, along with two other NUC submersibles, was "put on the blocks" until transferred to Southwest Research Institute in San Antonio. Shortly after the submersible's arrival, the prospective funding sources there "dried up" as well. In 1995, the submersible was on display in front of the Stennis Space Center near Bay St. Louis, Mississippi. Although never fully utilized to its potential, *Deep View* was a learning experience. The results of which further encourage the future use of glass and ceramic spherical configurations for deep sea exploration.

# *Significant Others*

MORAY

HIKINO/MAKAKAI

HUMAN POWERED SUB RACES

DEEP VOYAGER

TOURIST SUBS

PISCES/HYCO

SNOOPER 1

SNOOPER 2

SEA OTTER

DEEP ROVER

DEEP ROVER 2

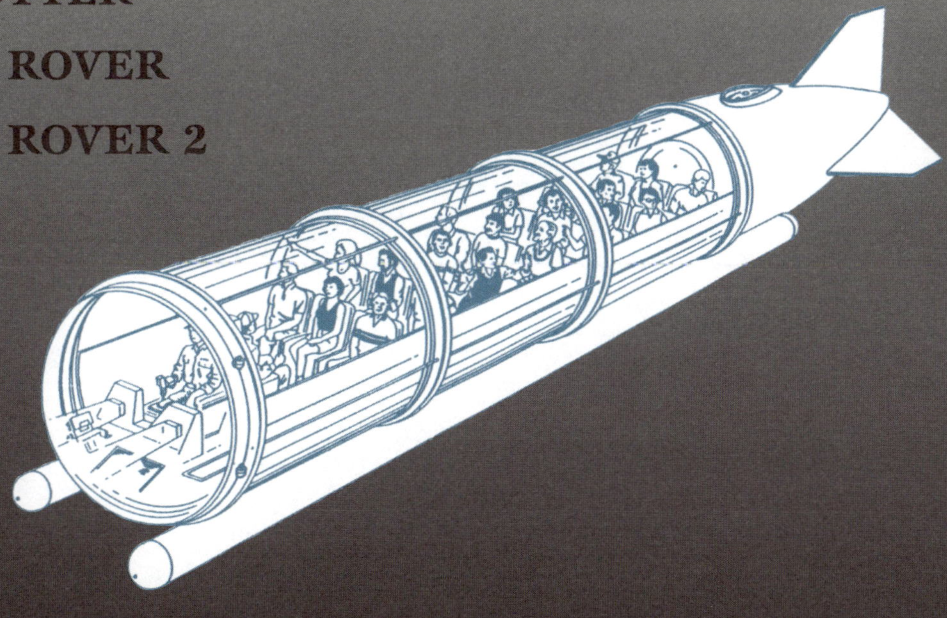

***1965–(      )***

# Significant Others

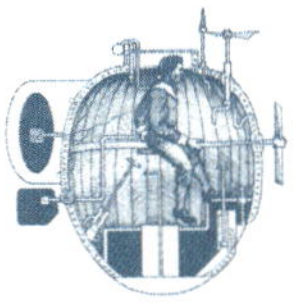

## Moray

The concept of small, fast submarines deployed by a support ship or submarine, like fighter planes from aircraft carriers, goes back a long time. Leading up to the Second World War, the Italian "Underwater Chariots" and the Japanese "Kaitens" developed what proved to be reasonably successful designs, once they were deployed. Ed Heineman, the famous aircraft designer, while employed by Douglas Aircraft Corp. in the 1950s, made studies and proposals to the Office of Naval Research (ONR) for "A Squadron of Research Submarines" and "Midget ASW (anti-submarine warfare) Submarines."

From whatever source the idea originated, the *Moray* manned torpedo, or mini-sub, was built at China Lake, and was launched at San Clemente Island, in the mid 1960s. The sub had an initial design depth of 6,000', a 15 kt. cruise capability, and was 64" in diameter by 33' long. The external flooded hull of the "torpedo" shape was made of ring-stiffened fiberglass and synthetic foam sandwich construction which provided a significant portion of the sub's buoyancy as well as contributing to the streamlined shape.

Two pressure resistant, windowless spheres, 5' in diameter with 1 $^{1}/_{2}$" walls of cast aluminum were located inside the forward portion of the fiberglass hull. The forward sphere housed two operators who "flew" the vehicle, totally on instruments, except at the surface where a TV periscope provided the visual means to see in any direction about the vessel.

***Name:*** *Moray* ***Builder:*** *Naval Ordinance Test Station*
***Length:*** *33'* ***Maximum Depth Rating:*** *1,500'*
***Capacity:*** *2 Man* ***Year Launched:*** *1964*

The second sphere contained the electronics and instrumentation necessary for the extensive sonar, guidance, and control systems used for quickly achieving underwater courses, headings and desired depths. A 90 H.P. Astro torpedo motor, turning a counter rotating propeller, was powered by a bank of 240 silver zinc rechargeable batteries. It was estimated that the cruise speed of 6 kts. could be maintained for 3.6 hours. A mercury trim system was used for pitch (nose up or down) trim with an automatic release feature for emergencies. (Mercury spills, dumped in an emergency, were almost impossible to salvage or decontaminate.) *Moray* had a displacement of 16 tons.

One of *Moray's* operational characteristics was its positive buoyancy, which required the sub to have forward motion to submerge. The first few operations were conducted in a storage reservoir at China Lake while the sub hung from a crane, using drag disks to simulate propulsion loads in place of propellers. Manned dives were all conducted at San Clemente Island. On one of these dives, Al Berryman, as pilot, made a 60 degree climb to the surface at close to full speed. Over half of the hull came out of the water and the sub came down like a breaching whale with a horrendous splash, which messed up the sonar in the nose of the sub. Operations of that kind were forbidden thereafter. Control was made easy by using a control stick with a turning wheel from a Cesna aircraft. The fins, which were the control surfaces of the sub were the same diameter as the hull, similar to submarine torpedoes.

About 25 manned dives were conducted at San Clemente Island, mostly in straight line courses with accurate depth regulation in shallow water of 250' or less. When the initial straight courses were completed, reverse turns were made to end up in the vicinity of the launch/recovery catamaran named *Catamoray*. The pilots entered or exited the sub while it was snugged up tight under the catamaran between the pontoons. Later, the catamaran became the *Sea See* boat when a transparent acrylic sphere was added. It could be extended anywhere up to 12' to 15' below the boat for viewing the ocean while underway or at anchor.

On about the 25th dive of the sub at San Clemente Island, the aged and/or unevenly discharged silver zinc batteries gassed off enough to exceed the benefits of the inert nitrogen gas they were housed in and the hydrogen/oxygen gas mix exploded, bursting the pressure housing they were contained in, and the sub sank to the bottom in 220' of water. The pilots released an emergency float and began pumping mercury ballast overboard using a pump powered by the emergency battery. At first the sub did not budge from the bottom. Slowly, the nose began to rise, but due to the loss of the buoyancy in the gas-filled battery pod, the stern didn't rise until it reached an angle of about 80 degrees. The nose of the sub just barely broke the surface in a slow manner when the Navy divers on hand – who were some of the best – quickly strapped the buoyant loose hull pieces together to prevent the sub sinking while keeping a lift-line on the sub until it was snugged up under the *Catamoray* and the pilots were out. Pilot Berryman stated later that the underwater explosion felt like he had been in a compact car at a stop light that got rear-ended by a large truck.

Up until the explosion occurred, underwater speeds of up to 20 kts. had been reached, which exceeded the original design goals. After the accident, the weakest battery cells were replaced with spares on hand, and approximately ten more dives were made at low power settings in shallow water. Once the feasibility of the concept had been proven and demonstrated, the project was discontinued.

***Name:*** *Hikino* ***Builder:*** *Naval Ordinance Test Station* ***Length:*** *16'*
***Maximum Depth Rating:*** *20'* ***Capacity:*** *2 Man* ***Year Launched:*** *1966*
Courtesy NOTS China Lake, California

## Hikino/Makakai

*Hikino* was an early attempt to demonstrate transparent hulled submersibles at China Lake, California. The 16' long, 5,700 lb. experimental mock-up was designed for omnidirectional underwater viewing using a thin, readily available, acrylic hull for operating to depths of 20' or less. Basically, it was a catamaran made of marine plywood covered with painted fiberglass with a 56" diameter by 1/4" thick acrylic sphere set between the pontoons as a capsule for the pilot and observer. Access and egress by a clam shell opening of the two acrylic hemispheres made flat water operations a requirement. Floodable sections of the pontoons regulated the buoyancy status for diving or surface operations.

Two, four-bladed cycloidal thrusters powered by 1 1/4 H.P. electric motors were mounted on the pontoons for propulsion and control. About 30 or 40 lake dives were made in the submersible.

*Makakai* was an upgraded version of *Hikino* launched in Hawaii on 14 April, 1972. The difficult and lengthy programs which would have been required to get an initially programmed glass sphere for its pressure hull certified, led instead to the use of a 5' diameter Nemo sphere for the pressure hull. The two pontoons were made cylindrical to house buoyancy chambers and contained 20 golf-cart batteries for propulsion, with three additional batteries for instrumentation power.

Two Kirsten Boeing Pi-pitch cycloidal propellers, mounted 45 degrees to the vertical axis of the submersible, provided thrust in any direction around the perimeter of the thrusters by altering the angle of the individual blades on each revolving thruster plate.

The 19' long, 5.3 ton submersible had a 600' operating depth for its two-man crew

***Name:*** *Makakai* ***Builder:*** *Naval Undersea Center* ***Length:*** *19'*
***Maximum Depth Rating:*** *600'* ***Capacity:*** *2 Man*
***Year Launched:*** *1972*
Courtesy of Naval Undersea Center, San Diego, California

with a payload of 870 lb. The experimental nature of the submersible included the development of the cycloidal propeller system, oil flooded motors, batteries, and the first digital multiplexing system used under water, as well as oil filled electrical conduits. The many developmental features made reliability a problem. In spite of these problems, the submersible was certified for 600' operations and made 40 or 50 dives in Hawaii at the Naval Undersea Center, before being shipped to San Clemente Island, for a brief period of training dives. The submersible was put on the blocks in 1973 and in 1995 sat on display at the Naval Undersea Museum at Keyport, Washington.

## Human Powered Sub Races

H.A. (HAP) Perry, president of the H.A. Perry Foundation, had dreamed for many years of a human powered submarine contest. He was partly influenced by the Kremer 5,000 lb. award offered in England, beginning in 1960, for the first human powered flight in a heavier than air type aircraft. (It took until 1977, after the contest had become international, for some Californians to fly a mile to win the prize of 50,000 lbs., followed in 1980 by the 22 1/4 mile wide English Channel flight by human power for $100,000.)

The progress in innovative aircraft design resulting from the Kremer contests confirmed Hap's dream of a human powered sub race. He felt that submersible design could use extensive improvement as well. He was also deeply concerned about the need to motivate and create a challenge for students that would give them practical experience in project management in the real world, to go along with their theoretical education and computer-oriented environment.

Hap shared his dreams of the races with Stanley Dunn, Chairman Ocean Engineering Department, Florida Atlantic University (FAU). Dunn's eagerness to help led to numerous brain storming sessions by Hap's industrial people and Dunn's University members. Consultation for underwater safety and operational guidelines was obtained from the Naval Surface Warfare Center at Panama City, Florida.

The first International Submarine Races (ISR) convened in West Palm Beach, on 22 June 1989. A human can generate approximately 1/2 H.P. under water for very short periods. The theoretical maxiumum speed possible is considered by some to be around 8 kts. The three top speeds in the first time

*Da Vinci and Crew*

trials were 2.7 kts., 2.5 kts., and 2.4 kts. Most of the submarines failed to go faster than 1.5 kts. Inability to complete the time trials was due to mechanical failure, buoyancy control, directional maneuverability, or seamanship.

A Naval Reserve diving unit provided most of the personnel for the underwater support team. On shore, the FAU students provided most of the support team on the beach as well as for the small boats used to tow the submarines back and forth from the dive sight.

The second I.S.R. in 1991 was also held off Palm Beach and had 36 entrants. After thirteen failed to finish the 100 meter time trials for various reasons, passing all the safety and water tests, the top speeds were 4.71 kts., 3.72 kts., and 3.40 kts. FAU won in the oval race track competition.

The third I.S.R. was held in 1993 off Ft. Lauderdale, Florida. Of the 50 original entries, six were withdrawn prior to the race while 24 qualified at the time trials with the top speed reaching 4.26 kts. by students of FAU. In addition to the U.S. contestants, there were entries from Germany, Canada, the Bahamas, and the U.K.

The 1995 ocean races were not held in Florida by Hap Perry and FAU for many reasons including the rising cost to Perry.

The first West Coast Invitational for Human Powered Submarines was held from 24-29 March 1994 at the Offshore Model Basin in Escondido, California. The event featured fourteen submarines and was open to submarines with either one or two crew members in three divisions — high school, college, or professional. Designs fell into two categories: propeller and non-propeller.

The West Coast races are more or less a spin-off from the Florida races. In contrast to Florida's open sea arena, an enclosed pool with 72 degree fresh water measuring 300' long, 50' wide and 15' deep is instrumented for 10 meter long, time trials after the submarines have attained full speed — in effect, similar to drag races. The first year's winner was a propeller entry from Florida Atlantic University (FAU) which went 5.9 kts. The West Coast Invitational is organized under the auspices of UCSD's Scripps Institute of Oceanography, and the April dates placed the events in the middle of spring break to encourage student participation.

During Christmas vacation in 1995, the fourth International Submarine Races (ISR), were held inside the David Taylor Model Basin (DTMB) operated by the Naval Surface Warfare Center, Carterock near Bethesda, Maryland. The fourth ISR consisted of time trials only, in a large enclosed pool. The pool used for the time trials at DTMB measures 1,186' long by 51'wide by 20' deep filled with Potomac river water. Records set by the twelve propeller driven entrees were:

- Two man, 10 meter run by Tennessee Tech. 5.656 kts.
- One man,10 meter run by Nicoloff 4.920 kts.
- Two man 100 meter run by Tennessee Tech. 5.359 kts.
- One man 100 meter run by Nicoloff 5.301 kts.

## Deep Voyager

At China Lake, California in the late 1960s, we were busy trying to get 82" diameter, one piece, glass spheres with hatch holes for submersible pressure hulls, fabricated by contract. As principle investigator, I was managing a $1,200,000 contract for the manufacture of several 16" diameter glass models and three 82" diameter full size glass spheres. The fabrication of the

16" models was a success. The full scale spheres had a 6" space between the male and female molds into which molten glass at 2,500 C (4,500°F) was poured and then cooled at a controlled rate, using heaters in the casting molds to prevent the glass surfaces from cooling faster than the core and cracking. It was a time consuming program.

***Name:*** *Deep Voyager* ***Builder:*** *W. Forman* ***Length:*** *70'*
***Maximum Depth Rating:*** *20,000'* ***Capacity:*** *3 Man*
***Year Launched:*** *Never Built*
***Description:*** *Artist's Rendition of Proposed Design of Deep Voyager*
Courtesy of Naval Weapons Center, China Lake, California

In the midst of the lengthy glass hull development program, I was approached by Tap Pryor who wanted to know if it was possible to build a submersible which could go under water at maximum depth from Hawaii to the Pacific Coast of the U.S without refueling enroute. (Tap's father had been a good friend of Charles Lindbergh, so the parallel between the benefits of long pioneering voyages was not lost.) The glass pressure hulls appealed to him greatly since their design depth was to be over 20,000' with omnidirectional viewing as an added bonus.

I suggested a technique for propulsion that was a bit unique. It encompassed the combination of two projects that had developed independent of each other. Torpedoes were recovered from the ocean floor after they had finished their test runs on the underwater range at San Clemente Island, when hydrazine inside the torpedoes was passed over a catalyst of palladium. That caused the generation of nitrogen gas which filled a balloon on board and lifted a float on a line to the surface for easy recovery.

At that time, General Dynamics/Convair was demonstrating in a San Diego swimming pool, a small submarine with wings that could glide downward when made heavy by ballasting, and glide upward when made light by adding compressed air to a buoyancy tank. The project was called "Whisper" for its lack of propulsion noises. Due to its compressibility, the use of air for submersible propulsion in a buoyancy/ballast system would be impractical for any significant depth, but using a gas that is less compressible and can be generated from a non-cryogenic source in place appeared feasible.

Combining the glass spheres, the hydrazine lift system and the gliding submarine concepts resulted in project "Deep Voyager." For maximum depths and minimum dive cycles, a course from Hawaii to San Diego would take the submersible into the 20,000' deep northeast Pacific basin between the Murray and Mendocino fracture zones at a glide angle of nine degrees. At 50 miles between the upper, near surface cycles during the trip, it would require a total of 50 cycles over the 2,500 mile course from Oahu to San Diego for a 16 day trip at 6 kts.

The proposed submersible was to consist of three glass spheres located astern of one another at the bow of a 70' long torpedo-shaped fiberglass fuselage. Aft of the third sphere were 30' of bellows filled with 8,000 gallons of hydrazine which extend into most of the teardrop-shaped afterbody (ogive). The lift from the 18' wings of the vehicle in the gliding mode was intended to augment the lift from the fuselage. The forward glass sphere housed the three submersible pilots and the cockpit with all of the operational controls and instrumentation. The middle sphere contained life support supplies.

Propulsion was to be provided by the lift in the buoyancy spaces between the spheres when filled with nitrogen gas during upward glides and flooded with sea water during the downward glides.

One thousand lbs. of water for drinking, or as an additive for freeze-dried foods, was to be supplied from sphere #3, as was 120 lbs. of freeze-dried food. One inch diameter stainless steel piping between the spheres was to transport, via pumps, all of the materials. Standard submersible instrumentation and equipment was included in the pilot's cockpit with a bed for the off-duty pilot to sleep in during the long trip.

The design of *Deep Voyager* was completed prior to the final attempt to cast an 82" diameter, glass pressure hull. On the next to last day of the controlled cool down of the third sphere, after two previous castings had failed, a heating element burned out resulting in the glass casting having a small crack through the entire thickness in one area. The operation was a success but the patient died. The large drain on the China Lake exploratory money supply prevented any further attempts at making glass pressure hulls, in spite of the data indicating that the process was practical. Without the glass pressure hulls *Deep Voyager* was not feasible and neither Tap — who had rounded up most of the necessary funds for the submersible and even its launching "hanger" at the Makai Range — was unable to find extra funding for the additional castings.

The gliding submersible concept could be extended to the transport of non-compressible or hazardous materials quite economically. Oil, wine, grains, etc., when freighted under water, beneath the surface storms, might in some cases compete favorably with existing surface ships. *Deep Voyager* as conceived, was workable. It offered the obvious pleasures of an adventure and probable scientific merit as it cycled from surface to sea floor in its long transit. The military applications of a vehicle that neither gives off heat nor sound during movement equally challenges a creative person's imagination.

## Tourist Subs

The tourist submarine business is generally considered to have had its start in 1964 at Lake Geneva, Switzerland, where "The world's first passenger submarine" named the *Auguste Piccard* was the star attraction at the Swiss National Exhibition held in Lausanne. The submersible was built by Jacques Piccard and named after his famous father. From July 1964 to October 1965, it carried some 33,000 tourists to as deep as 1,000' in the lake for what was then the equivalent of $10 U.S. per dive per person. Unfortunately, the lake was absent of

| ATLANTIS TOURIST SUBMARINE OPERATIONS AS OF DECEMBER 1995 | | | | | | |
|---|---|---|---|---|---|---|
| **Tourist** | **Seating Capacity** | **Subs Tonnage** | **Date of First Dive** | **Dive Location** | **Total Dives to Dec. 1995** | **Passenger Dives*** |
| Atlantis 1 | 28 | 49 | Dec. 1985 | Grand Cayman | 18,997 | 531,916 |
| Atlantis 2 | 28 | 49 | Feb. 1987 | Barbados | 22,380 | 626,640 |
| Atlantis 3 | 46 | 80 | Jul. 1987 | St. Thomas, V.I. | 15,399 | 708,354 |
| Atlantis 4 | 46 | 80 | Jul. 1988 | Maui, Hawaii | 20, 998 | 965,908 |
| Atlantis 5 | 48 | 80 | Oct. 1988 | Guam | 23,135 | 1,110, 480 |
| Atlantis 6 | 46 | 80 | Nov. 1990 | Aruba | 8,136 | 374, 256 |
| Atlantis 7 | 46 | 80 | Sep. 1989 | Kona, Hawaii | 12,584 | 577,2089 |
| Atlantis 9 | 48 | 80 | Jun. 1991 | Honolulu, Hawaii | 14,033 | 673,584 |
| Atlantis 10 | 48 | 80 | Jul. 1991 | Honolulu, Hawaii | 12,876 | 618,048 |
| Atlantis 11 | 48 | 80 | Feb. 1992 | Cayman Islands | 7,169 | 3444,112 |
| Atlantis 12 | 48 | 80 | Aug. 1994 | Cancun, Mexico | 2,529 | 121,392 |
| Aloha 2000 | 66 | 160 | May 1994 | Honolulu, Hawaii | 5,204 | 333,056 |
| **Total Dives** | | | | | **163,440** | **5,270,658*** |

*May be slightly lower due to occasionally unfilled seats

***Name:*** *Atlantis 14*
***Builder:*** *Atlantis Submarine International*
***Capacity:*** *64 Man*
***Year Launched:*** *1964*
***Description:*** *Interior and Exterior Views of Tourist Submarine Atlantis 14*
Courtesy of Atlantis Submarines
Vancouver, B.C., Canada

| ACTIVE TOURIST SUBMARINES AS OF DECEMBER 1995 | | | | | |
|---|---|---|---|---|---|
| **Year** | **Submarine Name** | **Capacity** | **Location** | **Operator** | **Designer Builder** |
| 1985 | Atlantis 1 | 28 | Brand Cayman | Atlantis | Atlantis |
| 1987 | Atlantis 2 | 28 | Barbados | Atlantis | Atlantis |
| 1987 | Atlantis 3 | 46 | St. Thomas | Atlantis | Atlantis |
| 1987 | Mareia (RS 250) | 46 | Saipan | Dosa Pacific Sub | Laiviat./Wartsila* |
| 1988 | Looking Glass (LG 50) | 48 | St.Thomas/France | Looking Glass Tours | Fluid Energy |
| 1988 | Golden Trout | | Canary Islands | Subtrek | Laiviat./Wartsila* |
| 1988 | Atlantis 4 | 48 | Maui, Hawaii | Atlantis | Atlantis |
| 1988 | Enterprise (LG 50) | 46 | Bermuda | Enterprise Submarine | Fluid Energy |
| 1988 | Mareia 3 (RS 250) | 48 | Cheju-do, Korea | Daekuk | Laiviat./Wartsila* |
| 1989 | Atlantis 5 | 48 | Guam | Atlantis | Atlantis |
| 1989 | Coral Adventure (RS 250) | 46 | Kagoshima, Japan | Time Associates | Laiviat./Wartsila* |
| 1989 | Atlantis 7 | 48 | Kona, Hawaii | Atlantis | Atlantis |
| 1989 | Moglyn | 46 | Okinawa | Jap Sub Tourism Ltd. | Mitsubishi Heavy Ind. |
| 1989 | Golden Salmon (RS 250) | 40 | Canary Islands | Subtrek | W-Sub |
| 1990 | Sinbad MK 3 | 48 | Hurghada, Egypt | Sinbad Tourism Co. | W-Sub |
| 1990 | Jacqueline (SM 100) | 48 | Eilat, Isreal | Scandive Coral World | Submarine Oy/Frabeco |
| 1990 | Atlantis 6 | 48 | Aruba | Atlantis | Atlantis |
| 1990 | Su Fun (MK 3) | 46 | Alicant, Spain | Sun Fun SL | W-Sub |
| 1991 | SPT-16 | 48 | Florida/Columbia | Ft. Apache Marina | Piccard/Sulzer/US Sub |
| 1991 | Beluga (Odyssey II) | 16 | Bali | PT Sub Safaris Asia | ISS/ISE |
| 1991 | Mundo Magico MK 3 | 36 | Marabella, Spain | Top Diving S. A. | W-Sub |
| 1991 | Atlantis 9 | 48 | Oahu, Hawaii | Atlantis | Atlantis |
| 1991 | Atlantis 10 | 46 | Oahu, Hawaii | Atlantis | Atlantis |
| 1991 | Atlantis 11 | 46 | Grand Cayman | Atlantis | Atlantis |
| 1992 | Seabus (Acrylic) | 48 | Monaco/Bahamas | Comex | Comex |
| 1992 | Hai KYan No. 1 (MK 3) | 48 | Taiwan R.O.C. | Jan-An Steamship Co. | W-Sub |
| 1992 | Nemo Primero (MK 3) | 48 | Mallorca, Spain | Nemo Submarines | W-Sub |
| 1992 | Tritone (MK 3) | 48 | Capri, Italy | Gorgonio, S.R.L. | W-Sub |
| 1992 | Tritone 2 (MK 3) | 48 | Genoa, Italy | Plankton, S.R.L. | W-Sub |
| 1993 | Seamaid TS 4 | 48 | Martinique | CBSP | Bruker Meerestechnik |
| 1993 | Atlantis 12 | 46 | Mexico | Atlantis | Atlantis |
| 1993 | SM-100/26 | 24 | Columbia | Barracuda Sub Invest. | Submarine Oy |
| 1993 | SM-100/26 | 24 | Fiji | Liesure Sub Holdings | Submarine Oy |
| 1994 | Atlantis 14 (Aloha 2000) | 66 | Oahu, Hawaii | Atlantis | Atlantis |
| 1994 | SM-100/50 | 46 | Oahu, Hawaii | Submarines Hawaii | Submarine Oy |
| 1994 | SM-100/50 | 46 | Oahu, Hawaii | Submarines Hawaii | Submarine Oy |

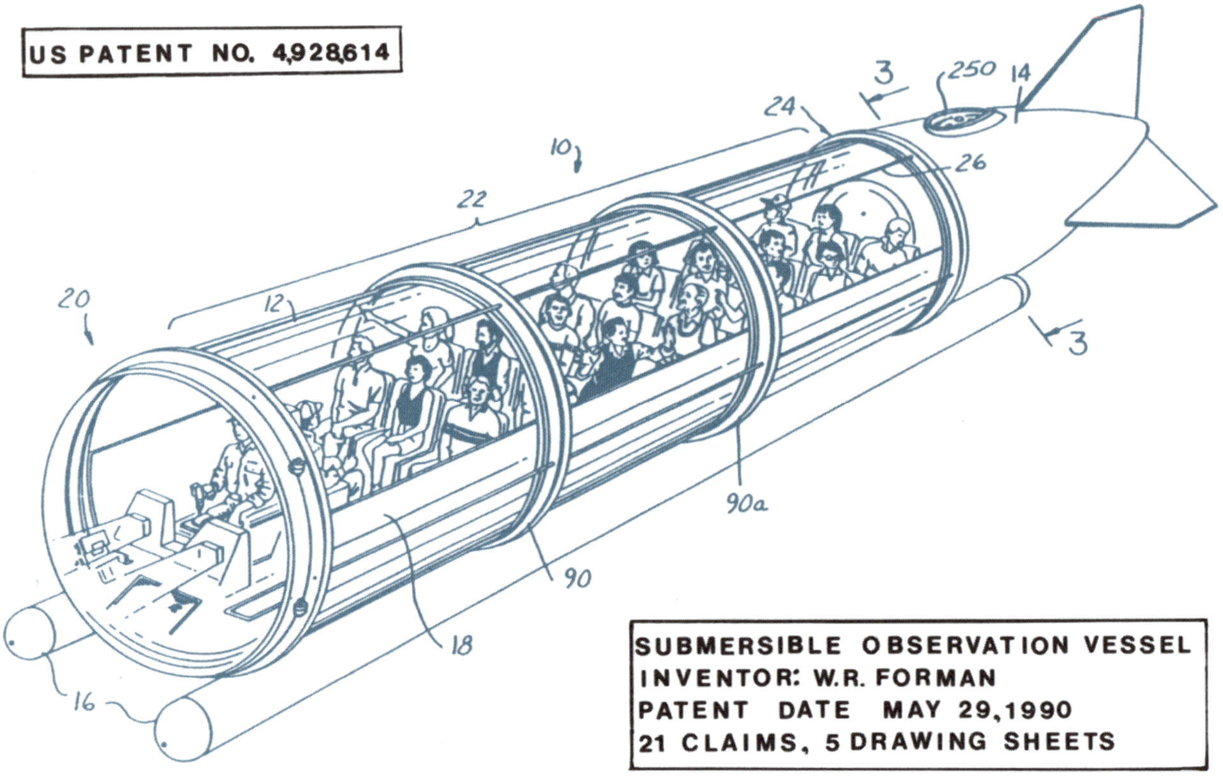


***Name:*** *Sea View* ***Builder:*** *W. Forman* ***Length:*** *35'*
***Capacity:*** *32 Man* ***Year Launched:*** *Never Built*
***Description:*** *Design of Sea View Transparent Hulled Submersible*

any life and uninteresting for tourists to look at. The submersible had seating for forty passengers, was 94' long vehicle and had a 75 H.P. propulsion motor with a pivotal stern propeller and enough lead acid batteries to provide a range of fifty nautical miles at 6 kts.

Atlantis Submarine International of Vancouver, British Columbia, a surviving pioneer in the tourist submarine industry, began the present day era in the business on Grand Cayman in the Bahamas on 18 December 1985 with *Atlantis 1*. As of December 1995, 12 of the 29 active submarines on the International Tourist Submarine list are owned and operated by Atlantis. Their latest version, *Atlantis 14* launched in May 1994, holds the current record for the largest passenger capacity at 64. The attached Atlantis operational summary of December 1995 shows a total of over 163,000 dives made by 12 submarines during an 11 year period in ten locations with more than 4.5 million passengers.

Deep submersibles with transparent hulls preceded tourist submarines by nearly two decades, but funding for acrylic tourist subs and the difficulties in certifying the new material for public transport daunted most efforts until recently. Designs were initiated for the *Sea View* tourist submarine in 1985 leading up to a U.S. patent issued in 1990. Full scale pressure hulls were fabricated in the late 1980s with mock-ups of the rest of the submarine, flown to the Tourist Oceanology International Exhibition and Conference in 1990 at Monte Carlo, Monaco, in an unsuccessful attempt to find investors. In March 1992, the acrylic hulled *Seabus* was launched at

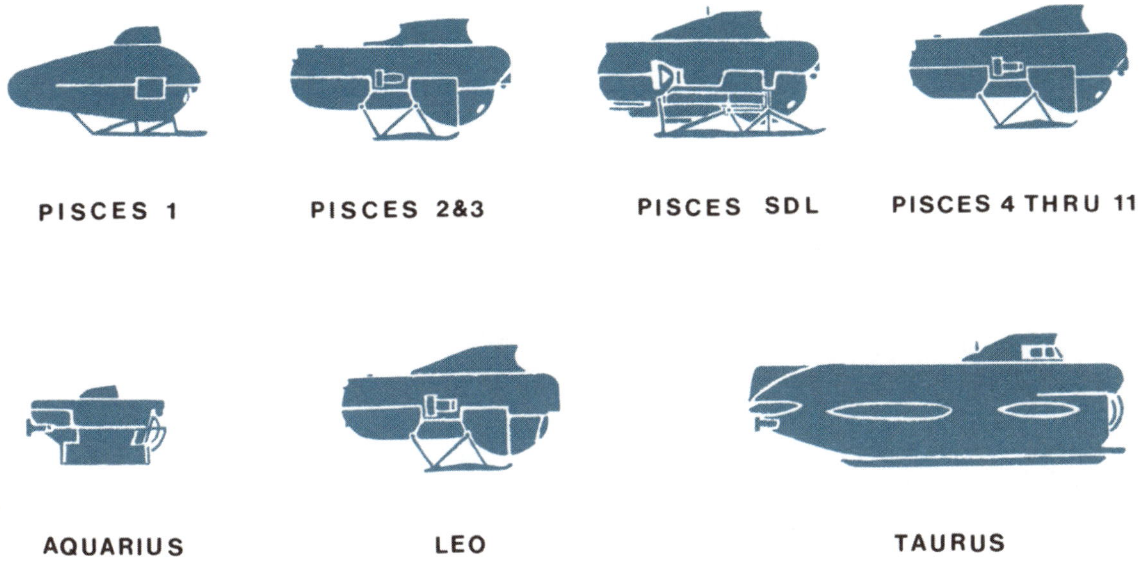


***Name:*** *Pisces/Hyco Submersibles* ***Builder:*** *Hydro-dynamics Co. Ltd.*

| *Name* | *Length* | *Capacity* | *Maximun Depth Rating* | *Year Launched* |
|---|---|---|---|---|
| *Pisces 1* | *16'* | *2 Man* | *1,500'* | *1966* |
| *Pisces 2 and 3* | *20'* | *3 Man* | *2,400' and 3,500'* | *1969* |
| *Pisces SDL* | *25'* | *6 Man* | *2,000* | *1970* |
| *Pisces 4-11* | *20'* | *3 Man* | *6,600 and 3,281* | *1972 and 1976* |
| *Aquarius* | *13'6"* | *3 Man* | *1,200* | *1973* |
| *Leo* | *21'* | *3 Man* | *2,000* | *1976* |
| *Taurus* | *34'* | *6 Man* | *1,100* | *1976* |

Monte Carlo by the French diving and submersible company Comex. In only a year, the submersible made over 1,000 commercial dives.

The 1990 Tourist Oceanology International convention has not been repeated in spite of several attempts, but new tourist submarines in new locations are being developed every year. Some submarine managers are choosing to have more than one location for a given submarine in the same year, depending on weather and the location of tourists during the year.

## Pisces/Hyco

*Pisces 1* was the first submersible built by International Hydro-dynamics Co. Ltd, later to be called *Hyco*, of North Vancouver, British Columbia. The original concept was to build a $20,000 vehicle in three months, but it took 26 months and $100,000 to complete the bare vehicle without ancillary equipment. *Pisces 1* was a two sphere vehicle, 16' long. The spheres were made up in six segments welded together, but not machined to an accurate spherical shape. The spheres were pressure tested by submergence to 2,300' in the Jarvis Inlet, 60 miles northwest of Vancouver to establish the operating depth as 1,500'. The air weight of the submersible was 7 1/2 tons. The droppable batteries had the largest battery capability of all the *Pisces* at 66 kw. hrs. Life support was provided for 72 hours. The 5 H.P. mid-ship, side mounted motors with 19" diameter propellers provided 2 to 4 kts. underwater speeds. Three plexiglass viewports 6" in diameter provided for viewing from the forward position of the front sphere.

Launched in 1966, *Pisces 1* made dive #500 in March 1970. During its career, *Pisces* was used to recover oceanographic devices under the arctic ice, retrieve off-shore drilling equipment, and serve as a salvage tool in the raising of a 90 ton tug from 670' of water. During 1969, along with *Pisces 2*, she successfully recovered over 120 torpedoes at the U.S./Canadian Nanoose underwater range. Vickers Oceanic, now British Oceanics, dismantled *Pisces 1* after its operations in the 70s, never to be reassembled again.

*Pisces 2* was launched in 1969. It was built for Vickers Oceanics, now British Oceanics of England. With an operating depth of 2,400', it was first used for general scientific and petroleum work in Europe which included a search for the Loch Ness Monster. It was used for extensive photographic coverage for Japanese TV of the Japanese 72,000 ton battleship *Yamato*, sunk in 1945 with 3,000 crewmen aboard. The submersible was used extensively on the underwater range in Canada for torpedo recovery. In 1987, shark research was conducted with the 20' long, 1,000 meter depth submersible in the Cayman Islands. Also in the Caymans, the three 6" diameter forward looking viewports were used by tourist observers beginning in October 1967. In the 1990s, the submersible was in storage in Florida.

*Pisces 3* was launched in 1969. It has a two sphere pressure hull and is of similar design as *Pisces 2*, but with an operating depth of 3,500'. Its first operations were for torpedo recovery in Canada, followed by survey work in the Bahamas. In 1971, the drain plugs in the after machinery sphere were accidentally left open before a dive. The submersible sank stern first shortly after launch in 600' of water with the stern down, almost in a vertical position. The Canadian submersible *S.D.L.-1*, which was undergoing sea trials nearby, attached a lift line to the downed submersible, and it was recovered eight hours later. On 29 August 1973, after making a routine nine hour dive installing the Canadian Transatlantic cable, 150 miles off Cork, Ireland, *Pisces 3* surfaced for recovery. During the recovery process, a towing line from its support ship fouled on the submersible's after machinery sphere hatch mechanism, which was missing a safety locking bar normally installed to prevent such an accident. The aft sphere flooded, with its hatch open, and the submersible sank rapidly. The crew in the forward sphere was safe but shaken when the submersible impacted on the bottom at 1,575' in a 70 degree bow up-angle. There was no heat in the vehicle with the outside Atlantic temperature 59 degrees and no help was immediately available.

After some delays, *Pisces 2* and *5* arrived on the site but a gale force seaway had blown up in the meantime. In an effort to connect a line to the disabled submersible, *Pisces 2* pulled off its manipulator, rendering it useless. *Pisces 5* was able to get a small line attached with a surface marker. Curve 3, a remote operated vehicle (ROV), arrived on September 1 (day 4), after a nightmare of delays, to attach a heavy lifting line to the downed submersible. After 3 1/2 hours of recovery operations, the *Pisces 3* was surfaced. The two member crew inside the submersible quickly threw open the hatch to get fresh air in a Hollywood movie type finish. They were down to 20 minutes of life support on their emergency face mask rebreather systems after an 83 hour dive, only nine of which were intentional. Five months later, *Pisces 3* was back operating in the same area with the same pilot, Roger Mallison, at the helm. The submersible is now in the museum in Waymouth, England.

The *S.D.L.-1*, Submersible Diver Lockout-1, launched in 1970 has a pressure hull comprised of two spheres joined by a tunnel. The hull was the back-up hull for *Beaver 4*. Like *Beaver 4*, three of the 11 small viewports in the bow were replaced with one large 36" diameter viewport in 1984 without decreasing the operating depth of 2,000' or its ability to make diver lock-outs as deep as 1,000'. It is operated by the Canadian Armed Forces and its home port is Halifax, Nova Scotia. A complete refit has been made of *S.D.L.-1* every four years at International Submarine Engineering Ltd. (ISE), since the 14 1/2 ton by 25' long submersible was launched in March 1970. The two 5 H.P. main thrusters were augmented by the addition of vertical and lateral thrusters in 1984. Dive series accomplished by the submersible include aircraft salvage, arctic research, and seabed studies off Nova Scotia, the Virgin Islands, and Bermuda. During 1986/87 many wreck dives were made in Canada and the Caribbean. An Arctic dive series took place in 1988. *S.D.L.-1* made its 900th dive in 1992. It remains one of Canada's most active submersibles.

*Pisces 4* was the first in the 2,000 meter or 6,600' Pisces series. Launched in 1972, it was originally scheduled for purchase by the U.S.S.R., until the Canadian government prevented the sale. It was sold to the Canadian government in early 1973 after completing sea trials and deep water tests. By 1975, she had made 300 dives for the Canadian Department of the Environment with another 133 dives made during the year for pollution studies, biological research, geological survey, etc. Dives in 1976, included seamount dives, plane crash surveys, missing undersea hardware searches, extensive diving for the Defense Research Establishment, the Institute of Ocean Sciences and Fisheries, as well as different universities. In 1983, she made dive #1,400 and in 1984-85 its operations included seamount dives and undersea chimney investigations. *Pisces 4* continues to be a busy submersible, now stationed at the Canadian Navy base in Halifax.

*Three Pisces on a Dock* Courtesy of Al Trice

*Pisces 5*, launched in April 1973, is similar to *Pisces 4* and was used by the Canadian Overseas Telecommunications Corporation to bury an offshore East Coast telephone cable. It also assisted in the recovery of *Pisces 3* in 1973. In 1974, *Pisces 5* dove on an A.T.&T. cable job off New York. After a brief stint in the North Sea on petroleum work, the submersible returned to Canada in 1976. *Pisces 5* was in storage in Canada for a while, sold first to France and then to the Cayman Islands. It was returned to active duty after ten years of inactivity, when it was sold to the Hawaiian Undersea Research Lab (HURL) in Hawaii in 1986. It continues to work actively through 1995 at HURL for investigations of seamounts, underwater volcanoes, cable routing, reef studies, marine life studies, and a great deal of diving on Loihi, the underwater mountain destined to be a new Hawaiian Island soon, in geological time.

*Pisces 6,* launched in 1975, is another in the 2,000 meter operational depth Pisces series similar to *Pisces 4*. The initial commercial dives were made off the Farallon Islands 30 miles west of San Francisco. The nuclear dump site there at 3,000' had been the repository of 55 gallon drums of nuclear waste from 1945-65. The Environmental Protection Agency used *Pisces 6* to make detailed observations of the site and to recover sample barrels. The submersible was then put to use in deep water oil fields solving problems on drill rigs, such as the deep water guiding of tool reentry and the reconnecting of tools. It set a record for drill inspection and deep work at 4,876'. The submersible was idle until 1987 when the multi-funded project Beebe was conducted in the Bahamas for undersea exploration, including a search for the giant squid, six gill sharks, etc.. Purchased by Andre Galerne's International Underwater Contractors in 1980, it remains part of their fleet of submersibles.

*Pisces 7*, built in 1975, and *Pisces 11*, built in 1976, are identical to *Pisces 4* and the other 2000 meter *Pisces*. Both were sold to the Soviet Academy of Sciences in Moscow. They were refurbished in Poland in 1989 after many years of successful operations.

*Pisces 8*, launched in 1975, has a 1,000 meter pressure hull. It was purchased by Vickers Oceanics, now British Oceanics, for European operations, including dives near Gibraltar and lengthy operations on the Trans Mediterranean pipeline from Sicily to Algeria. In 1991, it was taken out of ten years in storage and operated by Hallstrom Holdings of Singapore.

*Pisces 9* was never built. *Pisces 10* was launched in 1975 with a 1,000 meter pressure hull and sold to British Oceanics after launch and early sea trials for North Sea petroleum field work.

*Aquarius* was launched in 1973 and made its first dive in September. Its design depth is 1,200'. It differs from the *Pisces* submersibles by having a cylindrical pressure hull with hemispherical end caps. The bow dome is 39" in diameter and made of acrylic. The 13 1/2' long submersible has a 4 1/2 ton air weight. The stern mounted 5 H.P. driven propeller can turn 90 degrees to port or starboard. Bow planes up forward provide vertical control. The payload of 1,100 lbs., forward speed of 3 kts., and its small size make it a handy vehicle for a pilot and two observers. Originally built as an undersea oilhead workboat, it also made wreck surveys before being acquired by Can-Dive Services of northern Vancouver, Canada. It is currently used in conjunction with *Deep Rover* and *Sea Otter* for water tunnel inspections in the U.S. and Canada.

*Leo* was launched in 1976 and built for Sub Sea Oil Services of Italy. At 21' long, it has an operating depth of 2,000' and has a Pisces-like appearance with 5 H.P. thrusters mounted outboard for fore and aft movement. Two 1 1/2 H.P. thrusters provide vertical movement and two 1 1/2 H.P. thrusters, one forward, one aft, provide lateral motion as well as turning. *Leo* made some dives in the Gulf of Mexico for training purposes in the winter of 1978-79, but was placed in storage in 1991 and up for sale in 1992.

*Taurus*, launched in 1976, was originally a diver lock-out submersible with the pilot's section a 6' diameter by 6 1/2' long cylinder with hemispherical end closures. A diver lock-out sphere 7' in diameter is attached to the after portion of the pilot's section of the pressure hull. The operating depth is 1,100', weight 24 tons and *Taurus* has an overall length 34'. Early dives were made in the

***Name:*** *Snooper 1* ***Builder:*** *D. Siverts*
***Length:*** *14'6"* ***Maximum Depth Rating:*** *1,000'*
***Capacity:*** *2 Man* ***Year Launched:*** *1969*
Courtesy of Don Siverts

North Sea oil fields and for cable burial in the Bay of Biscay, France, in September 1980. Recently, it has gone into service after refitting as a tourist sub, carrying seven passengers at a time for ten one-hour dives per day for Silverton Ltd. of England in Loch Ness, Scotland.

## Snooper 1

In the early 1950s, Don Siverts was doing underwater inspection and photography of the outfall from the main ocean sewer pipes for Los Angeles County. The work at 200' was cold and, on scuba, the diving had to be limited to brief periods. When he was offered a dive as passenger in a friend's submersible to do the work he found that he was able to accomplish more in less time and in complete comfort which lead to his building his own submersible, *Snooper*.

The submersible has an overall length of 14 1/2' and weighs 2 1/4 tons. The cylindrical pressure hull has hemispherical end closures and an intersecting conning tower. The keel contains droppable batteries and the brow contains a manipulator and lights. The pilot sits upright with his head in the conning tower, while the passenger/observer lies prone on the floor. There are ten viewports in the pressure hull, four in the conning tower, one in the hatch and at the stern, with four in the bow. There are 72 hours of life support for the two person crew. Propulsion is provided by one stern mounted, pivotable 2 H.P. reversible electric motor and propeller. Surface support is provided by the 40' *Mother Goose* with a launching bay and overhead bridge crane. The twin diesel support vessel has a cruising speed of 9 kts. and a range of 1,000 miles.

*Snooper* was launched on 20 September 1969. Its first commercial dives were for the Los Angeles County Sanitation Department in 1969. In addition, there was part-time use of the combination submersible/support ship in 1970 for the Los Angeles County Sanitation Department for outfall photography, video taping, and general physical estimates of the conditions of the county's concrete outflow pipes at White Point, California. In 1971, the City of Los Angeles hired *Snooper* and Co. for the inspection of their outflow pipe system at Santa Monica Bay. From 1972 to 1975, both Los Angeles City and Los Angeles County hired the submersible and crew for outfall inspections. In 1977, *Snooper* made dives for ARCO Petroleum in the Santa Barbara Channel for geological survey purposes. Dives at Catalina Island were conducted for Motion Picture Marine, North American Rockwell, and the Avalon Sanitation Department. In 1979, *Snooper* was hired by Lloyds of London and Catalina Airlines to recover a sunken

***Name:*** *Snooper 2* ***Builder:*** *D. Siverts* ***Length:*** *14'4"*
***Maximum Depth Rating:*** *1,000'* ***Capacity:*** *1 Man* ***Year Launched:*** *1991*
Courtesy of Don Siverts

aircraft near Catalina Island. Dives were made for Paramount Pictures at Catalina the same year.

Routine work and outfall inspections for both Los Angles City and County continued in 1981. Another Catalina Airline and a private plane recovery were made near Avalon on Catalina Island. During 1982 dives were completed for Los Angles County outfall inspection, a marine biologist, the Shell Oil Company, Parker Diving at Catalina, and Motion Picture Marine at Santa Monica Bay. Dives in 1983 were made at Point Conception for Geo Cubic and at Catalina Channel for Parker Diving/Shell Oil. Dives in 1984 were made for inspection of the Encinitas, Goleta, and Los Angles County outfalls, with video coverage and eye-witness reports. Dives were also made for Geo Cubic in the Goleta area. The repeated outfall work for Los Angles County and City continued in 1985 with the addition of an unsuccessful two day search for a sunken racing boat in the 462' deep Lake Meade, Nevada. Diving in 1986 included outfall inspection for Los Angles County, Los Angles City, and Oceanside, California, a pipeline inspection job for Parker Diving at Aliso Creek, California, as well as dives for the University of Washington in an area of natural gas seepage. A total of 104 dives were made in 1986.

Most of the dives in 1988-91 were made in the Santa Monica Bay area for the Los Angles City outfall inspections. A total of over 1,000 dives during a 22 year period were made by *Snooper 1* between its launch on 20 September 1969 and its retirement on 20 September in 1991.

## Snooper 2

The second version of the Snooper submersibles holds only a pilot and weighs 1.6 tons. The operating depth of 1,000' is the same as for *Snooper 1*. The submersible was specifically designed for sewer outfall inspection with video recording and analysis.

The pressure hull is a 30" diameter cylinder with semi-eliptical forward and aft end domes and a conning tower. It has a total of seven viewports, two in the short conning tower and five in the forward dome, three are for direct viewing and two are video ports. Propulsion is provided by one main thruster driving a 3' diameter 3-blade, shrouded propeller. A rudder, aft of the propeller, directs the thrust. A vertical thruster and a bow thruster aid in maneuvering. A maximum speed of 3 kts. is attainable. The overall length is 14'-4". A distinctive feature is a curved pipe frame extending around the bow of the submersible to attach instruments, still cameras, lights, etc.

Launched on 9 June 1991, *Snooper 2* made outfall inspections of Los Angles City as well as Los Angles. and Orange Counties. In 1992, there were outfall inspection dives made for Los Angles City and Los Angles County as well as research dives for the Sandia Laboratories at Catalina Island. Dives for the Los Angles Sanitation Department made up the bulk of the 1993 dives. Dives made on the Los Angles City and the Orange County outfalls in 1994 and 1995 brought to a total 286 dives made since the launching of *Snooper 2* in 1991.

# Sea Otter

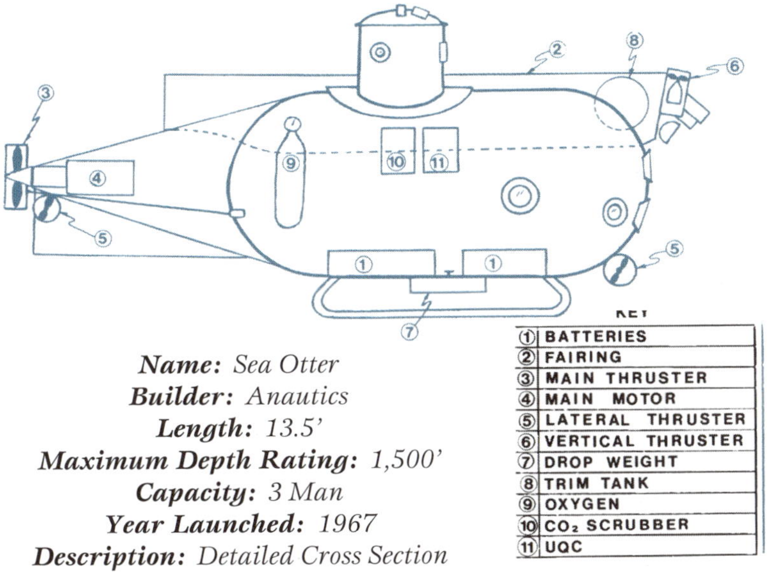


**Name:** *Sea Otter*
**Builder:** *Anautics*
**Length:** *13.5'*
**Maximum Depth Rating:** *1,500'*
**Capacity:** *3 Man*
**Year Launched:** *1967*
**Description:** *Detailed Cross Section*

*Sea Otter* was built for Anautics Inc. of San Diego, in 1967 as "Paulo 1." The pressure hull is a 4' diameter cylinder, 8' long with hemispherical end closures. The original configuration included a pivotable 4 H.P. electric motor and propeller, mounted on top of the flooded conical afterbody. Viewing was done through six viewports, in the forward end of the main hull and four in the turret and hatch. The original operating depth was designated as 600' with an estimated collapse depth of 3,650'.

After acquisition by Can Dive Services of Vancouver, British Columbia in 1971, the submersible was leased to Arctic Marine Ltd. Propulsion was changed to a 3 H.P. pivoting main motor driving a 9" diameter propeller in a shroud at the rear of the conical afterbody. In addition, two 1/2 H.P. lateral thrusters were added, one forward, one aft. A 1/2 H.P. vertical thruster

*Sea Otter* Courtesy of Mark Bradley

was also added at the bow. The pressure hull was eventually rated at 1,500'. Upgrading of the batteries increased the maximum propulsion speed to 3 kts. for 1 1/2 hours. The weight increased to 3.2 tons.

The first *Sea Otter* dive was made on 2 February 1972, at Burrard Inlet near Vancouver, British Columbia. Inspection of two new sections of the underwater power cables between Fidalgo, Decatur, and Lopez Islands in the San Juan Island group was made for Bonneville Power Administration in July, August, and September. A dive series on Cobb Seamont off Washington State was made in late September for the University of Washington, to investigate old beach terraces from the ice ages. In July and August 1973, a series of dives was made at and around Taku Inlet Alaska, near Juneau, for powerline inspections. *Sea Otter* was flown from Juneau to Crater Lake, Oregon by helicopter, for dives there in September, returning to Taku to finish another dozen dives the same year.

In 1975, a dive series was made in the Mackinak Straights of Michigan for power cable routing and pipeline inspection. A similar series was made early in 1976 in the Vancouver Island Area. In July, following an extensive refit program, *Sea Otter* was certified operational to 1,500'. In August, a series of dives was made in the oil fields off Louisiana. In September, she dove off Labrador on the sunken ferry William Carson. *Sea Otter* went through another refit program in 1978. A five function manipulator was added in 1982 during a major refit of the submersible. In September 1987, *Deep Rover* and *Sea Otter* made an inspection of the 20' diameter tunnel at Niagara Falls, Sanada where *Sea Otter* was on standby as *Deep Rover* dove over 1/2 mile into the flooded tunnel on a kevlar cable tether. A return to the Niagara tunnel was made in 1989 with the submersible *Aquarius* added to the group. *Sea Otter* did a few small jobs in the 1993-95 period for Can Dive of Canada as a standby for *Deep Rover.*

# Deep Rover

In 1980, Sylvia Earle made an unusual dive off Oahu, Hawaii. She was in a deep-ocean innerspace diving suit called a Jim Suit tethered to the submersible *Star 2.* On the bottom at 1,250', Sylvia was free to move about on the 18' line attached to *Star 2.* Inside the Jim Suit with the air at sea level pressure and temperature, the 2 1/2 hours passed "like 20 minutes" for Sylvia. After the dive she told her partner, Graham Hawkes, that the experience had whetted her appetite for a more perfect diving apparatus.

A few months after her critique, Hawkes showed her a new manipulator he called his "human equivalent arm." He then drew for her a sketch of small acrylic spheres with a person inside each one. He explained that the little spheres looked like submersibles, but that they behaved and controlled like individual diving suits, with manipulators that acted like the operator's arms and hands. The limitation, for the time being, would be 3,000', until such time as better transparent materials were available.

Funding for the new manned submersibles was non-existent until 1983, when their company, Deep Ocean Engineering (DOE), signed a contract with the Canadian company, Can Dive Services, to build an acrylic submersible called *Deep Rover.* This project was to be funded by the Government of Newfoundland and the Canadian company Petrocanada, who would use the submersible once it was constructed.

As constructed, the hull is made up of two 5" thick, cast-acrylic hypo-hemispheres, each 160 degrees in circumference, connected to aluminum rings which close together like a vertical clam shell to complete the 63" O.D. sphere-like one person pressure hull. Total length of the submersible is 8.2' with a weight of 3 1/2 tons.

***Name:*** *Deep Rover* ***Builder:*** *Deep Ocean Engineering* ***Length:*** *8'2"*
***Maximum Depth Rating:*** *3,281'* ***Capacity:*** *2 Man*
***Year Launched:*** *1984* Courtesy of Mark Brown

The operating depth of 1,000 meters (3,261') is based on a crush depth of 12,000'. The two 4' long manipulators provide human arm and hand-like control by sensory feed back technique. The lead acid batteries provide energy for the submersible lighting and the 4 electric 1 H.P. motors with propeller thrusters. Two of the thrusters are for main propulsion and are located aft of the pressure hull centerline. The other two are mounted near the fore/aft centerline of the acrylic hull and each can be rotated 360 degrees. The maximum forward velocity is 3.2 kts.

The acrylic hulled submersible was launched in Halifax Harbor, Nova Scotia, Canada in June 1984. The sale price of the prototype was $600,000. The project made its first research dive on 25 August 1984, for the Hawaiian Undersea Research Laboratory (HURL) at Waikiki. Working with the submersible *Makaliki (Star 2)*, a series of dives to 1,500' was made from the HURL Launch/Recovery/Transport (LRT) catamaran. In August 1985, a month long deepwater program was conducted involving National Science Foundation (NSF) funded bioluminescence measurements in the undersea canyon off Monterey. During November, the Coronado Canyon and the deep water slope offshore from San Clemente Island were explored. These explorations culminated in the first three dives to 1,000 meters.

After a year of planning, in September 1987, *Deep Rover* was launched inside the control gates of a dam at Niagara Falls, Ontario, Canada. The submersible dove 300' under the city of Niagara Falls for almost a kilometer inside each of two 9 km long by 46' diameter water-filled tunnels to make inspections. The ROV Phantom 300 was deployed from *Deep Rover* for close-up inspections and video recordings of the tunnel walls. A six hour dive was made on each of the two inspection days at a total cost of $230,000 for the inspection, instead of the $800,000 it would have cost to drain and inspect the system.

During 1988-90 a series of dives was made at Crater Lake, Oregon. The remoteness of the area required helicopter transport to the water-filled caldera, left when a mountain like Mt. St. Helens, erupted in about 4,600 B.C. A lightweight barge was constructed to dive into the six mile diameter lake, which is 1,932' deep with an average temperature of 38°F. The water is so clear that at times it is possible to look down the steep walls of the lake as deep as 115'. The first dive series took place from 2-23 August 1988.

Besides the general geological and biological research conducted during the first seventeen dive series, the water temperature was sampled throughout the lake in attempts to establish and locate the existence of geothermally formed currents. A conflict existed between factions supporting a proposed geothermal power plant only 7 $^{1}/_{2}$ miles from the lake and the Park Service that manages the National Park, that surounds the lake, with over $^{1}/_{2}$ million visitors a year. Contracts to Oregon State University (OSU) were let by the Interior Department to conduct the dive program over a three year period. Two OSU professors made most of the dives, along with several consultants who were brought in to give independent opinions after being taught how to operate *Deep Rover*. The investigators concluded from the dive data obtained in Crater Lake that the nearby geothermal sources of water were part of the Crater Lake circulation system. The power plant proposal was canceled and *Deep Rover* was returned to Can Dive in Canada. In late 1994, *Deep Rover* did a job for the New York Power and Light Utility. In order to inspect the water filled supply system near Niagara, it was necessary to take *Deep Rover* deep into its 20 foot diameter tunnel. A 1 $^{1}/_{2}$" diameter kevlar strength-member reinforced the fiber optic bundle was towed along for 2 $^{1}/_{2}$ miles to supply viewing by TV at the tunnel opening. It was the longest tunnel penetration ever done by a submersible at the time.

## Deep Rover 2

In September 1992, Deep Ocean Engineering was commissioned to build two, two-person submersibles similar to *Deep Rover 1*. Funding was provided by the French Cable Television Company, 'Canal'. The 66" O.D. spheres used were the ABS approved, cast acrylic "Nemo" type units approved for the Johnson Sea Link submersibles. The 1,000 meter operating depth, pressure hull assemblies, include two metal hatches, one on the top and one on the bottom of the spheres. In the *Deep Rover 2* design, the metal hatch topside is unopenable, used only for electrical as well as gas and hydraulic line pass throughs. The lower hatch is used for the normal access and egress to and from the sphere. The system requires a 20% increase in cabin pressure, above normal, to open the hatch when the submersible is at the surface of the sea due to the submerged depth of the hatch. The technique eliminates the ABS requirement for a sail or access trunk 30" above the sea surface. The technique requires that when normal access or egress is done on shore or ship, that the vehicle be elevated in some way, such as on blocks. The design, approved by ABS, is intended to eliminate accidental surface flooding.

The propulsion units are similar to those of *Deep Rover 1*, but the rotating side thrusters are fixed at a slight angle outboard from the vertical. The four thrusters are normally powered at 2 H.P. The environmental system was up-graded to 120 hours for each of two people for life support. Maximum speed is estimated at 2 kts. Overall length is 13'. The weight in air is 5 $^{1}/_{2}$ tons. Most of the other features are nearly identical to those of *Deep Rover 1*. The two-person *Deep Rovers* were purchased for $3 million before shipment from Oakland in June 1994. Shortly after they arrived in Southern France, training dives were initiated. American pilots from Deep Ocean Engineering instructed the French pilots who were going to work for the TV company. A scheduled around-the-world exposition was begun late in 1994, but stalled indefinitely in Africa in mid-1995. In 1995, Graham Hawkes and Sylvia Earle shared other more ambitious concepts for deeper diving and more manouverable versions of the *Deep Rover* as well as newer designs even more imaginative.

# *Epilogue*

# Epilogue

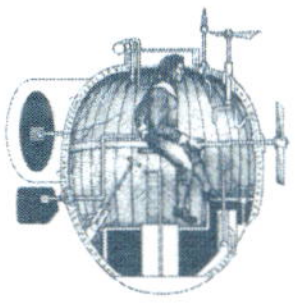

In researching the data for the 220 year period covered by the book, it was a little surprising to discover the extent that the history of submersibles was entwined with both American and world history. The *Turtle* of the Revolutionary War was well known and visited by Washington, Franklin, and Jefferson. Napoleon and Jefferson were directly involved with Fulton's *Nautilus*. Jules Verne said that he copied the name *Nautilus* from Fulton for his book 20,000 Leagues Under the Sea. That book was in turn an inspiration for Holland and Lake, let alone many of us in more recent years — with a little help from Walt Disney. President Lincoln was given a demonstration of the Union's submarine *Alligator* which was lost at sea before it could see combat. Simon Lake and John Holland virtually founded the present day submarine fleets of the world. Both had to go overseas to find interest and sales before the U.S. "battleship admirals" and their political allies could get interested in submarines. Beebe and Barton pioneered manned, undersea oceanography in the 1930s, hanging from a cable off a surface barge. Their pictures and stories in National Geographic still make enjoyable and inspiring reading.

In more recent times, *Alvin* and *Sea Cliff* have opened up a scientific cornucopia with their discoveries of hydrothermal vents and lava flows in nearly every oceanic sea floor area of the world. These areas, where life could have originated without the benefit of sunlight, contain life forms that live under heat and toxic conditions found nowhere else on earth — the stuff of science fiction that would defy even the imagination of Jules Verne. The 1977 discovery by *Alvin* of thermohydrovents near the Galapagos Islands is considered one of the most significant oceanic discoveries ever. Since hydrovents are now believed common to the edges of all submarine tectonic plates, and many other areas, both their known and potential frequency continue to escalate. These exciting areas represent an interface between our world and that planet just a thin crustal layer beneath the sea. This barely understood and newly discovered planet under the oceans has unlimited minerals and is an infinite heat source. Perhaps, it is this recent discovery that will stimulate the next technological and scientific thrust into the oceans. Never more than seven miles from the sea surface, the full value of this cornucopia will require extensive exploration by deep submersibles with trained scientists aboard.

Undoubtedly, there are new technologies at our fingertips or just out of reach that will make new explorations feasible. During the development of deep submersibles, hulls have gone from *Turtle's* wooden keg-like structure to *Alvin's* titanium to *Johnson-Sea-Link's* acrylic. Along the way, aluminum, steels of ever higher strength, glass, and composites have been progressively developed. Submersibles have progressed from human power to nuclear power. Candle light has yielded to various forms of electric light. The days when we mixed epoxy resin with glass micro-balloons in a trough using a hoe to form flotation units have progressed to factory-formed, graduated-sized hollow spheres embedded in ever stronger matrixes for use at maximum depths. As Auguste Piccard dreamed of acrylic-hulled submersibles, so do others now envision and anticipate glass-like or ceramic transparent hulls that will allow panoramic viewing while freeing mankind from depth limitations with unlimited access — perhaps with the power of an *NR-1* or "powerless" in the mode of *Deep Voyager*.

In the preface, submersible development was indicated to be at a "juncture." Indeed, many deep submersibles have become inactive or have been retired. Nevertheless, a few stalwarts prevail. The busiest of these are *Alvin*, two *Johnson-Sea-Links*, *Delta*, a few Perry subs scattered throughout the world and some Canadian *Pisces*. Most of the US Navy's small fleet is due for inactivation by 2000. By 1999, a skeleton crew will maintain *Avalon* on standby but inactive status while *Mystic* will remain active. While few regret the post cold war downsizing of the US defense budget, it surely marks the end of a once flourishing submersible era as well as most of the goals recommended by the committee formed after *Thresher* was lost.

There are some who point to the truly impressive advances in remotely operated vehicles as the advent of a new and more productive time in ocean exploration. The same argument is made against the use of man in space. And, yet, in either environment, there seems a place for both. Direct, unfiltered, 3-D viewing with the Mark-1 eyeball accompanied by on-site decision-making and management from manned deep submersibles is likely to remain the first choice for significant work under water — the more so as technology brings down cost and increases capability.

For one example, the industrial world that is baffled by how to rid itself of nuclear waste without contaminating the land, the ground water, or the oceans and its sea life may some day decide to use subduction zones of the sea floor. These areas where the tectonic plates come together return the sea floor to the earth's molten core where they originated. Manned submersibles will undoubtedly be needed in some parts of the process.

The standards and procedures of 30 years ago may seem crude today, but those of today will seem even more basic 30 years from now. Reflecting on past experience, there is one factor that is common to all work in the sea, that the oceans will do their best to humble the wise and destroy the unprepared or careless.

# Bibliography

# Bibliography

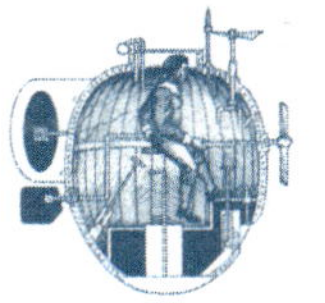


## CHAPTER 1 THE BEGINNINGS

### TURTLE (1775)

Abbott, Henry L. *Beginning of Modern Submarine Warfare Under Captain Lieutenant David Bushnell*. Hainden,CN: Archon Books, 1966.

Barnes & Noble. "The American Revolution." *NewAmerican Encyclopedia* Grolier Inc. 1991.

Burgess, Robert F. *Ships Beneath the Sea; A History of Subs & Submersibles*. McGraw-Hill Book Co. 1975.

Grant, Marion Hepburn. *The Infernal Machines of Saybrook's David Bushnell Patriot Inventor of the American Revolution*. The Bicentennial Committee of Old Saybrook, Ct. 1976.

Thomson D.W. *David Bushnell, The First American Submarine*. Naval Institute Proceedings Feb 42 Menasha WI.

Wagner, Frederick. *Submarine Fighter of the American Revolution*. New York. Dodd, Mead & Co. 1963.

### NAUTILUS

Philip, Cynthia Owen. *Robert Fulton: A Biography*. Franklin Watts, New York 1985.

Hutcheon, Wallace Jr. *Robert Fulton: Pioneer of Undersea Warfare*. Naval Institute Press, Annapolis, MD 1981.

Sutcliff, Alice Crary (Fulton). *Robert Fulton and the Claremont*. The Century Co. N.Y. 1909.

Parsons, William B. *Robert Fulton and the Submarine*. Columbia University Press, New York 1922.

Barnes and Noble, "The Napoleonic Wars, the War of 1812." *New American Encyclopedia*. Grolier and Co. New York 1991.

Thomson, David Whittet. *Robert Fulton and the Nautilus*, Naval Institute proceedings, Naval Inst. Press. Annapolis, MD Oct.1942.

Thomson, David Whittet. *Robert Fulton's "Torpedo System" in the War of 1812*. Naval Institute Proceedings, Naval Inst. Press. Annapolis, MD. Sept 1946.

Burgess, Robert F. *Ships Beneath the Sea: A History of Subs and Submersibles*. McGraw Hill, NY NY 1975.

### HUNLEY

Barnes and Noble. "The American Civil War." *New American Encyclopedia*, Grolier Inc.1991.

Blair, C.H. *Submarines of the Confederate Navy*. U.S. Naval Institute Proceedings, Annapolis, MD. Oct. 1952.

Burgess, Robert F. *Ships Beneath the Sea*. McGraw Hill, New York, 1975.

Lake, Simon. *The Submarine in War and Peace: Its Development and Possibilities*. J.B. Lippincott Co., Philadelphia and London1918.

Mazet, H.S. *Tragedy and Confederate Submarines*. U.S. Naval Institute Proceedings, Annapolis. May. 1942.

Polmar, Norman. *The American Submarine*. The Nautical and Aviation Publishing Co. of America. Annapolis. 1981.

Thomson, David W. *Three Confederate Submarines: Operations at New Orleans, Mobile, and Charleston1862-4*, U.S. Naval Institute Proceedings, Annapolis. Jan. 1941.

Von Kolnitz, Harry. *The Confederate Submarine.* U.S. Naval Institute Proceedings, Annapolis. Oct. 1937.

**ALLIGATOR**

Bolander, Louis H. *The Alligator: First Federal Submarine of the Civil War.* U.S. Naval Institute Proceedings, Annapolis, June 1938.

Mooney, James L. Editor. "Alligator, Submarine." *Dictionary of American Naval Fighting Ships.* Naval Historic Center, Washington DC 1991.

Winchester, J.F. Acting Master USS Sumpter. *Alligator, Loss at Sea.* Letter to the Secretary of the Navy, 8 Apr.1863. Office Naval Records and Library, Washington DC.

Worth, Richard A. *Alligator: Mr. Lincoln's Submarine.* The Submarine Review by Naval Submarine League. Annandale, VA.

**INTELLIGENT WHALE**

Anonymous, *Intelligent Whale*, A brochure from the Navy Museum Annex, Washington DC 1993.

Burgess, Robert F. *Ships Beneath the Sea: A History of Subs and Submersibles.* McGraw Hill,New York 1975.

Mooney, James L. Editor. "Submarine, Intelligent Whale." *Dictionary of American Naval Fighting Ships*, Navy Dept, Washington DC, 1991.

**JOHN HOLLAND**

Barnes and Noble. "John Holland." *New American Encyclopedia*, Grolier Corp USA,1991.

Cable, Frank T. *The Birth and Development of the American Submarine.* Harper Bros. Publishers, New York, 1924.

Cable, Frank T. *The Submarine Torpedo Boat Holland.* U.S. Naval Institute Proceedings. Annapolis, Feb. 1943.

Knowles, Morris. *The Story of the Holland Submarine.* U.S. Naval Institute Proceedings. Annapolis, Jan. 1960.

Knowles, Morris. *John P. Holland.* U.S. Naval Institute, Annapolis, 1966.

Lake, Simon. *The Submarine in War and Peace.* J.B. Lippincott. NY, 1918.

Lipscomb, F.W. *Historic Submarines.* Praeger Publishers, NY, 1970.

Morris, R.K. *Story of the Holland.* US Naval Institute Proceedings. Annapolis, MD Jan. 1960

Polmar, Norman. *The American Submarine.* Nautical and Aviation Publishing Co. of America. Annapolis, 1981.

Rucker, Colby. *The Fenian Ram.* U.S. Naval Institute Proceedings. Annapolis, Aug. 1935.

**SIMON LAKE**

Burgess, Robert F. *Ships Beneath the Sea*, Chapter 8, Accolade for an Argonaut. McGraw Hill, New York, 1975.

Hopkins, H.O. *Simon Lake and His Early Submarines: Discussion, Comments and Notes.* U.S. Naval Institute Proceedings, Annapolis, Mar, 1957.

Lake, Simon. *Submarine.* D. Appleton Century Co. New York, 1938.

Lake, Simon. "Modern Submarines in War and Peace." *International Marine Engineers Journal,* Jul. 1915 to Apr. 1916.

Lake, Simon. *The Submarine in War and Peace: Its Developments and Its Possibilities.* J.B. Lippincott Co. Philadelphia, PA, 1918.

**BEEBE/BARTON**

Barton, Otis. *Hunting Undersea.* The Atlantic Monthly Co. ,Boston, Jun. 1948.

Barton, Otis. *The World Beneath the Sea.* Thomas Y. Crowell Co. New York,1953.

Beebe, William. "A half mile down." *National Geographics Magazine.* Dec., 1934.

Beebe, William. "A Round Trip to Davy Jones's Locker." *National Geographics Magazine,* June, 1931.

Beebe, William. "A Wonderer Under Sea." *National Geographic Magazine* Wash, DC Dec. 1932.

Beebe, William. *Half Mile Down.* Harcourt, Brace and Co. New York, 1934.

Beebe, William. "The Depths of the Sea." *National Geographics Magazine.* Jan.1932.

Crandall, Lee S. *Charles William Beebe, In Memorium*. The Auk, NewYork Zoological Society. New York. Jan. 1964.

### CHAPTER 2 TRIESTE

Bascom, Willard. *The Crest of the Wave: Adventures in Oceanography*. Harper and Row, Publishers New York, 1988.

Bishop, Charles. Interview Re: Trieste acquisition. 23 Feb. 1993.

Cousteau, Jacques with Frederic Dumas. *The Silent World, the Submarine Dirigible*. Harper and Brothers, Publishers New York, 1953.

Cox, Donald W. *Explorers of the Deep, Pioneers of Oceanography*. Hammond Incorporated, Mapleton, New Jersey 1968.

Dietz, Robert S, Russell Lewis and Andreus Rechnitzer. "The Bathyscaph." *Scientific American,* Apr. 1958.

Keach, Donald. "Down to Thresher by Bathyscaph." *National Geographics* . June, 1964.

Keach, Don. Interview Re: Thresher search. 5 Apr. ,1993.

Michel, John E. Trieste presentation to the Deep Submersible Pilots Assn. Re: Sphere repairs prior to deep dive.13 Feb. ,1992.

Piccard, Auguste. *Earth Sky and Sea.* Oxford University Press. New York, 1956.

Piccard, Jacques & Thomas J. Abercrombie. "Man's Deepest Dive." *National Geographics* . Aug. 1960.

Piccard, Jacques and Robert Dietz. "Seven Miles Down." G.P. Putnam and Sons. New York, 1961.

Rechnitzer, Andreus. Interview Re: Trieste acquisition and early operations. 14 May, 1992.

Walsh, Don. *Conquering Inner Space.* U.S. Naval Institute Proceedings, Annapolis. Jan.1985.

Walsh, Don. Interview Re: Early Trieste operations. 18 Aug., 1992.

Walsh, Don. "Our 7-mile Dive to the Bottom." *Life Magazine,* 14 Feb, 1960.

### CHAPTERS 3 & 4 DEEP JEEP

American Society of Heating and Ventilating Engineers. *Heating Ventilating Air Conditioning Guide.* ASHVE. New York. 1960.

Bureau of Medicine and Surgery of the Navy. *Submarine Medicine Practice.* Department of the Navy, Washington 1956.

Forman, Will. "Deep Jeep: Design Through Operations" *Journal of Ocean Technology* July 1968.

Forman, Will. *Deep Jeep: A Developmental Research Vehicle.* AIAA. Los Angeles 21 Apr. 1966.

Forman, Will. *Deep Jeep: The Nots Deep Research Vehicle from Design to Operation at 2000 Feet.* Marine Technology Society, Washington 16 Jun. 1965.

Forman, Will. "Deep Jeep" *Undersea Technology* July 1966.

Forman, Will. *Evaluation of Deep Jeep Operations January 1964 to June 1966.* NOTS Report TS-67-130 Dec. 1966.

Gawn, R.W.L. *Effect of Pitch and Blade Width on Propeller Performance.* Transactions, Institute of Naval Architects, VOL. 95, 1953.

Hoerner, Sighard F. *Fluid Dynamic Drag.* Published by Author. Midland Park, NJ 1958.

Johnson, Phil. correspondence Re: Palomares,15 December, 1991.

Lewis, Flora. *One of Our Bombs is Missing.* Mc Graw Hill, New York.1967.

Messinger, Larry. Interview Re; Palomares, January 1992.

Roark, R.J. *Fomulas for Stress and Strain.* Mc Graw Hill, New York, 1950.

Rooney, Michael. correspondence Re; Palomares. 2 April, 1992.

Tsien,Hsue-Shew. "Theory for the Buckling of Thin Shells." *Journal of Aeronautical Sciences.* Vol. 9 1941-42.

Wendorf, Charles. interview Re; Palomares. 27, February, 1992.

Wenk, Edward. JR. *Feasibility Studies of Pressure Hulls for Deeply Submerged Submarines.* (NRC: CUW: 0256) National Academy Sciences, National Research council, Wash. DC Aug. 1958.

Woodson, Wesley E. *Human Engineering Guide for Equipment Designers.* University of California Press. Berkeley, CA 1954.

### CHAPTER 5 ALVIN

Ballard, Robert D. "A Long Look at Titanic." *National Geographic Magazine,* Dec. 1986.

Bland, Ed. Interview, Re: Alvin sinking accident,1 August, 1993.

Chandler, Rick. *DSV Alvin Dive Log 1964-1995.* Woods Hole Oceanographic Institute, Woods Hole, MA, 1995.
Corliss, John B. and Robert D. Ballard. "Oasis of Life in the Cold Abyss." *National Geographic Magazine* OCT, 1977.
Froehlich, Bud. Correspondence Re: Alvin's beginnings. 28 January 1992.
Froehlich, Bud. Interview, Re: Alvin's beginnings. 29 July, 1993.
Hollis, Ralph. Interview, Re: Thresher.12 September, 1993.
Hurtzler, J.R. "Project Famous." *National Geographic Magazine,* May, 1975.
Kaharl, Victoria A. *Water Baby; The Story of Alvin.* Oxford University Press, NY. 1990.
Kurtis, Bill. *The Secrets of Underwater Bolcanoes.* Bill Kurtis Productions, WTTW/Chicago. The New Explorers, Chicago, 1994.
Mc Camis, Marvin J. "Captain Hooks Hunt for the H Bomb." *Oceanus* Winter 88/89.
Wood's Hole Oceanographic Institute, Woods Hole, MA. Mc Camis, Marvin J. Interview, Re: Palomares. 24 August, 1993.
Momsen, Swede. Interview, Re: Alvins first funding. 26 March 1993.
Shumaker, Larry. Interview, Re: Alvin Pass thrus. 21 August, 1993.
Shumaker, Larry .Interview, Re: Project Famous. 23 August, 1993.
Walsh, Don. Interview, Re: Alvin's conception. 26 March, 1993.
Wilson, Val. Interview, Re: Palomares. 24 August.1993.

## CHAPTER 6 ALUMINAUT

Ellwood, E. E. The "Aluminaut." A Deep Submergence Vessel. Society of Automotive Engineers, paper 650661, NY 11 May, 1965.
Markel, Arthur. *Aluminaut: Historic Summary of Development and Operations.* Correspondence to Author, Reynolds Sub Marine Services Corp. Richmond VA 1 Jan, 1994.
Markel, Arthur. Interview, Re; Aluminaut 23 April, 1994.
Reynolds, J. Louis. *Aluminaut, The Deep Diving Aluminum Submarine,* Reynolds Sub Marine Services Corp. Richmond, VA 1968.
Sheets, H.E. and R.R.Loughman. *The Aluminaut.* American Institute of Aeronautics and Astronautics, AIAA paper 64-459 NY June, 1964.
Soule, Gardner. *The Ocean Adventure: Science Explores the Depth of the Sea,* Appleton-Century. NY 1966.
Soule, Gardner. *Undersea Frontiers: Exploring by Deep-Diving Submarines.* Rand McNally and Co. NY 1968.

## CHAPTER 7 THE DEEPSTARS

Bond, George. *Papa Topside: The Sealab Chronicles of Capt. George F. Bond USN.* Naval Institute Press, Annapolis, MD 1993.
Bradley, Robert. *Deapstar 4000 Dive Logs.* US Navy Electronics Lab. San Diego, CA 1970. Courtesy Mark Bradley, Vancouver, Canada. 1994.
Church, Ron. "Deepstar Explores the Ocean Floor." *National Geographic.* Jan. 1971.
Geyer, Richard A. *Submersibles and Their Use in Oceanography and Ocean Engineering.* Elsevier Scientific Publishing Co. NY.1977.
Jordan, S. A. *Gulfview: Deepstar Diving Log #106.* Westinghouse Undersea Division Westinghouse Corp. Baltimore, 1968.
Merrifield, R. *Undersea Studies with the Deep Research Vehicle Deepstar 4000.* Naval Oceanographic Office, Wash. DC. March,1969.
Shenton, Edward H. *Diving for Science, The Story of the Deep Submersible.* W. W. Norton & Co. Inc. NY NY. 1972.
Soule, Gardner. *Undersea Frontiers, Exploring by Deep Diving Submersible,* Rand McNally & Co. NY NY. 1968.
Thompson, Joe. Interview, Re; Deepstar 4000. 26 January, 1992.
Anon. *Deepstar (2000): Research Submersible System.* Westinghouse Research Labs. Baltimore, June, 1971.
Palmer, Hal. Interview, Re; Deepstar 2000 loss at sea, 19 June,1992.

Pritzlaff, John. Interview Re; Deepstar 2000. 5 April,1995.
Staehle, Charles & Hal Palmer & Robert Dill & John Pritzlaff. Interviews, Re; Deepstar 4000 & 2000. 15 March, 1995.

## CHAPTER 8 STAR 1, 2, 3 & ASHERAH

Bass, George F. *Archeology Beneath the Sea*. Walker and Co. New York.1975.
Bass, George F. "New Tools for Undersea Archeology." *National Geographic,* September,1968.
Beran, William T & Donald M Rosencrantz. *Asherah Design and Operations,* ASME Publication 65-UNT-3, American Society of Mechanical Engineers, NY NY. 1965.
Earle, Sylvia A. "A Walk in the Deep, First Open Ocean Use of Diving Suit "JIM" For Scientific Research." *National Geographic Magazine,* May, 1980.
Feed, Lou. *Logs of Star 1, 2, 3 & Asherah* correspondence 1993.
Geyer, Richard A. *Submersibles and Their Use in Oceanography and Ocean Engineering.* Elsevier Scientific Publishing Co. New York. 1977
Grigg, Richard W. "Hawaii's Deep Sea Jewels." *National Geographic* May, 1979.
Kirby, Terry. *Logs from Star 2 & Makalii.* correspondence; December,1993.
Walker, William W. Interview Re; Star 2 in Hawaii, 11 December,1993.

## CHAPTER 9 THE PERRY SUBS

Busby, Frank. *Manned Submersibles.* Office of the Oceanographer of the Navy, Wash. DC 1976.
Busby, Frank. *Undersea Vehicles Directory.* Busby Associates, Arlington, VA 1981, 1985 & 1990.
Fager, Irene. *One Half Fathom.* A Monthly Oceanographic Philately publication, Tampa, FL 1972 thru 1994.
Link, Marion Clayton. *Windows in the Sea.* Smithsonian Institution Press, Wash. DC. 1973.
Mac Leish, Kenneth. "A Taxi for the Deep Frontier; Project Man In The Sea goes Mobile." *National Geographic.* Jan 1968.
Perry, John, Interview, Re: Perrysubs history. 26 April, 1994.
Perry, John. *PC Specifications.* Perry Oceanographics, Riviera Beach, FL 1962-1982.
Perry, John. *Perry Submarine,* Update as of 11 March, 1994. correspondence, 28 March, 1994.
van Hoek, Susan with Marion Clayton Link. *From Sky to Sea; A Story of Ed Link.* Best Publishing Co, Flagstaff. AZ 1993.
Worthington, Robert, Editor *DSPA Newsletter.* San Diego CA 1974-1995.

## CHAPTER 10 DEEP QUEST

Anon. *Deep Quest: The Versatile Submarine.* Lockheed Ocean Systems, San Diego, CA 1982.
Saner, Don. *Deep Quest Dive Logs,* For 218 Dives, Over A 14 Year Period, 1994.
Saner, Don. Interview, Re: Deep Quest Operations, 17 Feb. 199
Wenzel, James. Interview, Re: Origins of the Deep Quest Project, 22 Nov.1993.

## CHAPTER 11 DOWB

Anon. *DOWB, Deep Ocean Workboat.* A.C. Electronics DRL,GMC. 1969.
Daubin, Scott C Interview Re; DOWB photos and reports. July, 1992.
Daubin, S.C. *The Deep Ocean Workboat (DOWB), An Advanced Deep Submersible Vehicle.* AIAA/SNAME paper no. 67-370. 24 May, 1967.
Fager, Irene. *DOWB Articles,* Half-Fathom Oceanic Philately Newsletter 1972-1985.
Megow, Larry. *Splash Up in the Pacific.* Slide Rule publication by Houston TX Engineering Society. 26 May, 1969.
Shenton, Edward H. *Diving for Science: The Story of the Deep Submersible.* W.W. Norton and Co. Inc. New York. 1972.
Worthington, Robert. *DOWB Articles,* Deep Submersible Pilots Association newsletter. 1972-1985.

## CHAPTER 12 TRIESTE 2

Anon. *History of the Bathyscaph Trieste 2.* Submarine Development Group One San Diego, CA 18 May, 1984.

Ballard, Robert D. *Explorations.* Hyperion publishers, New York 1995.
Ellis, George. Interview Re: Trieste 2 operations and dive logs. 27 Sep.1993.
Grenfell, E. W. (vice admiral) *USS Thresher* (SSN 593), 3 AUG. 61 TO 10 APR. 63. Naval Institute Proceedings. March, 1964.
Offley, Ed. "Lost in the Deep: Scorpion's Tale Hidden in a Web of Contradictions." *Navy Times* 22 Nov. 1993.
Parsons, Les. Interview Re: Trieste 2 operations and dive logs 26 Sep.1993.
Rean, Dan. Correspondence Re: Trieste 2 operations and dive logs.1994.
Scales, James. Interview Re: Trieste 2 operations and dive logs 6 Oct. 1993.
Tharp, Burt. Interview Re: Trieste 2 operations and dive logs 6 Oct.1993.

## CHAPTER 13 NEKTONS & DELTA

Anon. *Nekton Operations Manual.* General Oceanographic Inc, San Diego, CA 1982.
Ballard, Robert D. "Riddle of the Lusitania." *National Geographics* April 1994.
Caffarella, Tim. *Nekton: Alpha, Beta, and Gamma Dive Logs,* 1968 through 1987. (Correspondence 1993)
Lundberg, Eric. "A Crash, a Tremendous Crash: Free ascent from 240'." *Denver Post* 20 May 1980.
Privitt, Doug. Interview Re; History of Nekton and Delta Development. 3 Dec. 1990.
Slater, Richard A. "Commercial Submersibles Staying Busy with Delta." *Sea Technology* Dec. 1992.
Slater, Richard A. Delta Dive Summaries, 1983 through 1994.(Correspondence 1995).
Slater, Richard A. *Delta Operation,* As Reported to the Deep Submersible Pilots Association at 21 Nov. 1991 and 6 Oct. 1994 Meetings in San Diego, CA.
Slater, Richard A. *The Nekton Accident at Cataliina,* As reported to the Deep Submersible Pilots Association at 20 Sep.1990 meeting in San Diego, CA.

## CHAPTER 14 DOLPHIN

Anon. *Ships Historical Narratives,* Dolphin. 1971 through 1992, Submarine Development Group One, San Diego, 1993.
Anon. *Welcome Aboard,* Dolphin. Introduction and History. C.O. USS Dolphin, 1992.
Benya, John. *Dolphin,* Presentation to the Deep Submersible
Pilots Association. Luncheon/Meeting 18 May, 1995.
Sheets, Charles.(LT. USN). *Dolphin*, Presentation to Deep Submersible Pilots Luncheon/Meeting 21 May, 1992.
Sheets, Charles. (LT. USN). *Dolphin*, Submarine Tour 28 May, 1992.
Simpson, Charles. *Dolphin* (AGSS-555). Half Fathom Oceanographic Philately monthly publication, Tampa, FL. Aug. & Sep. 1992.

## CHAPTER 15 BEAVER MK 4

Black, D.L. *Beaver MK4. Submarine Work Boat,* Technical Description. Ocean Systems Operations of North American Rockwell Corporation July, 1968.
Cargile, Ed. Interview Re; Beaver 4 operations 7 June, 1992.
Gardner, W. *Beaver MK4 Submersible, Emergency Systems.* Ocean Systems Operations, North American Rockwell Corp. May 1968.
Rechnitzer, Andreus. Interview Re; Beaver 4 origins and designs. 29 Nov. 1994.
Simpson, Charles. *Beaver MK4.* One Half Fathom oceanographic philately publication. July, 1986.
Thompson, Joe. Interview Re; Beaver 4 operations 26 Jan. 1992.
Washington, Booker T. Interview Re; Beaver 4 operations by IUC 25 June, 1994.

## CHAPTER 16 BEN FRANKLIN

Barnes and Noble. Gulf Stream & Ben Franklin. Grolier Encyclopedia, 1991.
Emery, K. O. ,R.D. Ballard, and R.L. Wigley. "Dive Aboard Ben Franklin, Off West Palm Beach Florida." *MTS Journal Marine Technology Society* Wash. DC. Mar.-Apr. 1970.
Kazimir, Don. Interview Re: Ben Franklin operations. 30 Sep. 1994.
Piccard, Jacques. *The Sun Beneath the Sea.* Charles Scribners and Sons, New York, 1971.

Rand, Bill. Interview Re: Ben Franklin construction. 30 Sep. 1994.

Terrans, D. *Ben Franklin: Brief Systems Description,* Grumman Aircraft Engineering Corp. Beth Page, New York circa 1968.

## CHAPTER 17 SEA CLIFF

Anon. *Deep Submersible Vehicle Sea Cliff (DSV-4): Welcome aboard.* Submarine Development Group One, San Diego, 1992.

Anon. *Sea Cliff (1971-1992)* Record of Significant Events. Submarine Development Group One, San Diego.

Fager, Irene. *Sea Cliff.* Dive Reports by Crew Members, Half Fathom Oceanographic Philately publication. Tampa, FL.1973-1994.

Graham, David M. "Sea Cliff, A National Science Asset." *Sea Technology,* Arlington, VA. March 1988.

Laxo, Darryl E. "Navy's Sea Cliff to be Modified for Operations at 6100 Meters." *Sea Technology,* September 1980.

Simpson, Charles. *Sea Cliff History.* Half Fathom Oceanographic Philately, April 1992.

## CHAPTER 18 TURTLE (1968)

Anon. *Deep Submergence Vehicle Turtle (DSV-3): Welcome Aboard.* Submarine Development Group One, San Diego 1992.

Anon. *Turtle; Record of Significant Events 1977-1992.* Submarine Development Group One. San Diego. Burgess,Thomas F. "Eyes From Above Help Sub Escape Snare on Sea Floor." *San Diego Tribune* 9 Nov. 1989.

Fager, Irene *Turtle Dive Reports,* contributed by crew members. Half Fathom Oceanographic Philately publication. Tampa, FL. 1972-1994.

Murphy, Waldron J. "Turtle Dives Deep" *Underwater USA,* Fall 1987.

Northam, Don. *Turtle Dives San Juan De Fucua Ridge* 1994 Deep Submersible Pilots Luncheon/Meeting 20 Oct. 1994.

Owen, Thomas B.(Admiral) *Autec 1 and 2, Launch Remarks.* Office of Naval Research Wash. DC 11 Dec. 1968.

Simpson, Charles. *Turtle History.* Half Fathom Oceanographic Philately Publication, Dec 1992.

## CHAPTER 19 NR-1

Babb, Ivar G. and others. "Dual Use of a Nuclear Powered Research Submersible, The US Navy's NR-1." *Marine Technology Society Journal,* Winter 93-94.

Ballard, Robert D. "NR-1, The Navy's Inner-Space Shuttle." *National Geographic* April,1985.

Craven, John. Interview Re; NR-1 Origins 7 Oct. 1992.

Etter, Tom. *NR-1 Operations.* Presentation to the Deep Submersible Pilots Association at a luncheon/meeting 19 May,1994.

Grifith, D. O. (Admiral) Correspondence Re; NR-1 Characteristics and Operations, Office of the Chief of Naval Operations Wash. DC 3 June,1993.

NR-1 Nuclear Powered Research and Engineering Vehicle. Chief of Naval Operations Wash. DC 1968.

Polmar, Norman and Thomas B. Allen. *Rickover.* Simon and Shuster NY, NY. 1982.

*Welcome Aboard NR-1.* Chief of Naval Operations Wash. DC 1985.

## CHAPTER 20 KUMUKAHI

Cary, Cy. Interview, Re; Kumu fabrication and operations. 9 July, 1994.

Forman, W.R. *Kumukahi ASME* paper 70-WA/Unt-11, American Society of Mechanical Engineers Dec.1970.

Forman, W.R. *Kumukahi* Design and Operations Manual, Oceanic Institute, Makapuu HI June 1968.

Forman, W.R. *Remote Controlled Models as Submersible Simulators NOTS TP 4359* Naval Ordnance Test Station China Lake, CA May, 1967.

Lang, Tom Interview, Re; Utility Sub Origins and Design Details.11 July 1994.

Lang, T.G. and P.D. Scholz. "A Two Man Plexiglass Submarine for Oceanic Research." *Undersea Technology.* April 1963.

Stachiw, Jerry D. *The Origins of Acrylic Plastic Submersibles.* ASME paper 86-WA/HH-5. American society of Mechanical Engineers Dec. 1986.

Stachiw, Jerry D. *Acrylic Plastic Viewports for Ocean Engineering Applications.* TP 562 Naval Undersea Center, San Diego, CA. Feb 1977.

Walker, William W. Interview, Re; Kumukahi fabrication and operation 11 July, 1994.

### CHAPTER 21 DSRV-1 MYSTIC

*Command History 1977-1994* Submarine Development Group One. Coronado, CA.

*DSRV-1 Dive Record for Dives #1 to #306.* Submarine Development Group One. Coronado, CA.

Fager, Irene. *DSRV-1 Dive Reports,* Contributed by crew members. Half Fathom Oceanographic Philately Publication. Tampa, FL 1972-1994.

Forman, Will. *Preliminary Proposal for Interim Rescue Submersible.* NOTS Technical Paper TS 65-149 Naval Ordnance Test Station, San Clemente Island, CA Dec. 1964

Kermeen, Robert. Interview Re; *DSRV-1* Fabrication 3 March, 1995.

Wenzel, James. Interview Re; *DSRV-1* Fabrication 3 March 1995.

### CHAPTER 22 NEMO

Briggs, Ed. Interview Re: Construction and dive history of Nemo, May 1994.

Maison, Jack R. and Henning Ottsen. *Behavioral Stress Analysis of the Nemo Type Acrylic Hulls.* ASME paper 70-WA/Unt-8, American Society of Mechanical Engineers New York 3 Dec. 1970.

*Nemo: Twelve Plastic Pentagons Under the Sea.* Naval Civil Engineering Laboratory, Port Hueneme, CA. Circa 1970.

Piccard, Auguste *Earth Sky and Sea.* Oxford Univ. Press NY 1956.

Stachiw, J.D. *Acrylic Plastic Viewports for Ocean Engineering Applications* NUC TP 562 Naval Undersea Center San Diego CA. Feb.1977.

Stachiw, J.D. *The Origins of Acrylic Plastic Submersibles.* ASME paper 86- WA/HH-5 American Society of Mechanical Engineers NY NY Dec. 1986.

Trowbridge, Taylor *Optical Properties of a Spherical Plastic Underwater Observatory Nemo.* Transactions of the American Society of Mechanical Engineers New York May, 1971.

Tsuji, K. and R. Shelton. *Fabrication of Nemo Typoe Spherical Acrylic Capsules for Underwater Vehicles.* ASME paper 70-WA/Unt 4 American Society of Mechanical Engineers New York 3 Dec. 1970.

### CHAPTER 23 JOHNSON-SEA-LINK

Askew, Tim. Interview Re; Acrylic Hull History of JSL-1. 5 Aug. 1994.

Askew, Tim. *Johnson-Sea-Link: Cruise Planning Manual.* Harbor Branch Foundation, Fort Pierce, FL. Circa 1992.

Brooks, Robert A.(Chairman) and 7 others. *Entanglement of the Johnson-Sea-Link,* Expert Review and Report to the Secretary of the Smithsonian Inst. Smithsonian Inst. Wash. DC 21 Dec. 1973.

Ecklebarger, Kevin and Patricia A. Linley and Tom Smover (Editors) *Harbor Branch Profiles* Harbor Branch Foundation Inc. Fort Pierce, FL 1978-1984.

Fisher, Arthur. "See-All-Around Bubble Sub." *Popular Science Magazine* Sep. 1971.

*Harbor Branch Profiles* (1986-1992) Harbor Branch Inc. Fort Pierce, FL.

Link, Marion C. Windows in the Sea Smithsonian Institute Press. Wash. DC, 1973.

Maison, J. R.and J. D. Stachiw. *Acrylic Pressure Hull for Johnson-Sea-Link Submersible* ASME paper 71 WA/Un T-6, American Society of Mechanical Engineers NY 2 Dec. 1971.

Reed, John H. (Chairman) and 5 others. *Johnson-Sea-Link, Entanglement Off Key West FL 17 June 1973.* National Transportation Safety Board. Wash. DC. 15 Jan 1975.

Rogers, Michael. *The Sculpture Transparent, Bruce Beasley Turns 6 Tons of Acrylic Into Clear Art Work.* Science 83, American Association for the Advancement of Science. Dec. 1983.

Simpson, Charles *Johnson-Sea-Link Rescue.* One Half Fathom. Oceanographic Philately Soc. Tampa, FL. July 1989.

Stachiw, Jerry D. Interview Re; Acrylic Hull History from Beasley to JSL 10 Aug. 1994.

van Hoek, Susan with Marion Link. *From Sky to Sea.* Best Publishing Co, Flagstaff, AZ 1993.

### CHAPTER 24 DSRV-2 AVALON

Cooper, Kip. "Sunken Submarine Rescue Tested." *San Diego Union* 15 May 1979.

Fager, Irene. *DSRV-2 Dive Reports,* Contributed by Crew Members. One-Half Fathom Oceanographic Philately Publication. Tampa, FL 1972-1994.

Jones, Richard and John Drable. "U.K.'s Submarine Rescue Efforts." *Sea Technology,* Dec. 1993.

Phillips, George. Interview Re. DSRV-2 Fabrication by Lockheed. 4 Mar. 1994.

*Record of Significant Events 1979-1992.* Submarine Development Group One. Coronado, CA.

*Report on Exercise Latent.* Commander Submarine Force U. S. Atlantic Fleet. Washington DC. June, 1979.

Sorbet Royal 92. Commander Submarine Group Five. San Diego, CA.

Simpson, Charles. *History of Deep Submersible Rescue Vessel Avalon (DSRV-2).* Half Fathom Oceanographic Philately Publication. Tampa, FL

## CHAPTER 25 DEEP VIEW

Forman, Will and George Horn. "Deep View Glass/Metal Hull." *American Society of Mechanical Engineers* 16 Sep. 1971.

Forman, Will. "Glass Submersibles." *Journal of Hydronautics* Feb. 1969.

Forman, Will. and Robert De Hart. "Submersible Deep View Pioneers: Glass/Metal Bonding." *Undersea Technology* Dec. 1967.

Forman, Will. Submersibles "With Transparent Structural Hulls" *Astronautics and Aeronautics Magazine* April 1969.

Forman, Will. *Research Vehicle Deep Niew Operating Manual.* Naval Undersea Research and Development Center, San Diego, CA. Dec. 1970.

Frocht, Max M. *Photoelasticity.* John Wiley and Sons Inc. New York 1941.

Mazzone, Walter. Interview Re: Deep View Operations in Hawaii. 6 Jan.1995.

## CHAPTER 26 SIGNIFICANT OTHERS

### MORAY

Berryman, Al. Interview Re. Moray Operations 21 Mar. 1993.

Morse, Robert (Chairman) and Committee. *Undersea Vehicles for Oceanography.* Interagency Committee on Oceanography. Washington DC March 1965.

Terry, Richard D. *The Deep Submersible.* Western Periodicals Co. No. Hollywood, CA 1966.

### HIKINO/MAKAKAI

Mazzone, Walt. Interview Re. Makakai Operations 6 Jan. 1995.

Moore, Don. Interview Re. Hikino and Makakai Operations.12 Jan. 1995.

Murphy, Doug. Interview Re. Hikino and Makakai Operations 13 Jan.1995.

### HUMAN POWERED SUB RACES

Bigelow, Bruce V. "Sub Races Have Sunk to a Change of Venue" ( To Escondido, CA.) *San Diego Union Tribune,* 12 April,1994.

Britton, Peter. "Muscle Subs" *Popular Science* June, 1989.

Britton, Peter. "Technology Race Beneath the Waves." *Popular Science* July 1991.

Dunn, Stanley. Interview Re. FAU Involvement in the HPS Races. 20 December 1994.

Merry, Stephanie. "Human-Powered Submersibles: Generating New Technologies." *Sea Technology,* Dec.1989

*Official Race Guidelines and Rules Handbook* 1989-91-93.

Submarine Race Committee, Perry Foundation. Washington DC. Perry, Hap. Interview Re; Submarine Race Origins 20 Dec. 1994.

Reay, D.A. *The History of Man Powered Flight.* Pergamon Press, N.Y. 1977.

### DEEP VOYAGER

Hoyt, Richard. "This Sub Will Glide to L.A." *Honolulu Star Bulletin* 18 Feb. 1969.

Pryor, Taylor and Will Forman *Deep Voyager* ( A Proposal ) Oceanic Foundation, Makapuu Oceanic Center, Waimanalo, HI. 17 May 1969.

Rains, Dean A. *Ballast System & Trajectory Analysis, Project Deep Voyager.* TRW Systems. Redondo Beach, CA 7 Oct. 1969.

Shand, Errol B. *Glass Engineering Handbook.* MC Graw Hill Book Co. New York 1958.

**TOURIST SUBS**

Forman, Will. *The All Acrylic Tourist Submersible Sea View,* The Tourist Oceanology International 90 in Monte Carlo, Monaco 10 Oct. 1990.

Forman, Will. "The First Tourist Submersible with Acrylic Hull." ROV 89. *San Diego.* 16 March 1989.

Jones, Bruce L. *Tourist Submarines Operational.* 1994, U.S. Submarines Inc. La Conner, WA 28 Dec. 1994.

"New U.S. Built Acrylic-Hull Tourist Submanrine Steals the Show at Monte Carlo." *Waves Magazine* Jan/Feb 1991.

Piccard, Jacques *The Sun Beneath the Sea,* Charles Scribner's Sons, NY 1971.

Roberts, Tom. *Atlantis Submarine Fleet,* Atlantis Submarine International Inc. Vancouver, B.C. Canada. Correspondence and brochures 4 Jan. 1995.

**PISCES/HYCO**

Busby, Frank *Undersea Vehicle Directory.* Busby Associates Inc. Arlington, VA, 1985.

Chapman, Roger. *No Time on Our Side: The 83 Hour Dive In Pisces 3.* W. W. Norton and Co. New York 1975.

Clark, Eugenie. "Down the Cayman Wall (Pisces 2)." *National Geographic,* Wash. DC Nov. 1988.

Clark, Eugenie and Emory Kristof. "Sharks at 2000 Feet" (Pisces 6) *National Geographic,* Nov. 1986.

Galerne, Andre. "Pisces 6-Deep Eyes and Arms for Discovered Seven Seas." *Ocean Industry Magazine* Nov. 1979.

Simpson, Charles. *Pisces 1's 500th Dive.* One Half Fathom Oceanography Philately Publication, Dec. 1990.

Simpson, Charles. *Pisces 4.* One Half Fathom Oceanographic Philately Publication. Sep 1993.

Simpson, Charles. *SDL-1* One Half Fathom Oceanography Philately Publication. Nov.1993.

Trice, Al. Correspondence Re; Data On Hyco Subs. ISE Research Ltd Port Coquitlam, BC. Canada.

**SNOOPER**

Siverts, Don. *Dive Records, Snooper One and Two.* Correspondence, June 1995.

Siverts, Don. Interview Re; Snooper Operations. 12 Dec.1994.

*Snooper 1 & 2 Specifications.* Underwater Graphics Inc. Torrance, CA 1991.

Stewart, Jill. "Homemade Sub Makes Voyage to the Bottom of The Sea a Reality." *L.A. Times* 23 FEB. 1986.

**SEA OTTER**

Bradley, Mark. *Sea Otter Dive Logs;* 1972-76 Correspondence Oct. 1994.

Glover, Darrell. "Sub Checking San Juan Cable." *Seattle Post-Intelligencer* 3 Sep. 1972.

Hacking, Norman. "Sub Eyes Peak Under Ocean (Cobb Seamount)." *The Province* (No. Vancouver, BC) 16 Sep. 1972.

Hanson, Phil. "They'd Like Business to Go Down." Vancouver Sun 10 June, 1972.

Simpson, Charles. *Sea Otter* ( EX- PAULO 1). One Half Fathom Oceanographic Philately Publication Aug 1993.

**DEEP ROVER**

Ballou, Philip and Graham Hawkes. "Design Study: Two Person Research Submersible." *MTS Journal* Fall 1993.

Bouvier, John. "Deep Rover Submersible Features Force, Touch, Feedback." *Sea Technology* Feb. 1984.

*Deep Rover.* One-man Submersible. Deep Ocean Engineering Inc. San Leandro, CA Jan. 1985.

Earle, Sylvia A. "A Walk in the Deep" *National Geographic,* May, 1980.

Earle, Sylvia A. *Sea Change.* G.P. Putnam's Sons New York, 1995.

Hawkes, Graham S. "The Future of Atmospheric Diving Systems and Associated Manipulator Technology, With Special Reference to a New Microsubmersible, Deep Rover." *MTS Journal* 3rd Quarter, 1983.

Saunders, Alan. "Conquering Our Deepest Frontiers." *Popular Mechanics* Jan. 1985.

# Index

# A

# B



# C

# D

# E

# F



# G



# H

# I

## J



## K



## L

# M

# N

# O



# P

# R

# T

# U

# V



# W



# X



# Y



# Z